VOLUME ONE HUNDRED AND TWENTY SEVEN

Advances in
COMPUTERS
Edge/Fog Computing Paradigm:
The Concept, Platforms and
Applications

VOLUME ONE HUNDRED AND TWENTY SEVEN

ADVANCES IN
COMPUTERS

Edge/Fog Computing Paradigm:
The Concept, Platforms and
Applications

Edited by

PETHURU RAJ
*Edge AI Division, Reliance Jio Platforms Ltd,
Bangalore, India*

KAVITA SAINI
*School of Computing Science and Engineering (SCSE),
Galgotias University, Delhi, Uttar Pradesh, India*

CHELLAMMAL SURIANARAYANAN
*Government Arts and Science College, Srirangam
(Affiliated to Bharathidasan University), Tiruchirappalli,
Tamilnadu, India*

ACADEMIC PRESS

An imprint of Elsevier

ELSEVIER

Academic Press is an imprint of Elsevier
50 Hampshire Street, 5th Floor, Cambridge, MA 02139, United States
525 B Street, Suite 1650, San Diego, CA 92101, United States
The Boulevard, Langford Lane, Kidlington, Oxford OX5 1GB, United Kingdom
125 London Wall, London, EC2Y 5AS, United Kingdom

First edition 2022

ISBN: 978-0-12-824506-4
ISSN: 0065-2458

For information on all Academic Press publications
visit our website at https://www.elsevier.com/books-and-journals

Publisher: Zoe Kruze
Developmental Editor:
    Cindy Angelita Pe Benito-Gardose
Production Project Manager: James Selvam
Cover Designer: Greg Harris
Typeset by STRAIVE, India

Working together
to grow libraries in
developing countries

www.elsevier.com • www.bookaid.org

Contents

6. Microservices architecture for edge computing environments   183
Chellammal Surianarayanan

7. Edge data analytics technologies and tools   209
N. Jayashree and B. Sathish Babu

Contributors

J. Akilandeswari
Department of Information Technology, Sona College of Technology, Salem, TN, India

M.P. Anuradha
Department of Computer Science, Bishop Heber College, Affiliated to Bharathidasan University, Tiruchirappalli, Tamil Nadu, India

R. Anushia Devi
School of Computing, SASTRA Deemed University, Thanjavur, India

Vinay Aseri
Department of Forensic Science, Vivekananda Global University, Jaipur, Rajasthan, India

B. Sathish Babu
Department of Computer Science and Engineering, R V College of Engineering, Bengaluru, Karnataka, India

Kiranmai Bellam
Department of Computer Science, A & M University, Prairie View, TX, United States

Rushikesh L. Chopade
Department of Forensic Science, Vivekananda Global University, Jaipur, Rajasthan, India

A. Daniel
School of Computing Science and Engineering (SCSE), Galgotias University, Delhi, Uttar Pradesh, India

Swapnali Jadhav
Government Institute of Forensic Science, Aurangabad, Maharashtra, India

N. Jayashree
Department of Computer Science and Engineering, C Byregowda Institute of Technology, Kolar, Karnataka, India

S. Karthikeyan
VIT-AP University, Amaravati, Andhra Pradesh, India

Manju Khari
School of Computer and System Sciences, Jawaharlal Nehru University, New Delhi, India

N. Krishnaraj
Department of Networking and Communications, School of Computing, SRM Institute of Science and Technology, Kattankulathur, Tamil Nadu, India

Jenn-Wei Lin
Department of Computer Science and Information Engineering, Fu Jen Catholic University, Taipei, Taiwan

K. Lino Fathima Chinna Rani
Department of Computer Applications, Bishop Heber College, Affiliated to Bharathidasan University, Tiruchirappalli, Tamil Nadu, India

M. Marimuthu
Research Scholar, Sona College of Technology, Salem, TN, India

R.I. Minu
Department of Computer Science and Engineering, School of Computing, SRM Institute of Science and Technology, Kattankulathur, Tamil Nadu, India

Varad Nagar
Department of Forensic Science, Vivekananda Global University, Jaipur, Rajasthan, India

G. Nagarajan
Department of Computer Science and Engineering, Sathyabama Institute of Science and Technology, Chennai, Tamil Nadu, India

Swati Nigam
Department of Computer Science, Faculty of Mathematics and Computing, Banasthali Vidyapith, Banasthali, India

Uttama Pandey
School of Computing Science and Engineering (SCSE), Galgotias University, Delhi, Uttar Pradesh, India

Pritam P. Pandit
Department of Forensic Science, Vivekananda Global University, Jaipur, Rajasthan, India

Pethuru Raj
Site Reliability Engineering (SRE) Division, Reliance Jio Platforms Ltd. (JPL); Reliance Jio Cloud Services (JCS), Bangalore, India

Sunku Ranganath
Intel Corporation, Hillsboro, OR, United States

Partha Pratim Ray
Department of Computer Applications, Sikkim University, Gangtok, India

Kavita Saini
School of Computing Science and Engineering (SCSE), Galgotias University, Delhi, Uttar Pradesh, India

Mahipal Singh Sankhla
Department of Forensic Science, Vivekananda Global University, Jaipur, Rajasthan, India

R. Satheeshkumar
Department of Electronics and Communication Engineering, Galgotias College of Engineering and Technology, Noida, India

Prabha Selvaraj
VIT-AP University, Amaravati, Andhra Pradesh, India

Serin V. Simpson
Department of Computer Science and Engineering, Thejus Engineering College, Thrissur, Kerala, India

Anubhav Singh
School of Forensic Science and Risk Management, Rashtriya Raksha University, Lavad, India

P. Sivaprakash
VIT-AP University, Amaravati, Andhra Pradesh; PPGIT, Coimbatore, India

Swaroop S. Sonone
Department of Forensic Science, Dr. Babasaheb Ambedkar Marathwada University, Aurangabad, Maharashtra, India

Urvashi Sugandh
Department of Computer Science, Faculty of Mathematics and Computing, Banasthali Vidyapith, Banasthali, India

D. Sumathi
VIT-AP University, Amaravati, Andhra Pradesh, India

Chellammal Surianarayanan
Government Arts and Science College, Srirangam (Affiliated to Bharathidasan University), Tiruchirappalli, Tamilnadu, India

Veeramuthu Venkatesh
School of Computing, SASTRA Deemed University, Thanjavur, India

Preface

The faster adoption of several transformative and trend-setting technologies such as the Internet of Things (IoT), artificial intelligence (AI), edge computing, cloud-native and serverless computing models, 5G communication, etc. has set the ball rolling for the edge AI era. Centralized and expensive computing has become decentralized, distributed, and affordable as edge computing has surged in popularity. Market watchers estimate that billions of IoT edge devices and sensors are currently in use. The device ecosystem is growing fast with the plentiful production of sleek, fashionable, and useful devices. When these devices connect, communicate, collaborate, correlate, and corroborate directly or indirectly, the quantity of IoT data being generated and collected is massive, and IoT device data come in multistructured forms.

At present, IoT edge devices, through a plethora of technological innovations and disruptions, are being filled with more memory, storage, and networking capacities, and processing capability. In this way, edge devices are becoming part of mainstream computing. That is, centralized and consolidated computing is being gradually replaced by edge devices, which are decentralized and disaggregated. AI libraries are being deployed in resources-intensive IoT edge devices to carry out proximate and real-time data processing. This is described as on-device AI processing to create intelligent devices. Such a strategically sound shift is intended to produce a dazzling array of advancements and automations not only for businesses but also for individuals in their everyday activities. This book aims to articulate and accentuate how edge computing, analytics, and AI concepts are contributing immensely to visualizing, producing, and delivering a wide range of services that are state of the art, context-aware, people-centric, service-oriented, event-driven, mission-critical, and knowledge-filled.

Chapter 1, "Exploring the edge AI space: Industry use cases," explains the latest trends and transitions that are occurring in the edge AI space. Pioneering technologies and tools are emerging and quickly evolving to establish and sustain next-generation edge-native applications. This chapter illustrates the recent developments in edge computing and on-device AI processing domains.

Chapter 2, "Edge computing types and attributes," provides an introduction to various types of edge computing by broadly classifying it into four types, based on round trip latency requirements: IoT edge, wireless access edge, on premise edge, and network edge. The requirements and attributes of each of these edge computing types are discussed. The chapter then details the practical challenges across these edge deployments and explores how ETSI (European Telecommunications Standards Institute) Multi-access Edge Computing specifications help to address these challenges.

Chapter 3, "Industry initiatives across edge computing," focuses on various industry initiatives in the edge computing paradigm. To accelerate the evolution and adoption of edge computing, various standard bodies, open-source projects, and industry consortia have come together in recent times to revolutionize edge computing. This chapter discusses various initiatives around the world that have major traction in terms of collaboration, collateral produced, and industry impact.

Chapter 4, "IoT-edge analytics for BACON-assisted multivariate health data anomalies," explains anomaly detection, which in IoT-enabled systems can significantly improve the quality of the deployed systems. Although existing techniques can detect anomalies from a dataset, more efficient algorithms can be used to minimize the burden of excessive computational overheads on the resource-constrained IoT-edge device pool. In this chapter, the blocked adaptive computationally efficient outlier nominators (BACON) algorithm is implemented and illustrated along with the estimated-expectation/maximization method to improve the anomaly nomination for IoT-based health datasets. The weighted variant of the BACON algorithm package—"wbacon"—from the R repository is deployed to validate the utilization of anomaly nomination for an IoT-edge enabled health dataset.

Chapter 5, "The edge AI paradigm: Technologies, platforms and use cases," explains the various implementation technologies of the edge AI paradigm. Prominent industrial use cases are also discussed in this chapter.

Chapter 6, "Microservices architecture for edge computing environments," focuses on the importance of microservices architecture and event-driven architecture styles, as they help to visualize and implement edge-native applications that can be deployed and run on IoT devices.

Chapter 7, "Edge data analytics technologies and tools," explains the concept of edge analytics, the technologies and tools for enabling edge analytics, and provides various delectable use cases.

Chapter 8, "Edge platforms, frameworks and applications," observes how edge computing and analytics domains are receiving a lot of attention these days. Industries are exploring a variety of use cases that leverage integrated edge platforms and enable frameworks. These advancements enable enterprises to build a wide range of applications.

Chapter 9, "Edge computing challenges and concerns," discusses the various challenges and concerns of edge computing. There are several advantages associated with edge computing. However, there are also a few lacunae, and experts are working together to develop appropriate solutions and approaches that overcome the identified limitations of edge computing.

Chapter 10, "A smart framework through the Internet of Things and machine learning for precision agriculture," focuses on advanced techniques used in smart agriculture system based on IoT and machine learning algorithms. Several studies have been carried out on this system to offer smart services for real-time monitoring of any agricultural environment. IoT-based smart agriculture systems are an ideal approach to enhance the productivity of food items with reduced power and water consumption.

Chapter 11, "5G Communication for edge computing," highlights the concepts of edge computing and the working methodology. It also discusses in detail the importance and taxonomy of edge computing in 5G, and the functional components of edge computing. The evolution of 5G is summarized, and a brief explanation of the architecture of edge computing and 5G is also provided. Finally, the chapter explores recent advancements in edge computing for 5G.

Chapter 12 envisages "The future of edge computing." This chapter identifies the various technologies that are likely to be integrated with edge computing in order to realize next-generation edge applications and services.

Chapter 13, "Edge computing security: Layered classification of attacks and possible countermeasures," is about edge security. This chapter focuses on presenting layered classification of security attacks and identifies how these attacks can be overcome.

Chapter 14, "Blockchain technology for IoT edge devices and data security," explains the need for blockchain technology to ensure the tightest security for IoT edge devices and data.

Chapter 15, "EDGE/FOG computing paradigm: Concept, platforms and toolchains," focuses on edge concepts. The chapter also considers the emergence of various platforms and toolchains in facilitating the goals of edge computing.

Chapter 16, "Artificial intelligence in edge devices," discusses the implications of running AI algorithms and frameworks on edge devices.

The chapter also reviews the unique advantages and use cases when AI capabilities are embedded in IoT edge devices.

Chapter 17, "5G—Communication in healthcare applications," investigates the unique capabilities of 5G communication networks and how this next-generation cellular communication is useful in fulfilling a next-generation healthcare application.

Chapter 18, "The integration of blockchain and IoT edge devices for smart agriculture: Challenges and use cases," illustrates the architecture, challenges, and benefits of using edge computing to enable the industry 4.0 vision. This final chapter explains how the integration of blockchain technology and IoT edge devices is beneficial in visualizing and realizing smart agriculture applications.

The book aims to articulate and accentuate the significant contributions of edge computing and analytics technologies for the ensuing digital era. The book illustrates how the real digital transformation can be accomplished through the smart leverage of various innovations and disruptions in the edge AI space.

URVASHI SUGANDH
Research Scholar, Department of Computer Science,
Faculty of Mathematics and Computing, Banasthali Vidyapith,
Banasthali, India

DR. MANJU KHARI
Associate Professor, School of Computer and System Sciences,
Jawaharlal Nehru University,
New Delhi, India

DR. SWATI NIGAM
Assistant Professor, Department of Computer Science,
Faculty of Mathematics and Computing, Banasthali Vidyapith,
Banasthali, India

CHAPTER ONE

Exploring the edge AI space: Industry use cases

Pethuru Raj[a] and Jenn-Wei Lin[b]
[a]Site Reliability Engineering (SRE) Division, Reliance Jio Platforms Ltd. (JPL), Bangalore, India
[b]Department of Computer Science and Information Engineering, Fu Jen Catholic University, Taipei, Taiwan

Contents

Abstract

Now, edge devices, through a plethora of technological innovations and disruptions, are being stuffed with more memory, storage and networking capacities and processing capability. Thereby, edge devices are joining in mainstream computing. That is, the centralized and consolidated computing moves over to edge devices to be decentralized and disaggregated. AI libraries are being deployed in IoT edge devices to do proximate and real-time data processing. This is termed as on-device intelligence. Such a strategically sound shift is to bring forth a dazzling array of advancements and automations not only for business houses but also for common people in their everyday assignments. This chapter is to throw some light on the theoretical and the practical aspect of the edge AI paradigm.

Advances in Computers, Volume 127
ISSN 0065-2458
https://doi.org/10.1016/bs.adcom.2022.02.001
1

The faster adoption of a few transformative and trend-setting technologies such as the Internet of Things (IoT), artificial intelligence (AI), edge computing, 5G communication, etc., has set the ball rolling for the edge AI era. The centralized and expensive computing now becomes decentralized, distributed, and affordable with the surging popularity of edge computing. Market watchers forecast that there are billions of IoT edge devices and - sensors. The device ecosystem is growing fast with the plentiful production of slim and sleek, trendy and handy devices. When these devices connect, communicate, collaborate, correlate, and corroborate directly or indirectly, the size of the IoT data getting generated and collection is literally massive. Also, the IoT data comes in a multi-structured form is exponentially growing. With cloud computing, the IT becomes highly optimized and organized. Further on, the longstanding goal of IT industrialization and consumerization are seeing the reality with the steady growth of cloud applications, platforms and infrastructures. Now, with the growing solidity and sagacity of AI algorithms and frameworks being widely deployed on cloud environments, transitioning the fast-growing IoT data heaps into actionable insights in time is gaining the much-needed speed. Resultantly, there are insights-driven business workloads and IT services.

Now, edge devices, through a plethora of technological innovations and disruptions, are being stuffed with more memory, storage and networking capacities and processing capability. Thereby, edge devices are joining in mainstream computing. That is, the centralized and consolidated computing moves over to edge devices to be decentralized and disaggregated. AI libraries are being deployed in IoT edge devices to do proximate and real-time data processing. This is termed as on-device intelligence. Such a strategically sound shift is to bring forth a dazzling array of advancements and automations not only for business houses but also for common people in their everyday assignments. This chapter is to throw some light on the theoretical and the practical aspect of the edge AI paradigm.

1. The proliferation of IoT devices and sensors

There is a plethora of noteworthy breakthroughs in the field of artificial intelligence (AI). AI is succulently enabled through a host of machine and deep learning (ML/DL) algorithms. The prominent possibilities are computer vision (CV) and natural language processing (NLP) applications. Precisely speaking, the mesmerizing AI phenomenon is producing a dazzling array of sophisticated digital life applications. Not only business houses but also common people also started to experience the beauty and power of

AI technologies and tools. In this chapter, we are to focus on convolutional neural networks (CNNs) and recurrent neural networks (RNNs) and their unique contributions in visualizing and realizing game-changing use cases. These deep neural networks (DNNs), a critical part of artificial neural networks (ANNs), are guaranteeing in releasing and running vision-enabled and human-intractable business workloads and IT services.

Further on, the Internet of things (IoT) is being touted as the next-generation Internet comprising not only server machines, storage appliances, and networking solutions but also all kinds of digitized entities (All sorts of physical, mechanical, and electrical systems in our everyday environments will become digitized and integrated with the Internet). The size of the future Internet is bound to see an exponential growth due to the enormous and elegant participation of digital devices. Going forward, most of the enterprise-scale, service-oriented, process-aware, knowledge-filled, and event-driven business workloads and IT services are being modernized, migrated and executed in heterogeneous cloud servers (public, private, edge and hybrid). With the concept of cloud-native computing flourishing, IT infrastructures are being organized well and optimized deeply to run microservices-centric applications efficiently to achieve the much-needed portability, interoperability, findability, accessibility, availability, maneuverability, scalability, extensibility, and reliability.

There is another fledgling concept of cyber physical systems (CPS) gaining momentum in the recent past. Mission-critical physical assets at the ground level are being digitized and synchronized with cloud-hosted software applications and databases in order to be cognitive in their everyday operations. The concept of digital twins is also fast emerging and evolving for empowering ground-level artifacts. Especially sophisticated systems such as medical instruments, defense equipment, automobiles, avionics, space electronics, manufacturing assembly lines, etc. are empowered through their cloud-hosted digital twins, which continuously collect the data from their physical twins, leverage AI-driven data analytics in real time, and articulate their findings with the concerned in order to intrinsically automate a range of manual and error-prone tasks. Technically speaking, the IoT technologies and tools are for digitization, connectivity and integration of everything with everything else in the vicinity/neighborhood and with remotely held entities. Such a brewing trend is seen as a breakthrough transition for the future of the world.

In short, we are seeing an unprecedented growth of the IoT devices in our everyday environments. By fusing the distinct AI capabilities with these handy and trendy, slim and sleek handhelds, wearables, portables, and implantable, the world is all set to be adequately and artistically enabled through self, surrounding and situation-aware devices, systems, and environments. In other words, digitized and connected assets are to be empowered with vision, perception, decision-making and actuation capabilities. Also, these empowered devices are to interact with human beings through natural interfaces. This chapter is incorporated in this book in order to tell all about how technologies gel well to envisage advanced and automated capabilities.

2. Activating on-device intelligence

There are billions of IoT devices and sensors generating a staggering amount of multi-structured data at the network edge. That is, the IoT edge devices deployed across the globe could easily generate exabytes of data. To make sense out of exponentially growing edge device data, the concepts of edge computing and analytics are being seen as the trend-setting phenomenon. AI processing is being done directly on edge devices. Streaming data analytics is being performed on edge devices. Therefore, the next-generation capability of on-device intelligence is gaining prominence as edge devices are being instrumented with more processing capability, network, memory and storage capacities. Such a well-defined and strategically sound instrumentation facilitates interconnectivity and acquiring intelligence in an automated manner. Another noteworthy point is the much-discussed knowledge discovery and dissemination happen in real time. Thereby real-time intelligent applications are bound to flourish in the days to come and the world is to see real-time enterprises sooner than later.

Precisely speaking, the hugely popular IoT paradigm is for digitization. That is, all the ordinary things become extraordinary in their actions and reactions with the digitization-enablement process. The common and cheap things in our midst are designated to be digitized to join in the mainstream computing. Digital assets and assistants, when interacting with one another or when collaborating for accomplishing complex business processes, or correlating with one another for uncovering hidden patterns, are bound to emit out a tremendous amount of poly-structured data. The IoT devices and sensors generate a lot of data through interaction and collaboration. Now there is a widespread acceptance that data is a strategic asset for any organization to be competent and clever in their offerings. When the generated data gets collected, cleansed and crunched, the world is ready to receive and leverage actionable insights out of voluminous data. In other words, data gets transitioned into information and into knowledge. There are path-breaking technologies and tools to automate the arduous task of converting data into knowledge. That is, knowledge discovery and dissemination have become the talk of the town with the ready availability of big data. There are a number of digital technologies such as software-defined cloud centers, big, fast and stream data analytics, blockchain, artificial intelligence, microservices architecture, cybersecurity, etc., the IoT-generated data gets processed and analyzed to produce useful insights in time. The resulting insights are then looped back to business workloads, IT services, and IoT devices to empower them to be cognitive in their service deliveries.

Device to Cloud (D2C) Integration—The figure below vividly illustrates how cloud and enterprise applications get empowered through the seamless and spontaneous integration with ground-level devices and sensors. There are wider options for data transmission protocols, data representation, exchange and persistence formats, and network topologies.

Real-time Data Analytics at the Edge—With the accumulation of IoT edge devices, there is a massive amount of poly-structured data. The real-time analytics capability is gaining wider acceptance at the edge. In this big data era, IT experts understand that AI turns out to be a great game-changer especially in arriving at accurate inferences out of voluminous data. Also, there is a paradigm shift in proximate processing. Today IoT data gets collected, transmitted and stocked in hyperscale cloud centers to do casual and comprehensive data analytics. But due to latency, bandwidth, and security issues, increasingly data gets analyzed at the source itself. That is, one or more edge devices are being clubbed together to embark on real-time data analytics in an affordable, amenable, and artistic manner. The solidity and significance of edge analytics are being understood by business executives and technology professionals. There are technologies and tools galore for setting up and sustaining ad hoc, dynamic, purpose-specific and temporary clouds being formed out of edge devices in order to facilitate real-time edge data analytics at the source itself. With the general availability of lightweight artificial intelligence (AI) libraries, models, frameworks, and engines, the aspect of AI processing at the edge brightens and blossoms.

The formation of Edge Clouds for enabling Edge Analytics—The data processing slowly and steadily moves over from large-scale, centralized, clustered, consolidated, automated, and shared cloud environments to edge clouds. On one side, edge clouds are being formed out of a few server machines for enabling proximate data processing. On the other side, heterogeneous edge devices in a particular environment such as homes, hotels, hospitals, etc. can form ad hoc, small, and purpose-specific device clusters/ clouds quickly. There are enabling technologies and tools for automatic set up and sustenance of edge device clouds. Leading market watchers and analysts have predicted that there will be billions of connected devices in the years to come. Increasingly edge devices are being stuffed with more computing, storage and networking capacities and capabilities and hence they are intrinsically capable of finding, binding and collaborating with one another to solve bigger business problems. Thus, the edge power is growing rapidly with the proliferation of edge devices. The amount of data getting generated by machines is far higher than men-generated data. It is estimated that more than 850 zettabytes of data will be generated every year hereafter. Besides faraway cloud storage appliances, resource-intensive edge devices offer a better platform for stocking and processing edge data.

The Promise of Edge AI—The edge computing model is the recent phenomenon gaining a lot of interest among business executives and IT experts. As indicated above, sending all the IoT device data to faraway cloud environments to be stored and subjected to a variety of investigations is wasting network a lot of bandwidth and also increasing the network-induced latency. For low-latency and real-world applications, the concepts of edge computing, storage, and analytics are gaining momentum. In short, the local computing through edge devices and AI-inspired deep analytics of edge data generated by scores of IoT devices and sensors are gaining a lot of attention. Since the edge devices and services are closer to users than the cloud, the idea of edge computing is expected to flourish and fulfil hitherto unheard real-world and real-time requirements. The figure below vividly illustrates the macro-level edge computing architecture (Fig. 1).

The technologies and tools empowering edge computing and the faster maturity and stability of AI algorithms have concertedly resulted in a series of digital disruptions and innovations. The fusion of AI and edge computing is increasingly natural and beneficial. AI model creation, evaluation, optimization and deployment have been accomplished in cloud environments. Now with the general availability of highly miniaturized, multifaceted yet powerful processor architectures empowering edge devices and the surge of lightweight AI frameworks and libraries, the new era of edge AI has dawned. That is, running AI models directly in edge devices is gaining momentum for practical reasons. There is a bright and better scope for such an intersection. The point is that the distinct AI powers are being now realized through our everyday devices. In other words, the aspect of AI democratization is surging ahead.

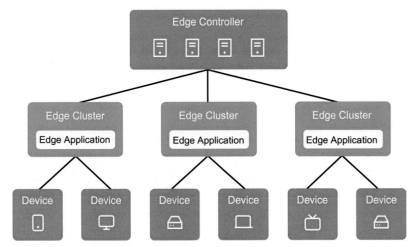

Fig. 1 Edge device to edge controller integration.

Through these transitions, there will be a plethora of people-centric applications. The enterprise IT has moved to cloud IT and now it is heading toward edge IT. In other words, not only business behemoths but also commoners are literally to enjoy a dazzling array of durable and distinct benefits out of the booming AI domain. The movement toward people IT is progressing well. There is a greater affinity between AI and edge computing paradigms. There are explanatory and exploratory articles articulating and accentuating long-term benefits of this disruptive and deft convergence. AI is all set to become penetrative, participative and pervasive. AI has turned out to be a key driving force behind emerging frontiers such as self-driving cars, smart diagnosis and treatment, intelligent transportation, cognitive applications, smart homes, hotels, hospitals, etc.

As indicated above, there will be billions of IoT edge devices including smartphones, consumer electronics, manufacturing machineries, robots, cameras, wearables, handhelds, portables, and fixed devices. The skyrocketing number of heterogeneous IoT devices lead to the generation of massive amounts of multi-modal data (still and dynamic images, audio files, etc.) of the edge devices' physical surroundings that are minutely monitored and sensed. There are big, fast and streaming analytics platforms for extracting actionable insights out of edge device and environment data. The data analytics platforms are being shrunken and deployed in edge devices to facilitate real-time data capture and crunching. In the recent past, AI models are being developed and deposited in model repository to solve a variety of problems. With the faster adoption of AI model optimization techniques, AI models are pruned to be run on edge devices. Thus, the distant and distinct goal of edge AI is seeing the grandiose reality with a series of noteworthy technology and tool developments.

The world is elegantly experiencing the power of AI algorithms. There are a litany of innovations and disruptions in the AI space. There came pioneering machine and deep learning (ML/DL) algorithms in order to enable everyday machines to automatically learn from data directly. Such knowledge-filled machines are capable of exhibiting real-time and real-world intelligence in their actions and reactions. Further on, the domains of computer vision (CV) and natural language processing (NLP) are seeing a lot of advancements with the steady stability of powerful ML and DL algorithms. However, due to the network latency issue, increasingly ML and DL frameworks are being installed in edge devices in order to democratize and demonstrate AI capabilities for every person and every group of people at any place. Edge-based AI processing feature is being keenly used for a wider range of applications such as smart factories and cities, face recognition, machine translation, human machine interfaces (HMIs), medical imaging, etc. Several industry verticals and business functions are exploring the possibility of leveraging the distinct advancements in the AI space to conceive and provide premium services to their consumers, employees, partners and other constituents.

3. The artificial intelligence (AI) processing at the edge

The faster maturity and stability of machine and deep learning (ML/DL) algorithms are setting a stimulating foundation for the huge success of the AI paradigm. Now there are miniaturized AI libraries to be deployed to facilitate running AI applications on IoT edge devices. Such a combination brings forth a number of personal, social and business use cases. For an example, it brings the ability to identify usable patterns and detect anomalies in the data points sensed and captured by the edge device, Sensors attached with edge devices are capable of monitoring population distribution, traffic flow, humidity, temperature, pressure, and air quality continuously and if there is any perceptible deviation, then the edge device will raise an alarm or alert to the concerned to ponder about the best course of remedial actions in time. Further on, the insights extracted from the sensed data are fed to automated systems for taking decisions and then for plunging into appropriate actions. There are automated systems such as traffic congestion avoidance systems, public transportation planning, driving and parking assistance systems, etc. The automation capabilities being introduced through AI processing at the edge fulfils the long-pending goal of operational efficiency, property and people safety, etc.

The aspect of edge AI is gaining prominence these days as there are requirements for producing and running real-time applications, which are prepared to succulently automate and orchestrate multiple manual tasks for industrial environments as well as peoples' every day environments. With the general availability of powerful processors, AI processing at the edge is seeing a good progress. The device ecosystem is on the growth path. There are multifaceted yet disappearing sensors and actuators being produced in large quantities. The emergence of data transmission protocols, data representation, exchange and persistence formats, the steady adoption of event processing middleware solutions and streaming databases, etc. are seen as a positive factor toward the ensuing era of edge AI. The direct or indirect integration of edge devices with faraway cloud platforms through intermediaries is also making it easier and popular for the edge AI phenomenon to gain the much-needed boost. Historical and comprehensive data analytics in consonance with cloud-based data lakes in hyperscale cloud environments is being touted as a key differentiator.

Digital twins are being constructed and made to run on cloud environments for all kinds of mission-critical and complicated appliances, machineries, rockets and their launchers, defense equipment, medical instruments, robots, drones, consumer electronics, gadgets and gizmos, etc. That is, these ground-level devices are having their virtual/cyber/software/logical versions running in clouds. Through the seamless and spontaneous connectivity and integration between physical entities at the ground and their equivalent constructs at virtualized and containerized cloud environments significantly, it is possible to visualize and realize additional competencies for edge devices. Digital twins also incorporate data analytics and AI capabilities in order to substantially empower edge devices. The performance and throughput levels, health condition, and operational quality of distributed and disparate edge devices can be minutely monitored and digital twins can integrate both current and historical data of edge devices to embark on comprehensive analytics to unearth actionable insights. Thus, the on-device intelligence capability coupled with cloud-based analytics facility is being pronounced as the way forward for the ensuing era of knowledge.

The traditional cloud architecture is primarily challenged on several fronts. Latency, centralization, cost, reliability, and security are seen as the

grave barriers for the thriving of the conventional cloud computing paradigm. With edge devices natively supporting the aspects of distribution
and decentralization, the widely articulated concerns of public cloud environments are being surmounted through edge cloud computing. Edge data
are environment-specific and people-centric. The self-, surroundings- and
situation-aware details are carefully collected, cleansed and crunched
quickly and efficiently to arrive at and articulate actionable insights in time.
Scores of real-time and context-aware services can be accurately decided and
delivered in time. Thus, the phenomena of data–driven insights and insights-
driven applications are set to be the new normal. There are two aspects:
Machine learning and Deep learning at the Edge.

4. Machine learning (ML) at the edge

Machine learning libraries are being taken to edge devices in order to
sufficiently enable them to do self-learning without any involvement,
instruction and interpretation. Primarily, blogs and articles illustrate two
main use cases for ML in the industrial IoT landscape.
1. Anomaly detection
2. Extract higher-valued features such as remaining uptime for industrial
   machineries
For getting these use cases, experts turn toward executing ML frameworks
(there are several lightweight versions) in edge devices, which are typically
situated near the data source. Where the action is, there edge devices are.
Even people carry and wear edge devices. The process of digitization is as
follows. All kinds of physical, mechanical and electrical systems in our personal, social and professional environments are enabled by externally and/or
internally embedding a variety of sensors and other enablers on them. There
are several powerful digitization and edge technologies fast emerging and
evolving to set up and sustain the digital world. Through this act of digitization, ordinary items become extraordinary. Casual things become smart
objects by applying noteworthy technological advancements. Dumb objects
become animated things and sentient objects. Common things become cognitive entities. In short, everything is destined to be digitized. Digitized
objects can be further enabled to be sensitive, perceptive, computational,
vision-enabled, communicative, responsive, decision-making, and active.
Natural interfaces can be attached with digitized elements to seamlessly
and spontaneously interact to complete everyday tasks with all the confidence and clarity.

Precisely speaking, digital and edge technologies and tools, on meticulous planning and execution, result in scores of connected and cognitive systems in and around us in large numbers. The sensors attached in these systems minutely monitor different parameters (physical, operational, and transactional details, health condition, performance level, and log data), capture and transmit them the ML frameworks and models, which, then, infer something unique and useful immediately. The ML frameworks and models can run in those devices or in nearby edge devices. The brewing idea is to do proximate data processing and local analytics, which emits out a lot of advantageous information and insights. Such an extracted knowledge can be disseminated to the concerned systems, devices, services and people in time.

The workflow for machine learning consists of two main steps: firstly, it is all about training and developing a ML model. Once the model reaches a state of stability and accuracy, then the model is destined to production environment. The first step is typically an off-line operation where stored data is used to train and tune a model. Then the trained, tested, and optimized model is ready to make inferences on real-time data. Increasingly IoT edge devices are used for executing matured ML models. However, ML model generation is an iterative process. That is, the model output is being checked and a proper feedback is created and shared to the original model architecture to bring in appropriate changes. Hence it is not a single and straightforward activity. Instead, the ML model is bound to go through multiple iterations to attain a reasonably good model. In this refinement activity, there can be a number of traversals between edge devices and the central cloud.

Machine Learning for Streaming Data—When preparing and producing ML models, data has to go through a variety of pre-processing steps such as data cleaning, deletion, augmentation, and addition, In other words, data has to be presented in such a way that the target environment can unambiguously understand and consume the data without any hitch or hurdle.

Generally, when training a ML model, data is stored in files or a database with all time-stamped sensor values. This is because the model gets the same set of data each time. In a streaming environment, the scenario is quite different. That is, the sensor data is getting received serially, with each sensor sending data at repetitive intervals but independent of all other sensors. Before we can deliver streaming data to a ML model, we must align the data on regular time boundaries. Also, sometimes it is necessary to adjust the time windows to get data from sensors that deliver data with a lower rate. Video cameras presented as multifaceted sensors are being stuffed with new-generation capabilities in order to enable them to be intelligent in their operations, offerings and outputs. Newer possibilities are being unearthed and fresh use cases are being articulated and accentuated. That is, video cameras are being deployed in industrial areas, advanced homes and office buildings, entertainment plazas, stadiums, eating joints, bus stations, nuclear installations, hospitals, and other important junctions. Experts have visualized the following use cases.

- **Vision inspection for yield optimization**—Intelligent video cameras can perfectly identify scrap material and faulty products as early as possible in order to ensure the highest quality.
- **Count and measure**—Empowered video cameras can easily count products, objects, people and measure position, alignment, color and other attributes.
- **Intrusion detection and people safety**—Enabled video cameras can quickly trigger an alarm or stop a machine when people or objects get too close or enter where they shouldn't be.

With intelligent cameras abounding in our everyday environments, a variety of physical and behavioral aspects are being monitored ceaselessly. Not only capturing the happenings in those places precisely, but also taking appropriate counter measures in time based on the gathered and gained understanding are being widely appreciated. By applying ML capabilities on sensor data locally, it is possible to transition ordinary devices into cognitive devices. Real-time actions can be initiated and implemented through streaming data analytics on edge devices. Simple as well as complex activities such as real-time classification, clustering, association, regression, vision, recognition, detection, and translation can be accomplished by running advanced data analytics tasks on edge platforms and infrastructures. With the ready availability of container orchestration platform solutions such as K3s, KubeEdge, and other lightweight Kubernetes versions, edge devices are clubbed together or clustered to form dynamic, purpose-specific, and ad hoc edge device clouds. Thus, with the capability of setting up and sustaining edge infrastructures quickly and easily is fast maturing and stabilizing, edge analytics is all set to become the new normal. Other prominent use cases are being explained below.

Edge devices in healthcare—The need for on-device intelligence is gaining ground as decisions and actions have to be taken quickly in order to save lives. Value-adding body parameters have to be collected and crunched locally to take correct and real-time decisions. Collecting and carrying them to faraway cloud environments for data storage and analytics is not a good sign especially for the healthcare sector. Even performing streaming data analytics using cloud-based streaming databases and analytics platforms is being seen as a risky thing. The network latency comes into the picture. If there is a slowdown in network speed or if there is a network breakdown, then the consequences could be unthinkable and irreparable. Thus edge-based real-time machine learning is emerging as the way forward to monitor, measure, and manage ICU patients safe.

We have features-rich sensors and they can form ad hoc smart networks to accomplish bigger and better things. The connectivity solutions are pervasive with 5G emerging as the boon for reliable indoor networking. There are lightweight libraries and frameworks to run AI models on edge devices. In the recent past, there are concerted research works and contributions to bring in highly optimized AI models. Thus, the technological power is facilitating real-time healthcare services. Neurological activity and cardiac rhythms can be monitored continuously and analyzed in real time to identify if there is any deviation. There is another term "Ambient intelligence (AmI)" coined and popularized some years back to fulfil the vision of intelligence everywhere. Now with the ambient communication, ubiquitous computing, pervasive sensing and perception, edge-based AI processing, etc., the goal of AmI is all set to see the grand reality soon. Every sneeze and activity of people under observation can be meticulously monitored and cared in case of any abnormality.

The device ecosystem grows fast. We have a variety of purpose-specific and agnostic devices. We have toasters, robots, smartphones, physical activity trackers, smart watches cameras, drones, and wearable gyroscopes or accelerometers. With the ready availability of a dazzling array of instruments, electronics, equipment, wares, gadgets, gizmos, appliances, etc. the era of edge devices joining in the mainstream computing is brighter than ever. Fresh use cases are being achieved with the surge of multifaceted devices. By applying the distinct ML capabilities, on-device intelligence is being enabled. That is, the realization of intelligent devices and services is being simplified and speeded up.

Mining, oil, and gas and industrial automation—The business value of edge-based ML becomes is gaining momentum in the oil, gas, or mining industry. Employees have to work in isolated, risky and faraway places. Therein, the connectivity is irregular. Therefore, the much-needed computation has to be taken to the devices and machines there. Sensors attached on edge devices and robots can capture large amounts of data and accurately predict things like as pressure across pumps. If there is any deviation in one or more operating parameters, then appropriate alerts get articulated in time. Preventive and predictive maintenance of industrial assets and artifacts are being facilitated through real-time prediction of ML-enabled devices.

Further on, embedded sensors of all machineries inside a factory or warehouse can capture and store all kinds of data (images, videos and audios) locally. By applying machine and deep learning algorithms on the data, it is automatically decided by the machines to have some self-rest. If there is any repair required, that details can be shared across. With the fast-growing AI power, machines can self-diagnose, heal, configure, defend, manage, etc. Machines become fault-tolerance and performant.

Streaming high-frequency data—In the increasingly streaming world, data is pouring in continuously in high speed. It is therefore important to extract and report knowledge continuously matching up with the data generation and ingestion speeds. As accentuated above, AI models play a vital role here in knowledge extraction. But because of high-frequency data, there is a need to accommodate a proven messaging system before sending the streaming data to a polished AI model. The figure below vividly illustrates the data flow and knowledge discovery and dissemination (Fig. 2).

With the confluence of the IoT, emerging computing paradigms such as cloud-native, serverless and edge, artificial intelligence, the goal of digitally transformed homes, cities, offices, retail stores, manufacturing floors, etc. is to be fulfilled quickly. AI-enabled data analytics go a long way hand in hand in setting up and sustaining digitally transformed systems, solutions, services and environments. Easily usable and manageable intelligence will become penetrative, pervasive and persuasive.

The adoption of ML capabilities is gaining momentum these days. There are several everyday applications getting immense support from ML algorithms and models. The majority of these ML models are process and data-intensive. Generally, ML models need cloud-like infrastructure to be run comfortably. For real-time computation and analytics, edge-centric computing is being presented as the way forward. Also, IoT device and

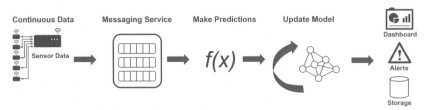

Fig. 2 The data flow toward knowledge discovery and dissemination.

sensor data are mainly streaming. In order to ensure real-time data analytics, ML models are being taken to edge devices. To bring ML libraries, frameworks and accelerators to the edge, a variety of innovations and transformations are being introduced and implemented. Processor architectures are going through a host of distinct disruptions. Highly optimized communication protocols are emerging and evolving. Data security at transit and persistence is being ensured through a host of powerful technologies and methodologies. Energy efficiency at the edge is being given extra thrust these days. Mitigating time and space complexities and achieving energy complexity are pronounced as the key barriers. Researchers are also equally working in a concerted fashion to surmount these constraints. Creating ML models with smaller memory footprint is the need of the hour.

There are an arsenal of tricks and techniques for ML model compression. The prominent ones among them are pruning, quantization, sparse modelling, knowledge distillation, transfer and federated learning, etc. Such breakthrough models can be easily run on a cluster of edge devices to perform edge-centric and ML-based inferences.

5. Deep learning at the edge

There arise new DL models to do image segmentation at the edge (https://bdtechtalks.com/2021/05/07/attendseg-deep-learning-edge-semantic-segmentation/).

There are a number of neural network (NN) architectures for fulfilling a variety of problems such as computer vision, speech recognition, etc. A new neural network architecture has made it possible to perform image segmentation on resource-constrained IoT edge devices. As we all know, image segmentation is the process of determining the boundaries and areas of objects in images. Image segmentation is a vital requirement for mobile robots, self-driving cars, and other vision-based systems that interact and navigate the real world. However, segmentation requires a heavy load of computing power besides huge network and storage capacities. That is, developing such deep learning models mandate cloud servers and storages. The scientists at DarwinAI and the University of Waterloo have created a neural network architecture that provides near-optimal segmentation. Also, it (is named as AttendSeg) is small enough to run on resource-constrained IoT devices.

Computer vision is one of the key applications of deep learning. The domain of computer vision includes image classification, object detection, and segmentation. Segmentation is a complex classification task. As indicated above, CNNs are primarily used for computer vision tasks. The CNN complexity is being measured by the number of parameters. More parameters mean more memory and computation are needed. For an example, RefineNet, a popular semantic segmentation neural network, contains more than 85 million parameters. At 4 bytes per parameter, any application using RefineNet requires at least 340 megabytes of memory just to run the neural network.

Due to the huge hardware requirements of neural networks, applications of image segmentation mandate for cloud infrastructures. That is, deep learning models are being run on cloud platforms for their versatility and resource needs. However, this poses a series of challenges. The channel between the image segmentation application running on robots, cameras, drones and smartphones and the faraway cloud server has to be wider and secure. The network bandwidth has to be on the higher side for transmitting images between ground-level devices with cloud-based image segmentation applications and services. Also, the problematic network latency comes into picture here. There are other limitations to be looked into. The strategic requirement is to empower IoT edge devices with much-needed vision capability to collect image and video data streams quickly and process them instantaneously to emit out actionable insights in time. Real-time image segmentation is needed in sensitive environments wherein AI agents are being deployed. Running AI models locally is definitely taxing energy-starved IoT devices. Thus, the way forward is as follows. Building and sustaining highly optimized AI models is insisted for the ensuing edge era. Edge-native applications ought to be empowered through sophisticated AI models, which consume less energy with small memory footprint. That is the main reason for the domains of edge AI and TinyML to flourish lately. As articulated above, highly miniaturized, generalizable, and transferable models have to be designed, developed and deposited in order to publicly be found and bound. There are public and private model repositories to kickstart and rekindle the long-pending AI era.

In short, the aspect of image segmentation goes to the edge to serve better. The industry 4.0 vision involves a plethora of industrial applications with segmentation capability. The authors of this paper claim that the AttendSeg deep learning model performs semantic segmentation at an accuracy that is almost on-par with RefineNet while cutting down the number of parameters to 1.19 million. The memory footprint of AttendSeg gets reduced to one fourth. The model requires a little above one megabyte of memory.

This is quite easy to manage with most of the IoT edge devices. Experiments clearly show that AttendSeg provides optimal semantic segmentation by sharply cutting down the number of parameters and by reducing the memory footprint.

Model	Accuracy (%)	MACs (G)	Parameters (M)	Weight memory (Mb)
RefineNet	**90.0**	202.47	85.69	343
EdgeSegNet	89.15	77.89	7.09	28.3
AttendSeg	89.89	**7.45**	**1.19**	**1.19**

AttendSeg leverages "attention condensers" to reduce model size without compromising performance. Self-attention mechanisms improve the efficiency of neural networks by focusing on information that matters. Self-attention techniques have been widely used in producing natural language processing (NLP) applications. As found out, recurrent neural networks (RNNs) had a limited capacity on long sequences of data. Transformers, a key deep learning (DL) architecture, use self-attention mechanisms to overcome the RNN limitations to expand their range. That is, deep learning models such as GPT-3 leverage Transformers and self-attention to churn out long strings of text.

AI researchers started to leverage the proven attention mechanisms to improve the CNN performance substantially. Attention condensers improve the performance of CNNs in a memory-efficient way. One of the key challenges of designing TinyML neural networks is finding the best performing architecture while reducing the computational requirements. To address this, the researchers have used the progressive concept of "generative synthesis". This is a potential ML technique to create neural network architectures based on specified goals and constraints. The idea is that instead of manually fiddling with all kinds of configurations and architectures, the researchers can bring forth a problem space to empower the ML model to zero down the best combination.

The application areas for the AttendSeg neural network are growing steadily. It is gaining a lot of attention as it is capable of getting deployed and executed on edge devices. It is impacting decisively on manufacturing applications such as parts inspection, quality assessment, etc. Fixed as well as mobile robots gain a lot. Medical applications such as cell analysis, tumor segmentation, etc. are also benefiting immensely out of this shrunken neural network. Remote sensing applications such as land cover segmentation are also hugely simplified through the smart application of this neural network solution.

Thus, computer vision and speech recognition capabilities are being embedded in IoT edge devices with the help of deep learning (DL) models. Edge devices, through the on-device intelligence feature, are all set to be intuitive, informative, and intelligent in their operations, outputs, and offerings. Performing deep learning tasks typically requires a lot of computational power and a massive amount of data. With big data, the decision accuracy of deep learning algorithms/models is higher. Today as per the estimates provided by various market watchers and analysts, there are billions of connected devices, which, when interacting purposefully, can generate a massive amount of multi-structured data. Thus, the data availability in humongous quantity has speeded up the adoption of deep learning methods. Now with cloud computing facilitating on-demand, online, on-premise/off-premise availability of enormous amount of compute, network and storage resources, problem resolution through the unique power of deep learning processes gains the much-needed momentum. IoT devices and sensors generate a lot of data, which gets collected, cleansed and crunched through cloud-based AI platforms to squeeze out actionable insights in time. Now, considering the network latency issue, AI processing is being taken to edge devices to guarantee real-time data processing. With edge devices joining in the mainstream computing, we will be bombarded with edge-native cognitive applications.

6. Digging into the paradigm of edge AI

The diagram below vividly illustrates how edge devices (generally resource-constrained and primarily used for data collection and transmission) are empowered through edge servers (resource-intensive and used for edge device data storage and processing). Edge servers are embedded with communication modules (wired as well as wireless) to take edge device data to faraway cloud environments for comprehensive and cognitive analytics. The Internet is the prime communication infrastructure. There are other dedicated and high-end communication infrastructure to guarantee high-quality communication.

With all-round advancements, the domain of edge AI is gaining momentum. A number of industry verticals are keenly exploring and experimenting with edge AI pilots and proof-of-concepts (PoCs) to gain a deeper and broader understanding about the power of edge AI. The IT teams and organizations besides cloud and communication services providers are readily equipping themselves with the necessary know-hows to jump into the edge AI bandwagon. This section is to dig deeper to discuss the advantages of edge AI.

The distinct Advantages of Edge AI—Edge computing enables bringing AI processing tasks from the cloud to near the end devices in order to overcome the intrinsic problems of the traditional cloud, such as high latency and the lack of security. As experts indicated, moving AI computations to the network edge has several advantages.

- **Lower data transfer volume**—Data is processed by the edge device and only a significantly lower amount of processed data is sent to the cloud for long-term storage and comprehensive data analytics. By sharply reducing data traffic between edge devices at the ground with cloud-based applications and data sources, the unnecessary wastage of precious network capacity can be reduced.
- **Real-time computing**—Enterprises are all set to become real-time in their offerings and operations. Real-time processing has become a new normal in the digital era. Real-time applications are becoming paramount and prevalent. Real-time data capture, storage, processing, analytics, knowledge discovery and dissemination, decision-making and action are vital for real-time enterprises. In short, edge computing is the way forward. Time-sensitive or low-latency applications are being mandated across industry verticals.

The physical proximity of edge devices to the data sources makes it possible to considerably reduce network latency. This improves real-time processing of edge device data. It supports delay-sensitive applications and services such as remote surgery, self-driving vehicles, park assistance systems, etc. With more resources being stuffed in edge devices, real-time analytics of edge data gets facilitated through the setting up and sustenance of edge device clouds.

- **Privacy and security** - Keeping data at the edge strengthens its privacy. Edge data, most probably, does not travel in the porous, public, open, affordable, and worldwide communication infrastructure (the Internet). Therefore, data is safe and secure and is not vulnerable for any kind of theft and distortion. Edge computing ensures data stays in its device(s) itself. If there is a need to send edge data to off-premise cloud platforms, confidential and critical information can be still kept locally to ensure its privacy and security.

- **High availability**—Edge AI intrinsically prescribes decentralization. Edge devices individually and collectively guarantee high availability. If there is a slowdown even breakdown between edge devices and the cloud environments, edge devices can function offline and deliver their designated tasks without any failure. The widely articulated offline capabilities make Edge AI more robust. Edge devices, thus, can provide transient services during a network failure or when the network is under cyberattacks. Therefore, enabling edge devices to perform AI processing ensures high availability for mission-critical AI applications.

6.1 The blend of edge AI and 5G rekindles state-of-the-art applications

High-throughput and highly reliable communication is the need of the hour. The faster adoption of the 5G communication standard across the world has stimulated many industries to visualize and realize a plethora of next-generation applications. There are certain domains such as self-driving cars, real-time virtual reality (VR) experiences, remote surgery, etc. yearning for 5G communication and proximate data processing capability. Edge-native applications are the most sought-after ones in the digital world. Thus, the combination of 5G and edge computing spells a bright future for the digital era. 5G cellular communication is now extensively used in industrial environments. Thus, factory automation and smart manufacturing for creating and sustaining the industry 4.0 vision are being enabled through the faster maturity and stability of 5G and edge analytics. Especially with the deployment and usage of machine and deep learning frameworks, libraries and accelerators at edge devices, producing people-centric, context-aware, real-time and intelligent applications becomes a simpler and simplified process.

Edge AI Applications—With Edge AI capability, it becomes possible to power up real-time, edge-native and intelligent applications.

- **Computer and Machine Vision**—With the flourish of highly optimized convolutional neural network (CNN) models, computer vision especially machine vision capability is becoming the new normal. Vision-enabled devices and machineries are being greatly used in different mission-critical environments. Video and audio data analytics tasks are technologically simplified and strengthened to produce and deploy path-breaking AI applications. Computer and machine vision applications will be thriving. Vision will become the core and central aspect for next-generation applications not only at the centralized cloud environments but also decentralized edge devices.

- **Smart energy** applications will be prevalent and paramount. Especially connected wind farms will thrive. Smart microgrids will become ubiquitous. The hybrid model of cloud and edge computing paradigms will become the new normal in building and running smart energy systems. The wind farm uses a variety of multifaceted sensors and actuators for video cameras, access control systems, etc. Also, wind turbines will be internally as well as externally fitted with features-rich sensors to minutely monitor and manage them. All kinds of right and relevant data get meticulously collected and processed to emit out actionable insights in time.

- **Smart healthcare** applications such as remote surgery and diagnostics will see the grand reality. Various body parameters are proactively and pre-emptively being captured and subjected to a variety of investigations in order to extract intelligence to act upon with all the care, clarity and confidence. Virtual diagnoses followed by correct medication can be realized with the blend of 5G and edge AI competencies.

- **Entertainment** applications—With the general availability of 5G communication, fresh gaming and entertainment applications such as virtual reality, augmented reality, and mixed reality are flourishing these days. Increasingly streaming video contents are sent to virtual reality glasses. The form factor of such glasses can be reduced sufficiently by offloading computation from the glasses to nearby edge servers.

- **Smart manufacturing and factory automation capabilities** are being realized through the combination of cloud environments, 5G, and edge device clouds. People and property safety and security requirements are fully met through the advancements happening in the technology space. Machineries are empowered to be vision, perception, knowledge discovery, decision-making and actuation capabilities. Localized AI processing is being facilitated by taking computation to edge devices and servers.

- **Intelligent transportation systems**—These systems are becoming important for next-generation travel, trip and transport. Edge devices and clouds are playing an extremely vital role here in shaping up the future transport. Driver assistance systems, traffic congestion avoidance systems, accident-avoidance systems, etc. are being given thrust these days with the faster maturity of edge AI technologies and tools. In addition, unmanned vehicles can sense their surroundings and move safely in an autonomous and artistic manner.

6.2 Person re-identification (Re-ID) (https:// towardsdatascience.com/why-we-need-person-re-identification-3a45d170098b)

This is a new initiative getting a lot of attention these days. The gist of this project is to correctly pinpoint a person of interest across multiple and non-overlapping cameras. With the advancement of deep neural networks (DNNs) and the increasing demand for intelligent video surveillance, this problem has gained significant interest in the computer vision community. The goal of Re-ID is to determine whether a person-of-interest has appeared in another place at a distinct time captured by a different camera, or even the same camera at a different time instant. While tracking allows us to receive all the trajectories of movement of anyone in the scene and identify one person from another, the problem surfaces if there are multiple cameras. If the same person moves across a shopping mall and, for example, takes off his jacket in-between cameras, he will not be recognized. Different poses, outfits, backpacks, and other details can mess up our AI model and recognize the same person as two different ones. Generally, building a person re-identification system requires five main steps (video data collection, bounding box generation, training data annotation, model training and evaluation, and pedestrian retrieval).

Re-ID with Deep Learning Methods—Many of the recent models use deep learning (DL) models to extract features and achieve good performance. Convolutional neural networks (CNNs) due to their automatic feature extraction and engineering ability are being proposed by AI researchers.

Thus, with a series of noteworthy advancements in the DL space, person re-identification has achieved amazing performance. Especially this is to solve the problem of pedestrian retrieval across multiple surveillance cameras. In recent years, video-based re-identification has made great advances, because video sequences provide visual and temporal information that can be obtained using tracking algorithms. However, correctly annotating large number of visual data is a time-consuming activity. This is the reason why unsupervised video re-identification is being insisted. As we all know, in the case of supervised learning, each data point has to be labelled (each datapoint gets annotated). An intuitive idea for unsupervised learning is to estimate re-identification labels as accurately as possible. The estimated labels are subsequently used for feature learning to train robust re-ID models.

Re-identification(reID) is the process of associating images or videos of the same person taken from different angles and cameras. The key to the issue is to find features that represent a person. There are research works and practical implementations (https://arxiv.org/pdf/1807.11042.pdf) to build accurate CNN models. Here are a set of best practices to build and release highly optimized and generalizable re-identification model.

- **Using batch normalization after the global pooling layer**—It is important to avoid overfitting during training. Therefore, the recommendation is to go for batch normalization. Herein, the task is to normalize the output of each neuron using some mean and variance. Due to this, certain features are getting generalized during training. Such generalization allows us to apply the same model on different datasets.
- **Use one fully-connected layer for identity categorization**—In a CNN, there generally are two fully-connected layers. The first one plays the role of a "bottleneck" to reduce feature dimensions. The second layer performs identity categorization. The authors suggest removing the "bottleneck" and use the second layer directly.
- **Use Adam for CCN optimization**—The **Adam optimization algorithm** is an extension to stochastic gradient descent (SGD) that has recently seen broader adoption for deep learning applications in computer vision and natural language processing. Compared to the most popular used SGD, Adam works on lower-order moments, which allows us to smooth the variation between gradients.

By using re-identification and tracking models, it is possible to follow the path that a person is taking and make sure nothing illegal or inappropriate is done. Additionally, vehicles and other objects could be tracked. In this way, the road situation can be analyzed and further improved. If the power of edge devices is used wisely and in a timely manner, crimes and other illegal actions can be prevented and the offenders could be easily tracked. Edge AI will be a game-changing technology as the application space for such tracking feature is expanding fast.

7. Edge AI for next-generation retail experiences

Brick-and-mortar retailing is facing tough competition from B2C e-commerce due to the pandemic situation prevailing across the globe. As the world is through the third wave, online e-commerce activity is gaining momentum. How to empower physical stores to compete with virtual stores is the challenge. Smart Retail technologies (https://hailo.ai/industry/smart-retail/) is helping physical stores to compete against online ones through a series of technological innovations and disruptions. Digitalizing everything, establishing deeper and extreme connectivity, and deploying AI for intelligent automation empowers customers to get extraordinary experience, which is easily comparable with digital stores. A number of herewith unheard facilities and features are being added to physical stores to attract shoppers and to explore fresh avenues to bring in additional revenue.

The shoppers can enjoy an improved in-store experience. Mobile robots assist customers in different ways and means. Automated checkout and personalized offers are being realized. All the features of online stores are being provided to entice buyers. Plus, the shoppers can touch and feel the products in person. For the retailers, intelligent automation and in-store real-time analytics could translate into operational optimization and cost savings. AI is extensively used to increase customer traffic. Inventory and replenishment management get automated. Actionable insights for succulently increasing profits, and scores of means for optimizing store layout, shelves and displays, etc. Further on, embarking on personalized promotions and nurturing personal and professional relationship with customers, etc. gain momentum.

Computer and machine vision functionalities come handy here. IoT devices, sensors and systems are enabled to have the much-needed vision power. Such vision-enabled systems and environments are capable of bringing forth a number of breakthrough facilities. The clear business benefits of AI-based data analytics have made retailers to keenly adopt the path-breaking AI technologies and tools. The business and technology benefits of edge AI technology are being articulated and accentuated across. AI-based video processing on clusters of servers in the cloud with high latency is performed at high cost. Also, the electricity consumption and heavy heat dissipation into our fragile environment are on the higher side. Considering the various limitations widely discussed, the attention gets turned toward running AI frameworks on multiple yet integrated devices locally. That is, video data streams are run in IoT edge devices and servers in order to facilitate real-time data capture, storage, and processing in order to extract insights in time.

Keeping things local and under the retailer's direct control is exciting shoppers to gain real-time personal experience. Through the breakthrough power of edge AI, the retail field is experiencing a renaissance of edge-native, people-centric, and intelligent device services. There are a number of enabling technologies such as cloud-native computing, serverless computing, DevOps, and reliability engineering. Cloud-native applications are being realized through microservices architecture (MSA), containerization-enablement and container orchestration platform solutions. Especially the Kubernetes platform emerges as the most sought-after platform for running containerized applications across different and distributed cloud environments that include on-demand, online, off-premise/on-premise clouds. With the lighter versions of Kubernetes, it is possible to create and manage edge clouds, which are formed out of edge devices (heterogeneous). Thus, edge device clouds hosting AI libraries collect and crunch device and environment data in order to produce real-time services.

Intelligent video analytics turn data into significant insights that translate into better performance for retail operations. There are powerful and parallel processor architectures and intelligent camera solutions to do deep learning tasks such as face recognition, image segmentation, object detection and tracking, and classification for a host of applications such as

- **Customer flow analysis**—This is for people counting and to gain a deeper understanding of customers visits and their preferences.
- **Queue detection**—This is for reducing waiting time and to empower service staff with all the information to do their tasks efficiently.
- **Smart advertising**—This is for enabling classification and association tasks to guarantee targeted and personalized product promotion.
- **Inventory and Replenishment management**—AI-inspired actionable insights for visual merchandising, shelf management and stock management.
- **Security & privacy**—This is for ensuring the tightest security of people and store assets. Access control and blockchain database for guaranteeing unbreakable and impenetrable security and safety for stocks, bundles, packets, etc. for

There are several innovations happening concurrently. Powerful process architectures including the classical CPUs, GPUs, TPUs, VPUs, etc. Breakthrough processors for edge devices have laid down a stimulating and scintillating foundation for sophisticated retail applications and services to attract customers to physical stores. The personal, social and industrial use cases of edge AI are steadily growing and drawing the attention of business executives as well as IT professionals.

8. Edge AI for smarter cities

The implications of edge AI are greatly diversifying. With the accumulation of multifaceted edge devices in our everyday environments such as homes, hotels, hospitals, etc., the computing is all set to expand beyond its jurisdiction. That is, computing moves from centralized cloud servers to decentralized edge devices. Besides proximate data processing, a number of other benefits are being accrued through such a strategically sound transition. AI-instigated data processing is happening in edge devices in order to build and deliver edge-native applications in time. The long-drawn automation and augmentation tasks are being neatly accelerated through the smart leverage of proven and potential data analytics platforms, products and processes.

As we all know, there is a growing population of people migrating to cities in search of several things. Therefore, our cities are getting thickly populated and therefore there are a number of challenges and concerns widely expressed. Resource utilization has to be efficiently done. More facilities have to be created with advanced social, cyber, healthcare, and connectivity infrastructures. Garbage, drainable, water and energy management activities have to be advanced in order to meet up the growing population. Cities have to be made livable, lively and lovely. All kinds of citizen-centric services have to be made available with all the care and clarity. Safety and security, mobility and transportation, air quality, noise pollution, employment opportunities, green spaces, etc. are all needed to sufficiently enhance the quality of living. With the surge of edge AI-enablement methods, intelligent and intimate services are being developed and delivered instantaneously. Concierge applications will abound in the days to come. City process automation will pick up and gain speed with the technological advancements.

With digitizing and edge technologies in plenty in and around us, everything becomes digitized. Devices are getting connected with one another and with digitized entities. All kinds of ground-level disposables, implantable, wearables, handhelds, portables, etc. are being digitized and integrated with remotely held business workloads, IT services and databases. Especially software applications deployed on on-demand, online, and off-premise/on-premise IT environments are empowering and energizing physical objects. Thus, digitized elements, connected devices, and cloud-hosted applications combine well to automate most of the manual activities in setting up and sustaining smart cities. With edge AI revolution, intelligent devices become the new normal accelerating the fulfilment of smart cities. Sensors fuse with other sensors and actuators in the vicinity and with remote entities through one or other networking options to create and sustain next-generation capabilities. With real-time AI processing, an arsenal of fresh services and applications will be unearthed and delivered to city residents.

Emergency services will get the technological boost. Incident or accident response times will be very minimal. Inland security will be sufficiently strengthened through the new breed of surveillance cameras, which are intrinsically enabled with advanced deep learning capabilities. Thus, security cameras can collect a lot of data and analyses it quickly and automatically to emit out actionable insights in time. Edge devices are being empowered to form dynamic, ad hoc, purpose-specific, and efficient clusters in order to tackle complex problems. Especially data analytics consume a lot of computation and storage. Through the quick formation of edge device clouds, complicated edge data analytics get performed in order to simplify and speed up knowledge discovery. As discussed elsewhere, edge devices fitted with computing, communication, data storage and analytics capabilities are laying down a strong and sustainable foundation for envisaging sophisticated applications. Edge analytics complements cloud-based data analytics. The speed, scope, and sagacity of future applications will be clearly understood by people.

The city traffic management will be digitally transformed. All kinds of traffic crises can be predictably and pre-emptively handled to smoothen the vehicle movement across city roads. IoT edge devices instrumented with AI capability are to contribute immensely for intelligent traffic management. With edge AI, self-driving cars are to hit the road soon. The noteworthy point here is that the technologies and tools are emerging for simplifying the transition of data to information and to knowledge.

Road signage is an essential tool for managing traffic in cities. Such an arrangement alerts drivers to hazards and blockages and offers alternate routes and updates on delays. Surveillance cameras can analyses road signage to enhance its accuracy. Edge analytics comes handy in automating several aspects associated with road traffic. Poorly managed traffic is disastrous for the fragile environment. Air quality is deteriorating, noise pollution is dangerous for city dwellers, precious time gets wasted due to the delays and traffic snarls, etc. With the random deployment of sensors along the road is helping out in solving the traffic problem. So, data collection and processing in real time using multiple and heterogeneous sensors and actuators go a long way in succulently surmounting the city problems. With edge AI, the people health and wellbeing are bound to go up significantly.

Citizen safety and security—Fire alarms, gas, pressure, temperature, and humidity sensors and the faster spread of CCTV cameras across city areas in conjunction with cloud-based city monitoring, measurement and management systems are promising to enhance the security and safety of people and properties. AI-inspired video surveillance and analytics play a vital role in appropriately enforcing law and order. The solidity of edge analytics and increasingly sophisticated sensors are delivering more value to those who are focused on keeping a city safe and secure.

Acoustic sensors can sense and send out alerts to road users in time. This is in relation to the specific noises such as vehicle accidents, the sudden application of car breaks, gunshots, glass breaking, etc. Edge analytics can monitor a video stream and spot anomalies, unusual patterns, specific objects, and suspicious behavior. The edge processing can then raise an alert or notification to the concerned in time in order to avoid any kind of untoward incidents. Public address systems can be integrated with sensors, cameras and other devices to broadcast any kind of emergency situations such as the intrusion of criminals into the city areas.

Edge analytics can enable a smart city municipality to better manage and conserve precious resources like energy, water and fresh air. Analytics on top of IoT sensors in water systems and waste management systems will enable better monitoring and management. Innovative electric grids will increase energy efficiency for businesses and consumers alike. Edge analytics will also help in the monitoring and controlling of building operations like HVAC, lighting, and security to enable best possible living environment to the building occupants.

Traffic flow, parking space availability, utility usage and public streetlight management can be monitored by using IoT sensors on 5G network. Authorities can leverage edge analytics to find practical solutions to conserve energy, optimize water and power resources, and reduce environmental impact. Gradually, minimum traffic congestions and improved waste management will entice new residents and hence increase economic opportunities within the community.

Edge analytics and Edge AI enable advanced and secure video, sensor and communication systems to proactively monitor public spaces and law & order. Sensors embedded in critical infrastructure such as bridges and power plants can monitor structural data to identify potential dangers, protecting citizens and assets. Sensor-equipped drones can monitor vehicular traffic, crowds, construction sites and disaster areas to help monitor conditions continuously and support first responders.

Edge analytics will be the key enabler for the autonomous vehicles revolution using the Internet of Vehicles. Vehicles on the road will communicate with each other and with road infrastructures, improving overall road safety. It will also lead to reduced traffic congestions and enhanced driver comfort. Cars enabled with cloud connectivity can get a number of newer features. Edge AI-enabled cars can exhibit several real-time capabilities. Inertial and environmental sensors in smartwatches and fitness bands with deep learning (DL) capability can respond to local events. Wearable devices collect a lot of data on human activity, location, body parameters, etc. and alert their wearers. Video analytics within automobiles help in detecting and alerting distracted and drowsy drivers through the eye position and the state of the eye.

9. Edge AI for telecommunication

The Internet of Things (IoT) has matured fast to support and become an inseparable part of the telecommunication industry. Self-driving vehicles and the industrial IoT applications are the most prominent ones for substantiating the beneficial fusion between the IoT paradigm and the telecom domain. The telecom industry, in fact, has adopted the IoT idea strongly in order to be right and relevant to their customers and consumers. The IoT has laid down a stimulating foundation for enabling machines and devices to interact with one another to bring forth next-generation services to the humankind. Automated vehicles, augmented and virtual reality (AR/VR) and industry 4.0 applications will see the grand reality with the aspect of edge intelligence gaining solid ground. The nexus of IoT devices, 5G communication and AI frameworks can lead to hitherto unheard applications for industry verticals and common people. On-device intelligence will become the new normal. Here are some global IoT trends that are going to influence the telecom sector.

Telecommunication unifies Big and Streaming Data Analytics— As indicated above, the IoT paradigm creates a massive amount of multi-structured data. Smart phones and connected devices generate a lot of data. There are big and streaming data. Further on, there are big data analytics (BDA) platforms in plenty from the open-source community. There are streaming data analytics and streaming databases to emit out. Cloud service providers are setting up cloud environments across the world and communication service providers are communication networks for integrating ground-level IoT devices and sensors with cloud-based analytics applications. Thus, streaming data emanating from IoT devices gets integrated cloud-based historical data to enable comprehensive data analytics to bring forth actionable insights for short-term as well as long-term decisions.

The arrival of 5G communication standard has clearly enhanced the power and the value of IoT devices and data. Advanced 5G technologies (CAT-M1 & NB-IoT) facilitate the integration of IoT devices and sensors with faraway as well as nearby cloud platforms for data storage and analytics. Long-Term Evolution for Machines (LTE-M), an LTE upgrade supports low complex CAT-M devices and ensures improved battery life. Narrowband IoT (NB-IoT) is a Machine-To-Machine focused Radio Access Technology (RAT) that enables huge IoT rollouts. It also ensures last-mile connectivity using its extensive-range communication at low power. The use cases of these 5G technologies are exploding. For example, NB-IoT is used for monitoring street lights, for waste management, and remote parking. NB-IoT seamlessly tracks all kinds of pollution including water, land, and air for up keeping the environment's health. NB-IoT scrutinizes alarm systems, air conditioning, and complete ventilation system.

Telecom equipment may get affected due to extreme-level weather, fire, or any other cause (natural or man-made). The physical security of telecom assets is very important for ensuring the continuous delivery of the communication facility. The cyber security implications have to be also calculated. With IoT sensors getting attached on these mission-critical artifacts, there is a huge scope for solid improvement. The blooming blockchain technology is also promising toward attenuating cyberattacks. By capturing all kinds of the IoT data and subjecting them locally and remotely, it is possible to extract value-added insights for preventive and predictive maintenance of telecom equipment. By installing smart cameras in and around the tower area and other locations housing telecom equipment, the physical security of telecom assets can be fully ensured. Edge AI contributes immensely for anything that requirement real-time decision-making and action.

Considering the huge impact of the path-breaking AI paradigm, lightweight AI frameworks and libraries are plentiful these days. Such a transition simplifies the process of installing AI software into all sorts of IoT edge devices and sensors. There are highly miniaturized and hugely improved chipsets to specifically enable native AI processing. All these noteworthy advancements ultimately result in real-time edge device data analytics. TinyML is emerging as a popular subject of study and research. Edge AI is all about creating, evaluating and optimizing AI models for a variety of systems and applications. Currently, modelling happens in cloud environments as it requires a lot of computational resources. The tested and refined model is then taken to IoT edge device to do the inferencing on fresh data. Creating and sustaining highly accurate prediction models is the most common aspect of AI. Any kind of anomalies/outliers in telecom equipment and networks can be proactively pinpointed to ensure their continuous functioning. By attaching appropriate IoT sensors, telecom assets can be remotely monitored, measured and managed. Remote access, theft exposure, and fraud examination can be enabled through AI-attached IoT systems.

New-generation gadgets, drones, digital assistants, robots etc. are adequately enabled through IoT sensors in order to be communicative and cognitive. Telecommunication plays a very vital role in shaping all kinds of devices, appliances, equipment, machineries, etc. Device integration (local and remote) is being empowered through connectivity and communication capabilities. AI-powered digital transformation is being speeded up through ambient, unified, and adaptive communication.

The telco industry is heavily using AI for a variety of improvements and improvisations. Monetizing the edge capability, saving costs through hyper automation and improving customer experience and engagement. AI has become an integral part in conceiving and concretizing breakthrough telecom services. AI contributes in providing cognition-enabled connectivity services.

1. **Monetizing at the edge**—Having understood the immense potential of edge computing, communication service providers are keen on exploring and exploiting this unique phenomenon for their benefits. IoT devices are emerging as the next-generation input/output (I/O) device. Delivering a spectrum of low-latency telecom services in time is enabled through a host of IoT devices. With the impending arrival of 5G communication, IoT edge devices can deliver their services in greater speed. For example, in the healthcare domain, with a mesh of IoT devices and sensors, patients' physical parameters can be collected during the ambulance ride and transmitted to the doctors' devices so that quick remedial actions can be contemplated and applied once patients reach the hospital premises. Speed is the differentiator. Local analytics to envisage correct and context-aware actions is being facilitated through AI running on edge devices.

Further on, first responders can use intelligent drones to rapidly and remotely assess emergency situations with real-time streaming video and audio even before they arrive at the place. ATM machines can pre-emptively alert banks if there is a fraudulent attempt. Thus, edge use cases are growing fast.

2. **Saving costs and boosting efficiencies**—There are a number of technological evolutions and revolutions such as containerization, container orchestration platform solutions such as Kubernetes, resilient microservices communication frameworks such as Istio, DevOps toolsets, etc. All kinds of log, operational, performance, security, health condition data of business workloads and IT services are meticulously collected and analyzed deeply and automatically through the smart leverage of delectable AI advancements. AIOps is a new field of study and research and for cleverly using AI for automating and accelerating information and communication technologies (ICT) operations. Such a technological empowerment brings in the much-needed productivity, affordability, speed, and sagacity for service providers as well as consumers. There is a bright scope for the new idea of NoOps to flourish in the days to come. That is, network infrastructure modules can be minutely monitored and if there is any deviation from the prescribed limit, then a proper notification will reach the concerned so as to plunge into appropriate counter measures in time.

Network virtualization is enabling worldwide network service providers remarkably. There is a consistent transition from physical network functions (PNFs) to virtual network functions (VNFs) and cloud-native network functions (CNFs). A number of advantages such as network flexibility, affordability, availability, reliability, scalability, and maneuverability are being accrued through such evolutions.

3. **Improving client engagement**—Customer-centricity and propensity become the new normal. Customer experiences and engagements go up sharply. The long-standing goals of customer delight and ecstasy are bound to see the grand reality with the tight integration with AI. Edge computing delivers real-time experience for users. AI-enabled chatbots can answer complicated questions correctly. Network disturbances can be proactively understood to avoid any hiccups. AI systems can process large amounts of data, in particular call detail records (CDR), in the case of the telecommunication industry, identify patterns, detect and predict network anomalies.

Virtual radio access network (vRAN)—As noted above, the concept of virtualization (divide and conquer) is acquiring greater significance these days. It has permeated into everything significant. Like the virtualization of network functions has enabled telecommunication networks to be modernized, the radio access networks (RANs) is also virtualized in order to avail significant credits and profits. Through open and disaggregated RANs, telecom companies can simplify network operations and improve flexibility, availability, and efficiency while serving a large number of end-user devices and bandwidth-hungry applications. Cloud-native and open RAN solutions often lower costs, improve ease of upgrade and modification, and scale horizontally. As per the research reports, 6G communication will heavily rely upon the unprecedented successes of AI to be hugely distinguished in their services.

Edge analytics and Edge AI add the data analytics capability to edge devices. Edge analytics use cases are fast growing: cognitive devices, smarter homes and cities, autonomous cars and industry automation.

10. Conclusion

Edge AI is all about bringing AI-inspired processing of edge device data in order to emit out actionable insights out of data heaps in real time. The extracted intelligence can be used for producing real-time and intelligent edge-native applications. Further on, edge devices are slated to become cognitive in their operations, offerings and outputs through the tight integration with AI. With intelligent edge devices abound in a system and environment, the era of firming up intelligent environments such as digitally transformed homes, offices, manufacturing floors, retail stores, railway stations, eating joints, entertainment plazas, shopping malls, etc.

Further reading

[1] Edge Analytics in 2022: What it is, Why it matters & Use Cases. https://research.aimultiple.com/edge-analytics/.
[2] Edge Analytics – The Pros and Cons of Immediate, Local Insight. https://www.talend.com/resources/edge-analytics-pros-cons-immediate-local-insight/.
[3] What Is Edge AI and How Does It Work? https://blogs.nvidia.com/blog/2022/02/17/what-is-edge-ai/.
[4] Edge AI – Driving Next-Gen AI Applications. https://viso.ai/edge-ai/edge-ai-applications-and-trends/.
[5] Edge-native applications: What are they and where are they used? https://www.edgeir.com/edge-native-applications-what-are-they-and-where-are-they-used-20210912.
[6] Edge-native development best practices. https://www.ibm.com/docs/en/eam/4.0?topic=reading-edge-native-development-best-practices.
[7] The role of cloud in edge-native applications. https://www.f5.com/company/blog/the-role-of-cloud-in-edge-native-applications.

About the authors

Pethuru Raj working as a chief architect at Reliance Jio Platforms Ltd. (JPL) Bangalore. Previously. worked in IBM global Cloud center of Excellence (CoE), Wipro consulting services (WCS), and Robert Bosch Corporate Research (CR). In total, I have gained more than 20 years of IT industry experience and 8 years of research experience. Finished the CSIR-sponsored Ph.D. degree at Anna University, Chennai and continued with the UGC-sponsored post-doctoral research in the Department of Computer Science and Automation, Indian Institute of Science (IISc), Bangalore. Thereafter, I was granted a couple of international research fellowships (JSPS and JST) to work as a research scientist for 3.5 years in two leading Japanese universities. Focuses on some of the emerging technologies such as the Internet of Things (IoT), Optimization of Artificial Intelligence (AI) Models, Big, fast and streaming Analytics, Blockchain, Digital Twins, Cloud-native computing, Edge and Serverless computing, Reliability engineering, Microservices architecture (MSA), Event-driven architecture (EDA), 5G, etc. My personal web site is at https://sweetypeterdarren. wixsite.com/pethuru-raj-books/my-books https://scholar.google.co.in/ citations?user=yaDflpYAAAAJ&hl=en.

Dr. Jenn-Wei Lin is currently a full professor in the Department of Computer Science and Information Engineering, Fu Jen Catholic University, Taiwan. He received the M.S. degree in computer and information science from National Chiao Tung University, Hsinchu, Taiwan, in 1993, and the Ph.D. degree in electrical engineering from National Taiwan University, Taipei, Taiwan, in 1999. He was a researcher at Chunghwa Telecom Co., Ltd., Taoyuan, Taiwan from 1993 to 2001. His current research interests are cloud computing, mobile computing and networks, distributed systems, and fault-tolerant computing.

CHAPTER TWO

Edge computing: Types and attributes

Sunku Ranganath
Intel Corporation, Hillsboro, OR, United States

Contents

Advances in Computers, Volume 127
ISSN 0065-2458
https://doi.org/10.1016/bs.adcom.2022.03.001

Abstract

The chapter on "Edge computing types and Attributes" provides introduction to various types of Edge computing by broadly classifying Edge in to four types, based on round trip latency requirements, (1) IoT Edge (2) Wireless Access Edge (3) On Premise Edge (4) Network Edge. Requirements and attributes of each of these edge compute types are discussed. Chapter then details the practical challenges across these Edge deployments and further explores how ETSI MEC specifications helps address these challenges.

1. Introduction

The paradigm of edge computing enables connectivity between plethora of devices and core of the telecom data center in a manner that provides low latency communication and enables multiple computing services to be brought closer to the consumers. This requires edge compute architecture to be highly distributed and customized to be suitable and adaptable for various device needs. Use cases such as immersive media, Augmented Reality (AR) and Virtual Reality (VR), etc., require low latency and high bandwidth connectivity without loss of frame data. Industrial applications such as autonomous robots, machine to machine interaction, etc., require low latency, real time computation and connectivity leveraging Time Synchronous Networks (TSN). Vehicular communication such as autonomous vehicles, UAV connectivity, drone communication, etc., require low latency & distributed compute power spread across multiple geographic points of interest. Use cases such as smart retail kiosks with facial recognition or smart cameras with scene intelligence or smart cities that provide ubiquitous connectivity, etc., require heavily distributed compute power that can simultaneously interact with multiple categories of devices and ability to process huge volumes of data in short time while protecting privacy and identity of the end user. Use cases such as smart hospitals, remote healthcare, intelligent medical devices, etc., require private wireless deployments with low latency communication while protecting patient confidentiality and adhering to federal laws.

To facilitate various use cases, some of which are mentioned above, the edge computing infrastructure cannot be a monolithic architecture with set of compute servers as in a datacenter model but needs to be disaggregated into multiple layers of access points based on some of these aspects below:

- physical location
- security and privacy requirements
- round trip latency
- connectivity management
- scale requirements
- on premises requirements
- data locality
- real time communication requirements
- proximity to the end user
- service capabilities and many more.

The dissection of edge computing based on said factors heavily impacts the architecture, design, and deployment considerations for practical application purposes. Its highly imperative that the end user considers these requirements first before upgrading their existing infrastructure.

One easy way to understand various types of edge compute is based on its proximity to the end device and round-trip latency to core of the datacenter, as indicated in Fig. 1 [1]. To broadly understand types of edge compute, consider the division in to following categories:

- Internet of Things edge
- On premise edge
- Access edge
- Network edge

2. Internet of Things (IoT) edge

This section of edge compute broadly categorizes plethora of devices that are interconnected via common network infrastructure (either public or private) that enables these devices to be smart enough to communicate with each other and possess possibility of enough intelligence to make autonomous decisions for its operations. Broad set of devices with connectivity fall under this category, such as, intelligent and/or autonomous vehicles, mobile communication devices, autonomous robots, factory equipment, retail kiosks, smart cameras, connected city devices (such as intelligent streetlamps, street sensors, smart parking meters, etc.), smart buildings, transportation and logistics devices, intelligent cameras, smart energy devices, intelligent health care equipment and so on, as shown in Fig. 2 [2]. This type of edge is usually called "Far edge" from the perspective of core of the telecom network.

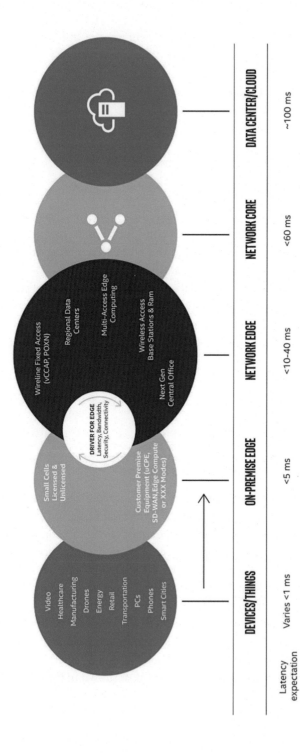

Fig. 1 Types of edge compute and corresponding latency expectation.

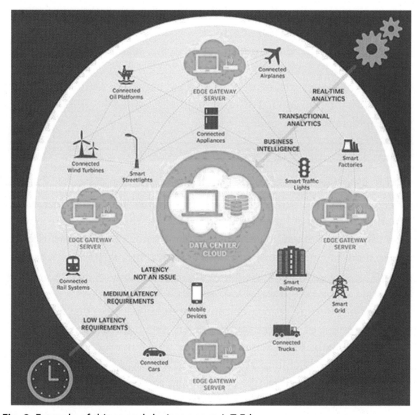

Fig. 2 Example of things and devices across IoT Edge.

One of the primary requirements for segregating multiple devices in to IoT Edge is its latency requirements. The typical round-trip latency for this type of edge computing is less than 1 millisecond (ms). Primary advantages of establishing IoT Edge:

- Accelerate development of innovative services that can interact and create value from data using millions of devices at the edge
- Provide capability to test, build, run and scale AI/ML models for various use cases
- Capability to operate locally to adhere to various city/state/country wide regulations as the data does not have to cross the pre-determined boundaries
- Option to utilize innovative hardware infrastructure such as GPUs, Infrastructure Processing Units (IPUs), Data Processing Units (DPUs), etc.
- Enables new service revenue models across industries as it brings together Operational Technology (OT) and Information and Communication Technology (ICT) companies

- Ability to remotely monitor and manage multiple edge devices through an intelligent and automated way
- Provides ability to root cause and contain any major security or privacy threats across the devices

Depending on the nature of the nature of the usage and nature of the business domains they are applied towards, following aspects become key part of their architecture and design:

2.1 Device type and mobility

The IoT Edge vastly extends to many of the day-to-day devices that human beings use or interact with. With the shift towards automating majority of the mundane tasks that humans perform, much of these devices are enabled with various levels of intelligence to augment them with enough capabilities to make them independent. While devices such as mobile phones, facial recognition devices, etc., are well known, the roll out of 5G is enabling multiple new classes of devices across various sectors.

Autonomous vehicles are an example of mobile devices that require tremendous amount of data transactions every second and very low latency requirements with V2X communications while managing the mobility handoffs across geographical regions. Smart city type deployments utilize multiple intelligent devices, like cameras that can detect riots on the streets or smart lamp posts that provide ubiquitous connectivity, etc., which are installed in a single location and don't have low latency mobility requirements and require aggregation points like gateways so that the data can be analyzed and acted upon. The architecture for edge deployments highly depends on the interaction among these devices and to the gateway node or edge server and communication protocols used between these entities.

2.2 Service capabilities

With the advent of 5G, devices today have ability to utilize frequency spectrum that provide high bandwidth and can utilize Ultra Reliable Low Latency Communication (URLLC). Not only that the devices can have intelligent capabilities, but new and innovative service models could be enabled Over The Top (OTT) on these devices. For example, new experiences such as immersive media, Artificial Reality (AR) and Virtual Reality (VR) could be enabled as services on mobile devices or intelligent vehicles or virtual reality goggles, etc. without constraints of limited set of geographic boundaries. Things and devices could be broadly classified based on the amount of service capabilities that they can enable across multiple business domains of industrial, retail, autonomous transport, hospitality, entertainment, health care, etc.

2.3 Security and privacy

Privacy becomes the key tenant in enabling services on the various devices and things across IoT edge. Due to scale of number of devices that are connected in thousands to millions at any given edge location, ability to protect identity and user information of each individual device across multiple classes of devices and the services that run on them is crucial. Security in terms of authorization and authentication across each of the layers of services consumed and services offered need to be baked into each of the devices during the lifetime of their connected existence. Edge deployments need to consider the scale at which the security and privacy needs to be enabled across the devices.

2.4 Latency requirements

The nature of applications and service models at the IoT edge generally is to aggregate and process the data from multiple devices and things and provide actionable insights using which necessary actions could be triggered. The nature of these data points generated could be once a second to minutes or hours or days. The crucial aspect of IoT edge is to bring compute closer to these devices and perform necessary analytics to provide insights in real time. Leveraging gateway-based aggregation points, efficient data transport protocols such as MQ Telemetry Transport (MQTT) and 5G front haul, latency requirement of 1 ms round trip latency could be enabled.

3. On-premises edge

Consider an industrial manufacturing floor that has mix or autonomous and pre-programmed robots along with various sensors on an assembly line that generate thousands of data points every second which needs to be aggregated, stored, processed to detect any anomalies or failures or to provide actionable insights to ensure smooth operations. Transporting the data across public networks or using centralized cloud data center to process the data would completely defeat the requirement of providing low latency and fast processing of data there by losing the value of the data that was generated. On-premises edge-based architectures help enable these type of use cases by having compute resources on customer premises and/or at various points of presence at customer site. Latency considerations for On-Premise Edge is generally targeted to be less than 5 ms.

Businesses such as large enterprises, industrial manufacturing floors, large retail operations, etc., benefit from these types of deployments to have ability to process the data close to point of its origin while still having proprietary rights of owning the necessary hardware. The business models could be that the on-prem equipment is operated by service provider or utilize a licensing model to utilize the necessary virtualized workloads or have necessary resources to life cycle manage the hardware and software stack. This type of deployments is usually popular for central office or branch office or point of presence type scenarios. Some of the important architectural considerations for deploying and utilizing an on-premises edge are share below. Primary advantages of On-premise edge:

- Reducing capital expenditure by reducing the need to buy purpose-built, vendor specific hardware
- Reducing operational expenditure through decreased running costs (less space needed to house all the equipment, less power needed to run them etc.)
- Saving time on long procurement processes with many different vendors
- Lowering the risk of rolling out new services by allowing providers to trial and roll back services, as the customer needs them
- Negate the need for engineer site visits through Zero Touch Provisioning
- Leverage universal set of APIs across multiple types of Edge clouds and Core Network Cloud for customers to develop value add applications on top
- Ability to utilize public cloud or hybrid cloud infrastructure and utilize any of the "as-a-Service" cloud models

3.1 Disaggregation

On-premise edge plays an important role in the distributed computing paradigm bridging the device ecosystem to the network edge and core data center in large scale enterprise/retail/industrial scenarios. An important associated with this type of deployment is that the sensitive and proprietary data is contained and secured on-premises while still can leverage connectivity and scalability of cloud computing. Traditionally network functions are run in a virtualized environment which are moving toward cloud native DevOps model of managing the on-prem network functions. Universal Customer Premise Equipment (uCPE) is a prime example of on-premise edge compute that provide combination of firewall, WAN optimizer, router within a single equipment that is deployed on-premise, as shown in Fig. 3 [3]. Due to advances in Network Function Virtualization (NFV) and ability for a service provider to remotely provision, update, upgrade and life cycle manage these network functions, on-premise edge is an enticing appeal.

3.2 Network requirements

Due to the nature of on-premise edge being the main interface between IoT edge and core of the datacenter, it needs the ability to interface with the IoT devices wirelessly and the ability to aggregate and transport the data securely to core the data center. The wireless connectivity requirements could span from leveraging private 5G wireless network or utilizing Citizens Broadband Radio Service (CBRS) spectrum or traditional Wi-Fi mesh network to connect to the corresponding set of devices, sensors, things, etc. as the data is aggregated, its packaged to be send to backend public or private cloud in most cases or utilize network edge infrastructure which in turn hauls the data to cloud environment. Software Defined—Wide Area Network (SD-WAN) is a prime example of software defined management of WAN services that allows enterprises to utilize combination of MPLS or LTE or broadband internet services. Figure 4 [4] provides high level connected architecture of SD-WAN with its open standard APIs providing capability to do zero-touch service lifecycle management.

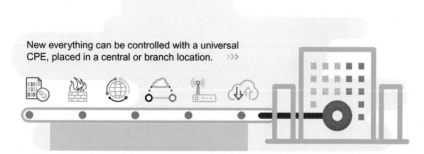

Fig. 3 Consolidation of multiple network functions on a uCPE.

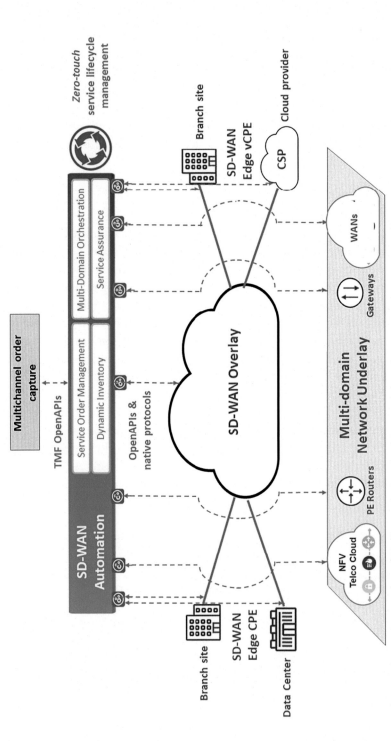

Fig. 4 High level architecture of SD-WAN with its Open Interfaces.

3.3 Scalability

The primary purpose of on-premise edge is to consolidate multiple network functions into limited set of hardware and install them at specific set of points of presence. The demand for data processing has to be gauged ahead of deploying these at appropriate locations in the overall network. While Scaling the network services on demand within on-prem edge is a challenge, much of the heavy lifting would be done using the connected cloud environment to handle unexpected spike in the end user activity.

3.4 Security

The ability to remotely deploy and manage network functions on the on-prem equipment opens the doorway for multiple security-based network functions to be configured and managed at run time. Network functions such as virtual Next Generation Firewall (vNGFW), virtual Deep Packet Inspection (vDPI), Denial of Service Attack prevention mechanisms, policy based secure access, etc., could all be delivered from cloud to the on-prem equipment on-demand basis to secure the traffic per use case basis. A combination of hardware-based security features, security network functions and communication security are necessary to secure overall deployment. Secure Access Service Edge (SASE) is a brand-new paradigm that is evolving the security at the edge to move towards Zero Trust based security solutions. SASE is the combination of network security functions, such as SWG, CASB, FWaaS and ZTNA, with SD-WAN capabilities to support the dynamic secure access needs of organizations. These capabilities are delivered using as-a-Service model based on the identity of the entity, real time context and security 5and compliance policies.

3.5 Life cycle management

One of the primary advantages of utilizing on-prem edge is its ease of managing the life cycle of network functions either in cloud native microservice based deployments or as virtualized deployment, by the service provider. The end user could offload all the necessary tasks of upgrading, updating, configuring, bootstrapping the infrastructure to the service provider and all of these can be done remotely once the equipment is installed. This appeal of providing low latency data management on-premises and ability to constantly update the network functions to the latest set of requirements, has increased the adoption of on-premise edge computing.

4. Wireless Access Edge

Using the advances in NFV, the once traditional Radio Access Network (RAN) that was available as a fixed function device has now been disaggregated to run as set of virtual functions in software. This reduces the cost by avoiding proprietary hardware and utilize Commercial Off The Shelf (COTS) servers. RAN is the crucial point that connects wireless devices to the core the operator's network. Concepts like virtual RAN (vRAN) and industry initiatives such as Open-RAN and ORAN have enabled interfaces to manage these virtualized deployments similar to managing any other edge device. Move towards cloud native instantiation of RAN function is easing the life cycle management of broad set of RAN deployments using Continuous Integration (CI)/Continuous Deployment (CD) constructs leveraging DevOps models. The set of infrastructures necessary to deploy and manage software RAN functions could be broadly termed as Access Edge. Some of the architectural considerations for designing and deploying Access Edge are as below.

4.1 Scalability

One of the most important benefit of dividing the RAN functions into individual software components is the reduction in costs when operating RAN deployments at scale. Few important aspects that help manage infrastructure using NFV for RAN:

- Ability to manage RAN infrastructure in an agile manner like infrastructure-as-a-service or platform-as-a-service models
- Utilization of less hardware as these network functions could be service chained and co-located in a performance and energy efficient manner
- Provides ability to instantiate, move or bring down workloads with minimal effort using suitable orchestrator
- Provides ability to elastically scale the necessary resources to address changing network demands

4.2 Distributed RAN functions

3GPP specifications for 5G have enabled RAN into a functional split as Radio Unit (RU), Centralized Units (CU) and Distributed Units (DU). CU takes of backhaul network connectivity to the 5G core network and connects to DU as midhaul, while DU connects to RU and takes care of fronthaul network. This has enabled network operators to deploy these units in a customized fashion to satisfy scale requirements, real-time performance management, optimize QoS for variety of service needs (for example, gaming, voice, video, etc.) and ease the operations & life cycle management. Operators can now satisfy variety of latency tolerance levels and customized the transport for different deployment scenarios like rural or urban that have different transport access like fiber.

ORAN consortium has laid out a standardize architecture that divides the RAN into following sections with relevant components:
- Radio side that includes Near-Real Time RAN Intelligent Controller (RIC), O-RAN Central Unit, O-RAN Distributed Unit and O-RAN Radio Unit
- Management side that includes service management and orchestration framework that contains Non-Real Time RIC

These entities are further interconnected using set of interfaces such as O1 or E2 or F1, etc.

4.3 Service models

The disaggregated RAN model has now enabled various service models that invites operators, service providers and third-party vendors to enable new service models providing value add on top of RAN deployments. O-RAN introduces concept of xApps that can run on Near Real-Time RIC utilizing the xApp APIs. Some of the very important functionality set could now be provided using xApps such as radio connection management, mobility management, QoS management, conflict mitigation among xApps, subscription management, security functions, user policy management, etc. This creates an ecosystem of vendors that can just provide unique differentiating xApps or combination of xApps and RIC functions or CU/DU to the operator. Testing the interoperability between these different vendors supplied components would also be a major service revenue model that can be utilized.

Another aspect of Near Real-Time RIC is the shared data layer with connectivity to appropriate database provides software vendors provide custom database software to satisfy huge set of data volume requirements with real time nature of the RIC. With move towards cloud native deployments, Non-Real Time RIC could now be implemented using industry's defacto choice of orchestrators such as Kubernetes or OpenStack enabling new service models to onboard, orchestrate, secure and life cycle manage the functions spanning Non-RT RIC and near-RT RIC. Figure 5 [5] provides a reference of how various O-RAN components can be deployed using Kubernetes based install.

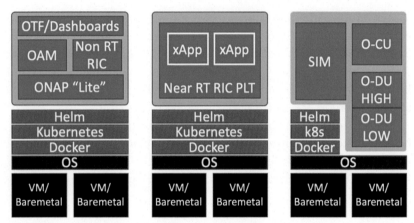

Fig. 5 Kubernetes based deployment of O-RAN components.

4.4 Intelligent RAN

One of the important tenets of moving to disaggregated RAN [R] is to imbibe the aspect of "intelligence" which is inherently provided by means of Artificial Intelligence (AI) algorithms. O-RAN approach not only opens interfaces between the distributed units but allows to reach management interfaces such as Radio Resource Management (RRM) and Self-Organizing Networks (SON) functions that can control radio resources and network operations. This enables service providers to implement AI models for radio network automation customized to various use cases across small cells and macro cells. Use cases such as intelligent Traffic Steering, QoE optimization, Massive MIMO optimization and QoS based resource optimization are considered for Phase 1 development to address immediate needs of the operators. Phase 2 of the enablement of intelligent RAN are focused on RAN sharing, RAN slice SLA Assurance, context based dynamic handover management for V2X, flight path based dynamic UAV resource allocation, and radio resource allocation for UAV applications.

5. Network edge

As the compute power is brought closer to the end users of devices and things, data needs to be aggregated from across multiple IoT edges, On-premises edge, and Access edge catering to the specific region before connecting to the centralized data center that can span across vast set of regions. This type of deployments could be broadly called "Network Edge" or near edge in relation to the core data center of the service provider.

The set of computes across network edge is usually called Edge data centers. These are deployed usually as nano Data Center (nano-DC) or micro–Data Center (micro-DC) consisting of few to many server racks. Key design and architectural considerations for managing these customized data centers and running services include, but not limited to:

- ability to support fixed function and dynamic functions that spawn over edge and cloud
- ability to utilize resources distributed across different pools in edge and cloud
- ability to onboard, execute and life cycle manage "Edge Native Services" (ENS) which are usually deployed in form of microservices that can adhere and adapt to constraints of edge compute
- ability to adhere to network requirements that are access agnostic and edge aware
- provide resilient fault tolerant platforms
- provide an easy service delivery vehicle using cloud native constructs such as distributed service mesh, containers, etc.

Edge data centers could be further classified based on deployment types to cater to needs of type of end user such as enterprise deployments, telecom providers, city, and public usage models, etc. The deployment types at network edge could be broadly classified as below.

5.1 Next Generation Central Office

The Central Office (CO) is usually the final node in the network managed by a carrier, or it can be a headend or hub from a cable company perspective or can be a set of baseband unit pools from a wireless carrier perspective. Since COs are where all the traffic is aggregated, it is usually the contention point and can be bottleneck that affects latency and throughput of corresponding services. Leveraging NFV, virtualization of CO addresses vendor lock-in issues and disaggregates traffic enabling value add services and controls costs. To effectively leverage the benefits from ability to deploy as network functions, COs can now expand towards utilizing edge data centers for network management, business & operation support systems (BSS & OSS) and analytics. Reference architectures such as Central Office Rearchitected as Datacenter (CORD) provides an integrated platform to create an operational edge data center with in-built service capabilities. Figure 6 [6] provides an example of a NGCO mini data center by Intel.

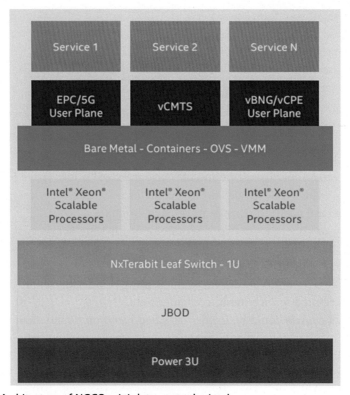

Fig. 6 Architecture of NGCO mini data center by Intel.

5.2 Wireline fixed access edge

The traditional wireline broadband that is governed by standards from Broadband Forum is undergoing industry shift towards convergence with rest of the 5G infrastructure by utilizing COTS servers and leveraging NFV. The wireline broadband edge broadly consists of networks that deliver broadband services to the subscribers such as IPTV, VoIP, internet services, etc. It is at the edge of the broadband network that Border Network Gateways (BNGs) are used to perform subscriber management such as session and circuit aggregation, Authentication Authorization & Accounting (AAA) management, policy and traffic management functions, multiplexing and demultiplexing of traffic to and from individual subscriber, etc. Multiple edge network models are available based on distributed nature of BNG.

Virtualizing cable headend and Cable Modem Termination System (CMTS) functions by dividing it into Control-Plane and User-Plane functions provides an easy way to migrate to edge cloud and deliver services from a data center enabling flexible, efficient, resilient, and scalable means of operating broadband architecture. Further architectural changes across wireline fixed access edge infrastructure towards cloud native include virtual Converged Cable Access Platform (vCCAP). vCCAP provides a shift towards Distributed Access Architecture (DAA) that combines headend functions of CMTS and Edge Quadrature Amplitude Modulation (EQAM) and implements these in a virtualized or cloud native microservice based infrastructure models, thereby increasing the capacity and throughput. vCCAP deployment models could vary across centralized and distributed model based on space availability, bandwidth consumption and server usage efficiency. The move towards cloud native Kubernetes based deployment models has further made it simpler to leverage unified set of life cycle management architectures.

5.3 Physical locations

The edge data centers are usually located at the bottom of a cell tower, telecom central offices, headend facilities of a cable company deployments, parking garages, building rooftops, etc. These are spread across closer to the end consumer of edge services and sized up based on the volume of services and traffic requirements.

6. Challenges in edge computing

As the 5G and IoT infrastructure are transforming toward unified set of architectures, they deliver brand new set of experiences and use cases using unprecedented compute power and distribute the intelligence across various devices. Across all the types of edge compute that have been shared earlier, one common architectural and deployment pattern is the move towards cloud native microservice based models. Traditionally limited set of hardware vendors and service providers, controlled end to end deployment of communication infrastructure. With the transformation towards distributed architecture and 5G advances multiple new business models and service delivery models are being enabled across different types of edges. Products and services from multiple infrastructure and software vendors could need to be working together providing a cohesive interoperability between various network functions and seamless scale up and scale out from across edge to cloud infrastructure. This brings forth various challenges to deploy, scale and manage the edge computing paradigm. Some of the challenges in utilizing edge computing are described below.

1. Software Infrastructure

   Transformation towards virtualized and cloud native models has essentially converted edge infrastructure to as-a-service deployment model using industry standard cloud orchestrators such as Kubernetes. Proprietary network functions now have a necessity to evolve towards microservice based architecture for service-oriented deployment models. These network functions need to be onboarded, tested for interoperability and scale and life cycle managed just like any other software application while maintaining the Service Level Agreement (SLA).

2. Unified Manageability Across Edges

   As we noted earlier that edge computing could be divided into multiple segments based on traffic type, applications being serviced, device connectivity and point of presence, there is an explosion of number of edge computing zones that span across geographical areas. Each of the edge computing zone that addresses specific area of geographical region or set of network bandwidth (that correlates to servicing specific set of devices) needs an interoperable mechanism with other edge computing zones to provide seamless connectivity. This arises the need for unified orchestration and life cycle management across these multiple clusters and cloud regions.

3. Public and Private Cloud

   The utilization of cloud native technologies across multiple edge scenarios essentially puts the service provider to enable private cloud clusters across these edges. However, hyperscale cloud service providers such as the likes of Microsoft Azure or Amazon Web Services, etc., provide the opportunity of hyperscale economics leveraging public cloud constructs. Infrastructure could now be scaled in an intelligent and cost-efficient manner while leveraging unified Application Programmable Interface (APIs) across the infrastructure provided by public cloud provider. Hybrid approach of leveraging public cloud for agile and flexible service offerings while being vendor agnostic and utilizing private cloud to utilize already existing in-built infrastructure is still an area that needs to be proven out in terms of efficiency and scale.

4. Security and Privacy

   Due to diverse nature of various types of edge deployments, the needs of a secure edge have evolved into multi-faceted set of approaches that needs to be customized per edge deployment. There is no one size fits all policy that can satisfy requirements across the edge types. Ability to provide Authentication Authorization and Accounting (AAA) across the distributed edge requires multiple levels policies to ensure every end user is accounted for. Privacy of an end user or end device is another critical aspect to maintain as the traffic flows across the edge. The security architecture needs to abstract out end user/device information as the data travels to the core of the network. Zero trust security architecture is one the latest paradigm with the belief that no aspect of data communication is secure as there are no trusted personas while providing secure abstractions starting from hardware root of trust, service to service communication, data security and end to end control of data access across the network. Implementing these with real time low latency requirements continue to prove to be a major challenge.

5. Hardware Abstraction and Utilization

Edge infrastructure's increased utilization of COTS servers has revolutionized each of the network function across the various edge types. However, this means abstracting out all the hardware features, accelerators, and any other enhancements available for virtualized network functions or cloud native microservices. The aspect of utilizing the underlying hardware features is left to cloud orchestrators that expose and maps the features to the network functions at the deployment time. With the advent of various as-a-service hardware models such as Graphical Processor Unit (GPU) as a service or Infrastructure Processor Unit (IPU) or in general x Processor Unit (xPU) as a service, additional intelligence needs to be enabled for latency sensitive network functions to fully utilize various hardware features with the ability to scale across the edge deployments.

6. Value of Data

Distributed and disaggregated edge computing paradigm brings out huge amounts of data across the end-to-end edge infrastructure. IoT edge, Access edge, On-premises edge, Wireless edge, Fixed Access edge all have connectivity to end users that provide gold mine of data to be analyzed to provide value added OTT type services customized for the end user. The challenge however is that the value of data decreases as the latency increases farther from the origin point. Thus, an efficient and low latency data processing and analytics mechanisms are needed at the distributed edges that are closer to the end users.

7. AI & ML Models for Edge

To derive the value of the data generated across the various edges, huge amounts of data points associated with a single end user or a single device need to be processed and analyzed at a constant time interval (for example every second in an Industrial automation use case). This calls for customized ML and AI models that need to be tailored for each of the edge type and management system that can apply appropriate model for the necessary use case. There is a huge scope for innovation and development in this space for various use cases across the edges.

8. Life cycle management

Software based network functions are deployed across edge infrastructure necessitates the operator to deal with packaging, onboarding, deploying the network functions; storing, updating, testing for interoperability, ensuring high availability with zero downtime, error free upgrading on the fly, chaos testing, demand-based scaling, supplying adequate infrastructure resources, detecting anomalies at run time and so on, which all constitutes the aspects of life cycle management. Due to the nature of complexity at each of the edge type, move towards microservice based containerized deployments instead of virtual machines has significant benefits over managing virtual machines. Some of the following practices are required to be customized towards each type of the edge computing:

○ DevOps practices: to deploy software across 100s to millions of devices automation is the key. DevOps provides methodologies for CI/CD necessary to test, install, upgrade, and manage at scale.

○ DevSecOps: Security must play an important role in the development and operations lifecycle. Automating security gates in the DevOps cycle ensure secure development and onboarding of the applications.

9. Operational Telemetry

Telemetry consists of various set of metrics from applications and infrastructure that can be exported to a database for them to be analyzed and derive meaningful insights. With the distributed nature of edge computing, the traditional telemetry collection and analysis models don't apply as they add to the latency in data collection and data processing in a centralized location. Metrics generation, telemetry storage and processing need to be distributed across the individual set of edge types before the data loses value because of latency. Combination of application telemetry and infrastructure/hardware telemetry enables new set of use cases such as closed loop automation, provide service assurance, QoS and QoE management, detect application and resource anomalies, etc. Analytics models are to be developed that can be customized and scale across different types of edge deployments.

10. Policy Management

Policies at the edge are set of rules and constraints that control over edge services deployment by different personas, such as administrators, service providers, application developers, service owners, operation personnel, etc. These policies help cloud based orchestrators understand the constraints for each of the application type such as hardware constraints, latency tolerance, application priority, run to completion models, security requirements, scale requirements and so on. Policies differ heavily based on type of edge. For example, policies at IoT edge need to consider low latency limitations while servicing large amount of user data in a resource constrained environment, while policies at Network edge need to ensure adequate resources are available for the heavy network functions and consider the ability to scale between the data center cloud and network edge. Ultimately a centralized policy manager is required that can interact with individual policy managers and enforcers at each of the edge types.

11. Network Automation

Managing and maintaining the distributed compute capacity and varying nature of network function requirements in edge environments requires 100s of operations at any given instant. This gets a lot more complex in a 5G based architecture where the intelligence in the network is widely distributed. Automation of network operations is a huge differentiating factor in owning, maintaining, and operating the network at scale. Zero-touch automation is an emerging area that aims to provide human free and interaction free automation of network issues leveraging AI by taking preemptive actions against a predetermined set of objectives.

Some of these universal challenges across the edge continuum necessitate a simplified architecture that interconnects the interfaces between IoT edge, Wireless Access edge, Fixed Access Edge, On-premises and Network Edge in a seamless manner for cloud native application functions to be operated across the infrastructure and in turn enable data movement across these networks to core of the network. One such architectural concept is Multi-Access Edge Computing.

7. Multi-Access Edge Computing

The concept of Multi-Access Edge Computing (MEC) moves disaggregates the centralized computing of traffic and services in a central cloud to the edge of the network and closer to the customer. As indicated in the ETSI's definition of MEC [7], it offers application developers and content providers cloud-computing capabilities and an Information Technology (IT) service environment at the edge of the network. This environment is characterized by ultra-low latency and high bandwidth as well as real-time access to radio network information that can be leveraged by applications. The MEC initiative is an Industry Specification Group (ISG) within ETSI whose is to create a standardized, open environment which will allow the efficient and seamless integration of applications from vendors, service providers, and third parties across multi-vendor Multi-access Edge Computing platforms. ETSI MEC provides set of standards, specifications and API definitions that help enable various edge service models across the different types of edge deployments. Figure provides high level framework for MEC. Figure 7 [7] provides high level framework for MEC as described in ETSI GS MEC 003 v2.2.1.

To simplify the nomenclature and most importantly indicate adherence of an edge deployment with MEC standards, the edge platform is usually referred to as a MEC platform. Features of a MEC platform include, but not limited to:

- The importance of MEC is highlighted by the fact that it unites IT and cloud computing capabilities with telecom networks across the types of edges described earlier.
- Enables new business models for applications and services to be hosted on top of mobile network elements such as RAN functions, gateway elements, customer premise equipment and so on.
- Creates set of unified APIs that application developers could leverage across both edge and cloud compute environment
- Provides mechanisms to reduce network congestion and accelerate network and application performance enabling low latency type services to IoT and mobile platforms
- Enables data processing to be done at the network edge closer to the user location there by reducing the burden on network and cloud resources
- Provides new types of data security constructs with ability to process sensitive user data locally adhering to government or policy requirements
- Provides cost-effective way to deploy and manage private wireless and wireline networks through an intelligent allocation of assets
- Real time analytics using heuristics or AI models with a lower latency turn around
- Ability to adhere to various compliance requirements in a distributed manner

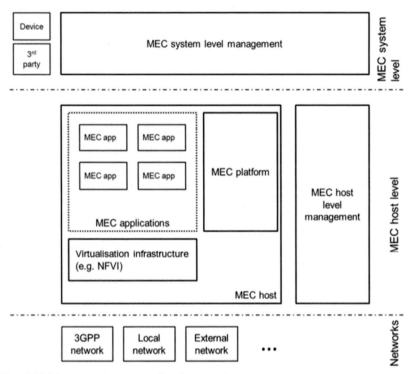

Fig. 7 Multi-access edge computing framework.

To provide standard set of terminologies, interoperable interfaces, and requirements around each of the MEC component, ETSI has published a reference architecture as indicated in Fig. 8 [7].

For application developers and service enablers, ETSI MEC has provided few considerations for application to be deployed at the edge of mobile networks. The nature of requirements varies based on service, placement, mobility requirements and so on and hence the requirements are classified under varying modes of operations. Applications at the edge need to be rearchitected to fit into edge requirements thereby making them Edge aware. Using the Domain Name System services, it allows several application requirements to be addressed causing few limitations in cases where support is required for high user mobility in conjunction with the need to frequently transfer application context between servers. In this case, there is typically a need to inform, through notifications, the application about a change of the server IP address. This calls for a change in the application

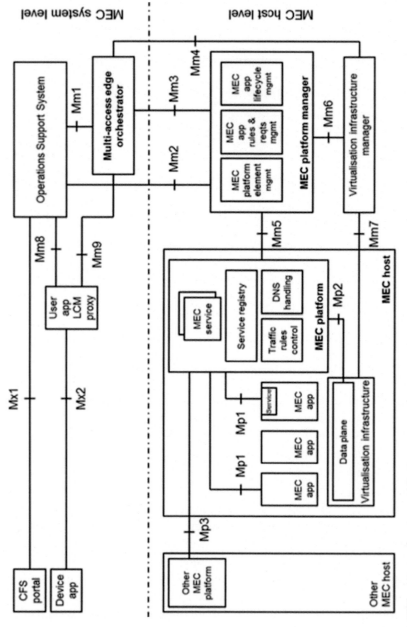

Fig. 8 MEC system reference architecture.

logic, therefore requiring applications to be Edge aware. Few aspects of edge-aware application enablement, as described in Harmonizing Standards for edge computing—A synergized architecture leveraging ETSI ISG MEC and 3GPP specifications [8]:

- Edge awareness by server applications:
  - ○ Applications can offer enhanced user experience by if network services such as locations, QoS or traffic influence could be used in conjunction with mobile networks
  - ○ Edge application may need to register and discover each other's services or have capability to access common set of services at the edge
  - ○ To realize low latency benefits, highly mobile devices need to be connected to most suitable edge cloud requiring edge cloud platform to assist in context migration to target edge cloud for stateful applications
- Discovery mechanism between application client and server application
  - ○ DNS based: An authoritative DNS can, through IP address resolution, perform optimal routing to an edge cloud. However, the DNS options have limitations for applications running on devices that are highly mobile or when the edge cloud is highly distributed. The enhancements to support mobility when DNS is used are still under study in 3GPP.
  - ○ Device based: a device hosted client may also be used to facilitate the discovery of server applications and the optimal edge cloud, termed an Edge Enabler Client (EEC) in 3GPP SA6 nomenclature. The same client can also assist service continuity because it can subscribe and receive information about mobility and possibly the decision to perform context migration during an application-level session. Edge aware application clients on the device can directly interact with this client to benefit from all these advantages.
- Mobility and context transfer
  - ○ Context migration can utilize network that provides underlying mobility support
  - ○ Over the top migration through communication between edge cloud applications is the usual model
  - ○ Examples include, but not limited to, network exposure of mobility events, network offered ability to influence traffic steering, etc.

The above said capabilities of a MEC platform are derived from set of specification published by ETSI MEC ISG. MEC ISG has produced over 30 specifications establishing and describing APIs across multiple edge to device connectivity and edge to network core scenarios and more. Some of the specifications published are listed as below:

- WLAN Access Information API
- Study on Inter-MEC systems and MEC-Cloud systems coordination
- General principles, patterns, and common aspects of MEC Service APIs
- UE Identity API
- API Conformance Test Specification; Part 1: Test Requirements and Implementation Conformance Statement (ICS)
- API Conformance Test Specification; Part 2: Test Purposes (TP)
- API Conformance Test Specification; Part 3: Abstract Test Suite (ATS)
- Framework and Reference Architecture
- Edge Platform Application Enablement
- Multi-access Edge Computing (MEC) MEC 5G Integration
- General principles, patterns, and common aspects of MEC Service APIs
- Traffic Management APIs
- WLAN Information API
- V2X Information Service API
- Device application interface
- Application Mobility Service API
- Radio Network Information API
- Support for network slicing
- Edge Platform Application Enablement
- MEC Management; Part 2: Application lifecycle, rules, and requirements management
- Study on MEC support for alternative virtualization technologies
- Location API
- Proof of Concept Framework
- Fixed Access Information API
- MEC Testing Framework
- UE application interface
- Support for regulatory requirements
- Framework and Reference Architecture
- General principles for MEC Service APIs

The viability of any new technology and its interfaces are put to test via building a Proof of Concepts (PoC). ETSI MEC ISG has laid out set of PoCs that provide a reference on possible use cases using MEC platforms.

The following PoCs are in various stages of development by its members using the PoC Framework laid out by the ISG:
- Video User Experience Optimization via MEC - A Service Aware RAN MEC PoC
- Edge Video Orchestration and Video Clip Replay via MEC
- Radio aware video optimization in a fully virtualized network
- FLIPS—Flexible IP-based Services
- Enterprise Services
- Healthcare—Dynamic Hospital User, IoT and Alert Status management
- Multi-Service MEC Platform for Advanced Service Delivery
- Video Analytics
- MEC platform to enable low-latency Industrial IoT
- Service Aware MEC Platform to enable Bandwidth Management of RAN
- Communication Traffic Management for V2X
- MEC enabled OTT business
- MEC infotainment for smart roads and city hot spots

References

[1] Intel Corp, A smarter network: Creating a Platform for Innovation with Edge Computing, intel.com.
[2] Martel Innovate, Together on the Edge, 2020. materl-innovate.com.
[3] Colt, What is Universal CPE, colt.net.
[4] Blueplanet, SD-WAN Automation, Blueplanet.com.
[5] L. Ji, Architecture, Integration and Testing, O-RAN, O-RAN wiki, 2020.
[6] R. Browne, P. Mannion, E. Walsh, Creating the Next Generation Central Office with Intel Architecture CPUs, whitepaper, intel.com.
[7] ETSI GS MEC 002 v2.2.1 (2020−12), Muti-access Edge Computing; Framework and Reference Architecture, ETSI ISG MEC, RGS/MEC-0003v221Arch, 2020.
[8] N. Sprecher, et al., Harmonizing standards for edge computing - A synergized architecture leveraging ETSI ISG MEC and 3GPP specifications, ETSI ISG MEC, 2020. ISBN: 979-10-92620-35-5.

About the author

Sunku Ranganath is a Solutions Architect for Edge Compute at Intel. For the last few years, his area of focus has been on enabling solutions for the Telecom domain, including designing, building, integrating, and benchmarking NFV based reference architectures using Kubernetes & OpenStack components. Sunku has been an active contributor to multiple open-source initiatives. He serves as a maintainer for CNCF Service Mesh Performance & CollectD Projects and participated on the Technical Steering Committee for OPNFV (now Anuket). He is an invited speaker to many industry events, authored multiple publications and contributed to IEEE Future Networks Edge Service Platform & ETSI ENI standards. He is a senior member of the IEEE.

CHAPTER THREE

Industry initiatives across edge computing

Sunku Ranganath
Intel Corporation, Hillsboro, OR, United States

Contents

Advances in Computers, Volume 127
ISSN 0065-2458
https://doi.org/10.1016/bs.adcom.2022.03.002

Abstract

To accelerate the evolution and adoption of Edge computing various standard bodies, open-source projects and industry consortia have come together in recent times to revolutionize Edge compute. This chapter goes through various initiatives across the world that have major traction in terms of collaboration, collateral produced and industry impact. Architecture, collateral produced and details of the projects involved are described.

As we explore various types of edges, their constraints, challenges and opportunities, various industry consortia have been formed to address some of the emerging challenges across these edge computing environments. This section describes these initiatives and the nature of work as it relates to type of edge computing detailed earlier, apart from the ETSI MEC ISG described earlier. There are multiple ongoing initiatives across edge ecosystem and new alliances are forming as we write this book. Some of the major projects that are creating impact across the industry are shared. They are at various stages of development in terms of maturity, from standardization, reference implementation and live trials. Together these standard bodies, open-source software and initiatives are shaping the future of 5G and Edge computing.

1. Linux Foundation Edge

Linux Foundation (LF) Edge has been formed to foster cross-industry collaboration across IoT, telecom, enterprise and cloud ecosystems and enable organizations to accelerate adoption and innovation for edge computing [1]. LF Edge provides broad umbrella of governance and organization for various Edge projects to uniquely stand out against other projects in the ecosystem while providing necessary framework to interoperate with overall edge ecosystem of projects. LF Edge divides edge compute types broadly under following types:

- Constrained Device Edge: the set of devices and things usually fall under Internet of Things category consisting of devices with very limited set of compute resources and power. Microcontroller based devices, embedded devices, low power sensors, etc., fall under this category.
- Smart Device Edge: the set of devices that are usually mobile and consists of relative intelligence and/or compute capacity to perform complex tasks. Smartphones, PCs, IoT gateways, etc., fall under this category
- On-Prem Data Center Edge: this edge is the connecting point between Device Edge and Access Edge and consists of micro-Data Centers or on-prem data centers.
- Access Edge: this edge consists of wireless and wirelines access networks, COs, aggregation hubs, etc., that have server-based compute capacity.
- Regional Edge: multiple Access Edges feed into Regional Edges and these edges have set of data centers that caters to a specific geographical region.

As shown in Fig. 1 [1], LF Edge hosts multiple projects at various stages of maturity and evolution, at the time of writing this book. Brief description of each of the projects is provided below.

1.1 Akraino

Considered to be a stage 3 impact project, it [2] provides open-source software stack that supports high-availability cloud services optimized for edge computing systems and applications. It provides autonomous turn-key solutions for service enablement, low latency placement, zero-touch provisioning, etc., in the form of blueprints and automated deployment stacks. Akraino is a set of open infrastructures and application blueprints for the Edge, spanning a broad variety of use cases, including 5G, AI, Edge IaaS/PaaS, IoT, for both provider and enterprise edge domains. These Blueprints have been created by the Akraino community and focus exclusively on the edge in all of its different forms. What unites all of these blueprints is that they have been tested by the community and are ready for adoption as-is or used as a starting point for customizing a new edge blueprint. Few of the blueprints and blueprint families provided by Akraino are:

- 5G MEC System Blueprint Family
- AI/ML and AR/VR applications at Edge

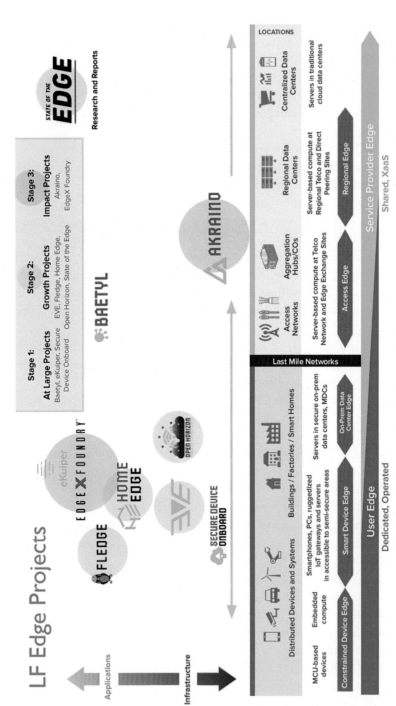

Fig. 1 LF Edge Projects categorized against type of Edge deployment.

- Automotive Area
- Edge Video Processing
- Integrated Cloud Native NFV/App stack family (Short term: ICN)
- Integrated Edge Cloud (IEC) Blueprint Family
- IoT Area
- KubeEdge Edge Service Blueprint
- Kubernetes-Native Infrastructure (KNI) Blueprint Family
- MicroMEC
- Network Cloud Blueprint Family
- Public Cloud Edge Interface (PCEI) Blueprint Family
- StarlingX Far Edge Distributed Cloud
- Tami COVID-19 Blueprint Family
- Telco Appliance Blueprint Family
- The AI Edge Blueprint Family
- Time-Critical Edge Compute

1.2 EdgeXFoundary

Also considered to be a stage 3 impact project, EdgeXFoundary [3] it provides flexible open-source software framework that facilitates interoperability between heterogenous devices and applications at the IoT Edge with consistent framework for security and manageability. It collects data from sensors and things at the Edge and acts as a dual transformation engine sending and receiving data to and from enterprise, cloud and on-premises applications. EdgeX enables autonomous operations and intelligence at the Edge as show in Fig. 2. Top use cases provided by EdgeXFoundary [3]:

- Manufacturing use case: Remote monitoring of production equipment, get data from multiple sources and filter/transform it to react at edge before sending to the cloud for aggregation, analysis and to optimize production and maintenance.
- Retail use case: the Open Retail Initiative (ORI) promotes the EdgeX framework in retail to ingest data from cameras (OpenVino), POS systems, RFID, etc., and use it at the edge for use cases like Loss Prevention and Inventory Management.
- Building Automation use case- Edge Control (control devices via a common API), use edge data to control building environment (HVAC, lighting, access). Connect to the cloud to optimize power consumption using ML.

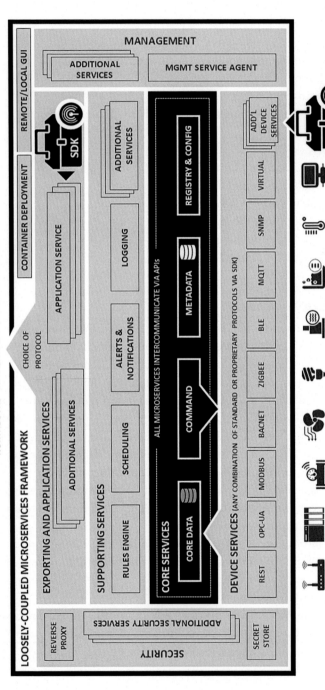

Fig. 2 Platform Architecture of EdgeXFoundary

1.3 EVE

Considered stage 2 growth project, EVE [4] provides an open abstraction engine that simplifies the development, orchestration and security of cloud-native applications on distributed edge hardware. Supporting containers, VMs and unikernels, EVE provides a flexible foundation for Industrial and Enterprise IoT edge deployments with choice of hardware, applications and clouds. Fig. 3 shows high level architecture of EVE. The goal of Project EVE is to enable IoT edge computing deployments with the following capabilities [4]:

- Access to hardware root of trust
- "Secure by default" deployment profile
- High efficiency and usage of device resources including remote control of CPU, memory, networking and edge device I/O ports
- Hosting of any combination of apps in virtual machines, containers and Kubernetes clusters
- Hosting of any guest operating system deployable in a virtual machine
- Ability to assign CPU cores and co-processing (e.g., GPU) to specific apps
- Ability to block unused I/O ports to prevent physical tampering
- Remote updates of entire software stack with rollback capability to prevent bricking
- Automated patching for security updates
- Automated connectivity to one or more backends (cloud or on premises)
- Distributed firewall to securely route data over networks per policy

1.4 Fledge

Considered stage 2 growth project, Fledge [4] provides an open-source framework and community for the Industrial Edge. Architected for rapid integration of any Industrial IoT (IIOT) devices, sensor or machine all using a common set of application, management, and security REST APIs with existing industrial "brown field" systems such as DCS (Distributed Control Systems), PLC (Program Logic Controllers) and SCADA (Supervisory Control and Data Acquisition) and the clouds. Fig. 4 provides high level overview of Fledge architecture. Some of the benefits of Fledge include [4]:

- Industrial Equipment Vendors—can build Your Next Generation Machines that can learn, maintain themselves, integrate with new cloud services and data systems

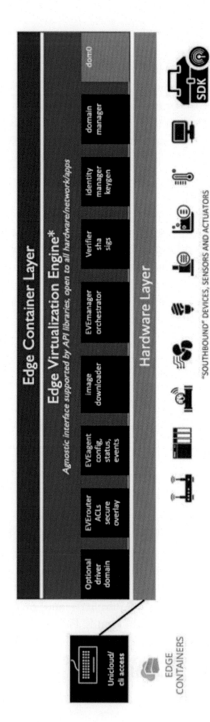

Fig. 3 Project EVE high level architecture.

FLEDGE Architecture

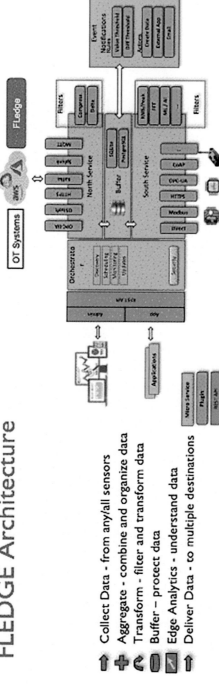

↑ Collect Data - from any/all sensors

✛ Aggregate - combine and organize data

✎ Transform - filter and transform data

▯ Buffer – protect data

◢ Edge Analytics - understand data

↑ Deliver Data - to multiple destinations

Fig. 4 High level architecture of Fledge.

– Industrial Operators can condition and perform predictive maintenance on all machines
– Industrial System Integrators can leverage a framework for over IIoT business

1.5 Home Edge

Considered stage 2 growth project, it [4] provides interoperable, flexible, and scalable edge computing services platform with a set of APIs that can also run with libraries and runtimes. It provides device and service management in the home network, service offloading, scoring manager as the major set of services. All the devices (TVs, fridges, washing machines, etc.) connected into a Home Edge Network are considered Home Edge devices and are assigned to Home Edge Nodes by the Home Edge Orchestrator. The Orchestrator continuous scans the home network looking for new devices and when it finds one, the device is assigned to a node or a new node is created by the Orchestrator. Fig. 5 shows high level view of Home Edge. Couple of top use cases of HomeEdge [4]:

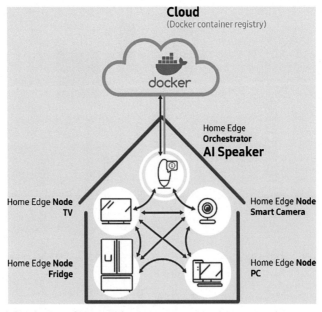

Fig. 5 High level view of Home Edge.

- Service offloading in a home environment when device doesn't have required capabilities
- Distributed computing framework to maintain low latency and high data privacy

1.6 Open Horizon

Considered stage 2 growth project, it [4] provides a platform for managing the service software lifecycle of containerized workloads and related machine learning assets. It enables management of applications deployed to distributed web scale fleets of edge computing nodes and devices without requiring on-premises administrators. Open Horizon simplifies the job of getting the right applications and machine learning onto the right compute devices, and keeps those applications running and updated. It also enables the autonomous management of more than 10,000 edge devices simultaneously—that's 20 times as many endpoints as in traditional solutions. Fig. 6 shows main components of Open Horizon. Key benefits of Open Horizon [4]:

- Add new capabilities to a single-purpose device
- Enable your device to use other services (both nearby and cloud-based) to enhance its existing capabilities
- Automate the hands-free management of workload lifecycle on the device
- Automatically deploy applications to all devices where policies match and an agreement is negotiated

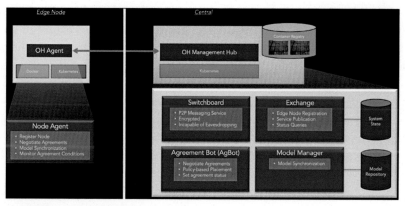

Fig. 6 Main components of Open Horizon.

1.7 State of the Edge

Considered stage 2 growth project, it [4]is an open-source research and publishing project with an explicit goal of producing original research on edge computing, without vendor bias. The State of the Edge seeks to accelerate the edge computing industry by developing free, shareable research that can be used by all. State of the Edge believes in four principles [4]:

– The edge is a location, not a thing
– There are lots of edges, but the edge we care about today is the edge of the last mile network
– This edge has two sides: an infrastructure edge and a device edge
– Compute will exist on both sides, working in coordination with the centralized cloud.

State of the Edge project produced following assets under the LF Edge umbrella:

– State of the Edge reports
– Open glossary of edge computing
– Edge computing landscape

1.8 Baetyl

Considered stage 1 at-large project, Baetyl [4] offers a general-purpose platform for edge computing that manipulates different types of hardware facilities and device capabilities into a standardized container runtime environment and API, enabling efficient management of application, service, and data flow through a remote console both on cloud and on prem. With modern container and serverless design concepts and engineering tools optimized for stand-alone and small multi-machines, Baetyl enables edge hardware and cloud native applications to work better and more efficiently together. It helps deliver stronger processing power to edge devices like smart home appliances, wearables and other IoT devices. Fig. 7 shows the components of Baetyl project and interaction between them.

   Baetyl is committed to help build an open-source framework for the edge that [4]:

– Abstracts different forms of hardware to a unified container environment, from IOT devices to distributed clusters, even embedded devices.
– Supports open application models, including plain Open Container Initiative (OCI) container and serverless modes such as FaaS and streaming.
– Provides a standardized remote management model with compatibility to k8s primitives.

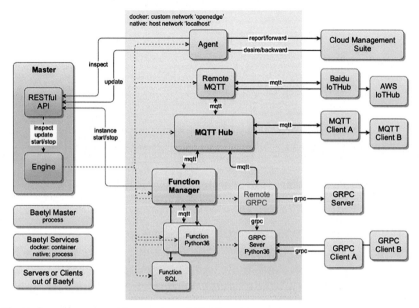

Fig. 7 Components and Interaction within Baetyl.

1.9 eKupier

Considered stage 1 at-large project, it [4] is an edge lightweight IoT data analytics/streaming software to run at all kinds of resource constrained edge devices. One goal of eKuiper is to migrate the cloud streaming software frameworks to edge side. eKuiper references these cloud streaming frameworks, and also considered special requirement of edge analytics, and introduces rule engine for developing streaming applications at edge side. It migrates cloud real-time cloud streaming analytics frameworks such as Apache Spark, Apache Storm and Apache Flink to the edge. eKuiper references these cloud streaming frameworks, and also considers any special requirements of edge analytics, to introduce rule engine, which is based on Source, SQL (business logic) and sink; rule engine is used for developing streaming applications at the edge. Fig. 8 shows interaction among project's components and their interfaces.

1.10 Secure Device Onboard

Considered stage 1 at-large project, Secure Device Onboard (SDO) [4] is an automated "Zero-Touch" onboarding service. To more securely and automatically onboard and provision an edge device, it only needs to be

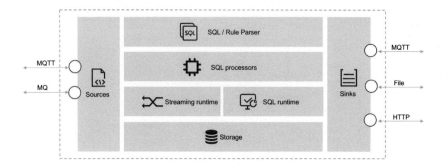

Fig. 8 eKupier Project Components and Interfaces.

drop-shipped to the point of installation, connected to the network and powered up. SDO does the rest. This zero-touch model simplifies the installer's role, reduces costs and eliminates poor security practices, such as shipping default passwords. Secure Device Onboard provides easier, faster, less expensive, and secure onboarding of devices. It expands TAM for IOT devices, and in turn accelerates the resulting ecosystem of data processing infrastructure. Most "Zero touch" automated onboarding solutions require the target platform to be decided at manufacturer. Fig. 9 [4] shows an example of provisioning with SDO. Key benefits of Secure Device OnBoard include [4]:

– Enables Build-to-Plan Model—ODMs can build identical IOT devices in high volume using a standardized manufacturing process. Reduces inventories, supply cycle times, and costs.
– SDO "Late Binding"—allows the device's target platform to be selected "late" in the supply chain, at first power-on.
– It's Open—means its service & cloud independent. Devices are bound to target the ecosystem at install. Works with existing cloud services, it does not replace them.

2. Linux Foundation for Networking

Linux Foundation for Networking (LFN) [5] aims to foster collaboration and innovation across the open networking stack from the data plane to the control plane enabling the necessary orchestration, automation, end-to-end testing and more. LFN is building the required infrastructure

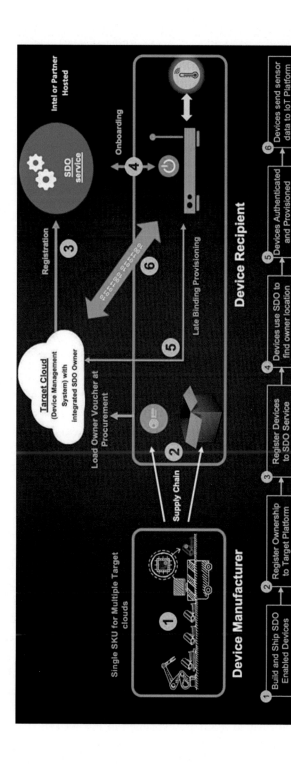

Fig. 9 Provisioning with SDO.

enablement and interoperability required for 5G leveraging hybrid approach of both VNFs and CNFs. LFN integrates with edge compute leveraging the following projects:

2.1 Open Network Automation Project (ONAP)

ONAP [6] is a comprehensive platform for orchestration, management, and automation of network and edge computing services for network operators, cloud providers, and enterprises. ONAP orchestrates edge cloud onboarding from distributed edge locations. Fig. 10 provides example of how ONAP helps orchestrate, analyze, and provide closed loop control in a centralized way across the multiple types of Edges and Core Cloud. Following enhancements are in various stages of planning and execution for ONAP releases to better address the needs of Edge compute:

- Optimization to address large number of edge-clouds
- Provide Mutual TLS with Edges, Secrets/keys protection (HSM/DHSM), Hardware rooted security, Verification of Edge stack (Attestation), and centralized security for FaaS
- Adhere to geographical regulations such as GDPR
- Enhancements to performance using containerized deployments
- ONAP to expose API for Application providers—to create MEC Service, instantiate MEC Service, provide MEC status, MEC analytics, etc.
- Ability for Zero Touch Provisioning (ZTP) systems to interact with ONAP to provide information about Edge inventory
- Aggregation of statistics & ML analytics for various edge deployments and provide infrastructure information to application providers.

2.2 Anuket

Anuket delivers a common model, standardized reference infrastructure specifications, and conformance and performance frameworks for virtualized and cloud native network functions. Anuket defines Edge Cloud Reference Architecture and provides a path to its realization using Reference Implementation. Anuket addresses a wide range of use cases from Core all the way to the Edge. Anuket artifacts include integrated, tested, and validated open software reference infrastructure used to design a conformance framework and verification program. The correlation between various industry standard groups and Anuket's Reference Models (RM), Reference Architecture (RA), Reference Implementation (RI) and Reference Conformance (RC) are explained in Fig. 11 [7].

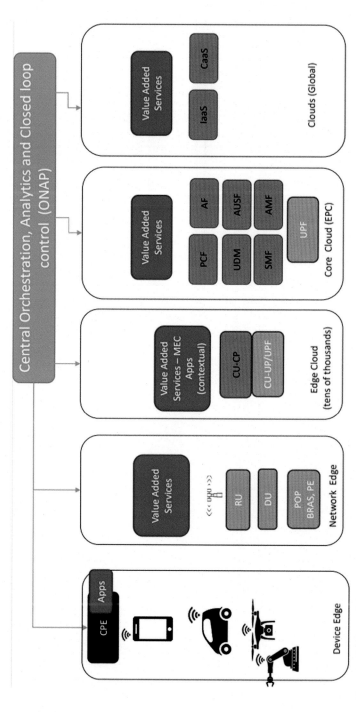

Fig. 10 ONAP enabling centralized control across Edge and Cloud.

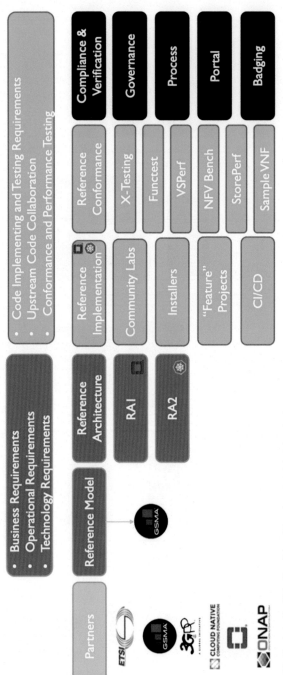

Fig. 11 End to End Technology Development by Anuket.

3. O-RAN alliance

O-RAN alliance [8] has been founded by operators across the world, AT&T, China Mobile, Deutsche Telekom, NTT DOCOMO and Orange and have become a world-wide community of mobile network operators, vendors, and research & academic institutions operating in the Radio Access Network (RAN) industry. O-RAN ALLIANCE's mission is to re-shape the RAN industry towards more intelligent, open, virtualized and fully interoperable mobile networks. To achieve this, O-RAN ALLIANCE is active in three main streams:

- The specification effort $\geq$ extending RAN standards towards openness and intelligence
- O-RAN Software Community $\geq$ development of open software for the RAN (in cooperation with the Linux Foundation)
- Testing and integration effort $\geq$ supporting O-RAN member companies in testing and integration of their O-RAN implementations

O-RAN focuses on technical aspects of the RAN and stays neutral in any political, governmental, or other areas of any country or region. O-RAN does its innovation and work through many of the technical working groups. The O-RAN specification work has been divided across technical workgroups, all of them under the supervision of the Technical Steering Committee. Each of the technical workgroups covers a part of the O-RAN Architecture. Brief description of technical workgroups is shared below. Each of the Working Groups have published a set of specifications that are listed below, at the time of writing this book.

- **WG1: Use Cases and Overall Architecture Workgroup**. It has overall responsibility for the O-RAN Architecture and Use Cases. Work Group 1 identifies tasks to be completed within the scope of the Architecture and Use Cases and assigns task group leads to drive these tasks to completion while working across other O-RAN work groups. Specifications produced by WG1:
  - o O-RAN Architecture Description 4.0
  - o O-RAN Operations and Maintenance Architecture 4.0
  - o O-RAN Operations and Maintenance Interface 4.0
  - o O-RAN Use Cases Detailed Specification 5.0
  - o O-RAN Use Cases Analysis Report 5.0
  - o O-RAN Slicing Architecture 4.0
  - o O-RAN Study on O-RAN Slicing 2.0
  - o O-RAN Information Model and Data Models Specification 1.0

- **WG2: The Non-real-time RAN Intelligent Controller and A1 Interface Workgroup**. The primary goal of Non-RT RIC is to support non-real-time intelligent radio resource management, higher layer procedure optimization, policy optimization in RAN, and providing AI/ML models to near-RT RIC. Specifications produced by WG2:
  - ○ O-RAN A1 interface
  - ○ O-RAN AI/ML Workflow Description and Requirements 01.02
  - ○ O-RAN A1 interface: General Aspects and Principles 2.02
  - ○ O-RAN A1 interface: Application Protocol 3.01
  - ○ O-RAN A1 interface: Transport Protocol 1.01
  - ○ O-RAN Non-RT RIC & A1 Interface: Use Cases and Requirements 3.00
  - ○ O-RAN Non-RT RIC: Functional Architecture 1.01
- **WG3: The Near-real-time RIC and E2 Interface Workgroup**. The focus of this workgroup is to define an architecture based on Near-Real-Time Radio Intelligent Controller (RIC), which enables near-real-time control and optimization of RAN elements and resources via fine-grained data collection and actions over E2 interface. Specifications produced by WG3:
  - ○ O-RAN Near-RT RAN Intelligent Controller Near-RT RIC Architecture 2.00
  - ○ O-RAN Near-Real-time RAN Intelligent Controller Architecture & E2 General Aspects and Principles 1.01
  - ○ O-RAN Near-Real-time RAN Intelligent Controller, E2 Application Protocol 1.01
  - ○ O-RAN Near-Real-time RAN Intelligent Controller E2 Service Model 1.0
  - ○ O-RAN Near-Real-time RAN Intelligent Controller E2 Service Model (E2SM), RAN Function Network Interface (NI) 1.0
  - ○ O-RAN Near-Real-time RAN Intelligent Controller E2 Service Model (E2SM) KPM 1.0
- **WG4: The Open Fronthaul Interfaces Workgroup**. The objective of this work is to deliver truly open fronthaul interfaces, in which multi-vendor DU-RRU interoperability can be realized. Specifications produced by WG4:
  - ○ O-RAN Open Fronthaul Conformance Test Specification 3.00
  - ○ O-RAN Fronthaul Cooperative Transport Interface Transport Control Plane Specification 2.0
  - ○ O-RAN Fronthaul Interoperability Test Specification (IOT) 4.0

- o O-RAN Fronthaul Control, User and Synchronization Plane Specification 6.0
- o O-RAN Management Plane Specification 6.0
- o O-RAN Management Plane Specification—YANG Models 6.0
- o O-RAN Cooperative Transport Interface Transport Management Plane Specification 2.0
- o O-RAN Cooperative Transport Interface Transport Management Plane YANG Models 2.0
- **WG5: The Open F1/W1/E1/X2/Xn Interface Workgroup.** The objective of this work is to provide fully operable multi-vendor profile specifications (which shall be compliant with 3GPP specification) for F1/W1/E1/X2/Xn interfaces and in some cases will propose 3GPP specification enhancements. Specifications produced by WG5:
- o O-RAN O1 Interface specification for O-DU 1.0
- o O-RAN O1 Interface specification for O-DU 1.0
- o O-RAN O1 Interface for O-DU 1.0 - configuration Tables
- o O-RAN Interoperability Test Specification (IOT) 02.00
- o O-RAN Transport Specification 1.0
- o O-RAN NR C-plane profile v4.0
- o O-RAN NR U-plane profile v4.0
- o O-RAN NR C-plane profile for EN-DC 2.0
- o O-RAN NR U-plane profile for EN-DC 2.0
- o O-RAN EN-DC C-Plane Table 1.0
- **WG6: The Cloudification and Orchestration Workgroup.** The cloudification and orchestration workgroup seeks to drive the decoupling of RAN software from the underlying hardware platforms and to produce technology and reference designs that would allow commodity hardware platforms to be leveraged for all parts of a RAN deployment including the CU and the DU. Specifications produced by WG6:
- o O-RAN Acceleration Abstraction Layer General Aspects and Principles 1.0
- o O-RAN Orchestration Use Cases and Requirements for O-RAN Virtualized RAN 2.0
- o O-RAN O2 General Aspects and Principles Specification 1.0
- o O-RAN Cloud Platform Reference Design 2.0
- o O-RAN Cloud Platform Reference Design for Deployment Scenario B 1.01
- o O-RAN Cloud Architecture and Deployment Scenarios for O-RAN Virtualized RAN 2.01

- **WG7: The White-box Hardware Workgroup**. The promotion of open reference design hardware is a potential way to reduce the cost of 5G deployment that will benefit both the operators and vendors. The objective of this working group is to specify and release a complete reference design to foster a decoupled software and hardware platform. Specifications produced by WG7:
  - O-RAN Hardware Reference Design Specification for Indoor Pico Cell with Fronthaul Split Option 6-1.0
  - O-RAN Hardware Reference Design Specification for Indoor Picocell FR1 with Split Architecture Option 7-2 2.0
  - O-RAN Hardware Reference Design Specification for Indoor Picocell FR1 with Split Architecture Option 8 2.0
  - O-RAN Indoor Picocell Hardware Architecture and Requirement (FR1 Only) Specification 1.0
  - O-RAN Deployment Scenarios and Base Station Classes for White Box Hardware 2.0
  - O-RAN Hardware Reference Design Specification for Outdoor Micro Cell with Split Architecture Option 7-2 1.0
  - O-RAN Outdoor Micro Cell Hardware Architecture and Requirements (FR1) Specification 1.0
  - O-RAN Hardware Reference Design Specification for Fronthaul Gateway 1.0
  - O-RAN Outdoor Macrocell Hardware Architecture and Requirements (FR1) Specification 1.0
- **WG8: Stack Reference Design Workgroup**. The aim of this workgroup is to develop the software architecture, design, and release plan for the O-RAN Central Unit (O-CU) and O-RAN Distributed Unit (O-DU) based on O-RAN and 3GPP specifications for the NR protocol stack. Specifications produced by WG8:
  - O-RAN Stack Interoperability Test Specification 1.0
  - O-RAN Base Station O-DU and O-CU Software Architecture and APIs 3.0
- **WG9: Open X-haul Transport Work Group**. This workgroup focuses on the transport domain, consisting of transport equipment, physical media and control/management protocols associated with the transport network. Specifications produced by WG9:
  - O-RAN Synchronization Architecture and Solution Specification 1.0
  - O-RAN Management interfaces for Transport Network Elements 1.0

- o O-RAN WDM-based Fronthaul Transport 1.0
- o O-RAN Xhaul Transport Requirements 1.0
- o O-RAN Xhaul Packet Switched Architectures and Solutions 1.0
- **WG10: OAM Work Group**. This workgroup is responsible for the OAM requirements, OAM architecture and the O1 interface.

Further, O-RAN alliance has focus groups that deal with topics that are over-arching the technical workgroups or are relevant for the whole organization.

- **SDFG: Standard Development Focus Group**. SDFG [8] plays the leading role on working out the standardization strategies of O-RAN ALLIANCE and is the main interface to other Standard Development Organizations (SDOs) that are relevant for O-RAN work, for which SDFG also coordinates incoming and outgoing Liaison Statements.
- **TIFG: Test & Integration Focus Group**. TIFG [9] defines O-RAN's overall approach for testing and integration, including coordination of test specifications across various WGs. This may include creating end-to-end test & integration specifications; profiles to facilitate O-RAN productization, operationalization and commercialization; approaches to meet general requirements; and specifications of processes for performing integration and solution verification. The TIFG plans and coordinates the O-RAN ALLIANCE PlugFests and sets guidelines for the 3rd party Open Test & Integration Centers (OTIC). Specifications produced by TIFG:
  - o O-RAN End-to-End Test Specification 1.0
  - o O-RAN Certification and Badging Processes and Procedures 1.0
  - o O-RAN End-to-End System Testing Framework Specification 1.0
  - o O-RAN Criteria and Guidelines of Open Testing and Integration Centre 2.0
- **OSFG: Open-Source Focus Group**. The biggest task that OSFG [9] has accomplished was the successful launch of the O-RAN Software Community. As most of open-source activities are happening directly in the O-RAN Software Community, the OSFG remains in a dormant mode.
- **SFG: Security Focus Group**. SFG [9] focuses on security aspects of the open RAN ecosystem. Specifications produced by SFG:
  - o O-RAN Security Threat Modeling and Remediation Analysis 1.0
  - o O-RAN Security Protocols Specifications 1.0

4. Open Network Foundation

The mission of Open Network Foundation (ONF) [10] is to transform access and edge networks by cross collaboration between academia and industry to build next generation mobile and broadband infrastructure. ONF provides solutions leveraging disaggregated and white box hardware, open-source software for SDN, NFV and cloud technologies. Fig. 12 [10] indicates current set of ONF Focus projects at the time of writing this book. A brief description of these projects is provided below. ONF also provides set of reference designs leveraging these projects to build a deployable platform.

4.1 VOLTHA

Virtual OLT Hardware Abstraction (VOLTHA) [10] is a project transforming broadband infrastructure by providing a common, vendor agnostic, GPON control and management system, for a set of white-box and vendor-specific PON hardware devices using open-source software. Highlights of VOLTHA, as indicated in Fig. 13 [10] below:

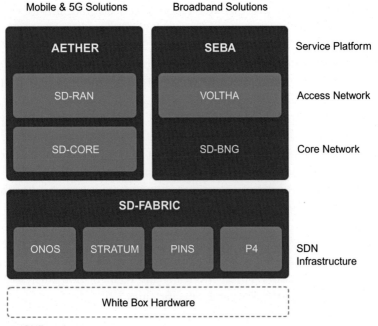

Fig. 12 ONF projects.

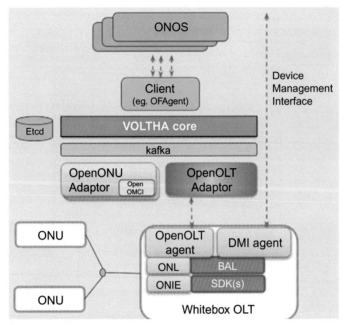

Fig. 13 VOLTHA Component Diagram.

- Makes an access network look like an abstract programmable switch
- Works with legacy as well as virtualized devices. Can run on the device, on general purpose servers, or in a virtualized cloud
- Provides unified, vendor/technology agnostic management interface
- DevOps bridge to modernization

4.2 SEBA

SEBA [10] a platform that utilizes open-source components to build a virtualized PON network to deliver residential broadband and mobile back-haul, as shown in Fig. 14. It supports a multitude of virtualized access technologies at the edge of the carrier network, including PON, G. Fast, and eventually DOCSIS and more. Highlights of SEBA:

- Kubernetes based
- High speed
- Operationalized with FCAPS and OSS Integration

4.3 Aether

Aether [10] is the first open source 5G Connected Edge platform for enabling enterprise digital transformation. It provides mobile connectivity

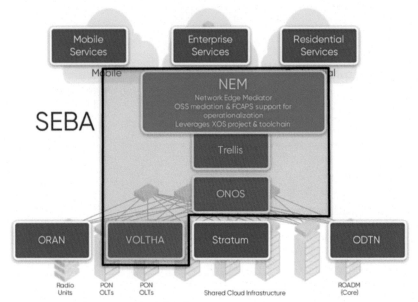

Fig. 14 SEBA Architecture—high level overview.

and edge cloud services for distributed enterprise networks as a cloud managed offering. Aether is an open-source platform optimized for multi-cloud deployments, and simultaneous support for wireless connectivity over licensed, unlicensed and lightly licensed (CBRS) spectrum [R]. As indicated in Fig. 15 [10], some of the highlights of Aether project include:

- Provides predictive end-to-end performance with reliability and security optimized for multi-edge sites
- Leverage CBRS spectrum and easy to deploy as WiFi
- Support AI/ML driven IoT, OT applications and mission critical edge applications
- Optimized end-to-end slicing with spectrum management, allocating dedicated slices for mission critical applications

4.4 SD-RAN™

It is a 3GPP compliant software-defined RAN that is consistent with the O-RAN architecture that is cloud-native and built on ONF's projects. SD-RAN™ [10] is developing a near-real-time RIC (nRT-RIC) and a set of exemplar xApps for controlling the RAN. This RIC is cloud-native and builds on several of ONF's well established platforms including the ONOS SDN Controller. The architecture for the SD-RAN nRT-RIC will leverage the O-RAN architecture and vision. Fig. 16 [10] provides

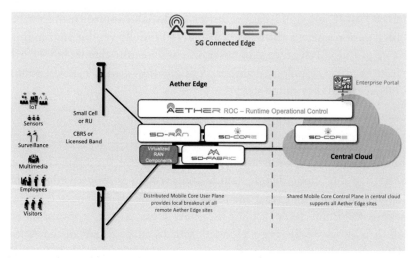

Fig. 15 Aether architecture for 5G Connected Edge.

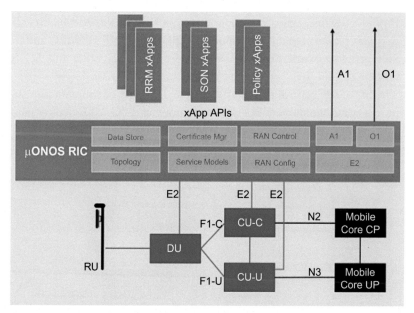

Fig. 16 Architecture of SD-RAN implementation with μONOS RIC.

reference of high-level view of SD-RAN architecture. Highlights of SD-RAN project includes, but not limited to:

– As the SD-RAN™ project creates new functionality, all extensions and learning that come from building the system will be contributed back to the O-RAN Alliance with the intent that these extensions can help advance the O-RAN specifications

- ONOS RIC is a cloud-native, carrier-grade SDN controller that enables high performance, availability, scalability in a multi-vendor environment
- The μONOS RIC uses a microservices architecture that includes certificate manager, topology manager, configuration manager, RAN control manager and distributed store

4.5 SD-CORE™

The SD-Core™ [10] project is a 4G/5G disaggregated mobile core optimized for public cloud deployment in concert with distributed edge clouds and is ideally suited for carrier and private enterprise 5G networks. It exposes standard 3GPP interfaces enabling use of SD-Core as a conventional mobile core. It is also available pre-integrated with an adapter (part of the Aether ROC subsystem) for those deploying it as a mobile core as-a-service solution. Fig. 17 [10] provides high level overview of SD-CORE architecture, some of the project's highlights include:

- Provides flexible, agile, scalable, and configurable dual mode 4G/5G core network platform that builds upon and enhances ONF's OMEC and free 5GC core network platforms to support LTE, 5G NSA and 5G SA services.
- The SD-Core control plane provides the flexibility of simultaneous supports for 5G standalone, 5G non-standalone and 4G/LTE deployments.

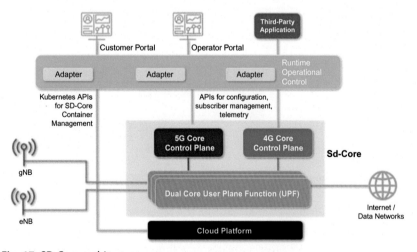

Fig. 17 SD-Core architecture.

- SD-Core provides a rich set of APIs to Runtime Operation Control (ROC). Operators can use these APIs to provision the subscribers in the mobile core; control runtime configuration of network functions; and provide telemetry data to third party applications. Third party applications can leverage telemetry data to create applications for closed loop control.

5. 3GPP

3rd Generation Partnership Project (3GPP) [11] unites seven telecommunications standard development organization known as "Organizational Partners" and provides their members with a stable environment to produce the Reports and Specifications that define 3GPP technologies. The project covers cellular telecommunications technologies, including radio access, core network and service capabilities, which provide a complete system description for mobile telecommunications. The 3GPP specifications also provide hooks for non-radio access to the core network and for interworking with non-3GPP networks. 3GPP specifications and studies are contribution-driven, by member companies, in Working Groups and at the Technical Specification Group level. The three Technical Specification Groups (TSG) in 3GPP are below and the product:
- Radio Access Networks (RAN),
- Services & Systems Aspects (SA),
- Core Network & Terminals (CT)

As a global initiative, 3GPP is well placed to build on its strong relationships and collaborations with ETSI MEC and GSMA. 3GPP Release 17 is foundational for edge computing but more will come in future releases given its importance in mobile communications and as we gradually move beyond 5G. Artificial Intelligence and edge computing can both serve as building blocks but in different ways:
- Network layer perspectives: AI can further optimize edge computing applications.
- Application layer perspectives: Edge computing can be a building block for AI, e.g., offloading limited capabilities from the device to the network.

SA2 currently responsible for [12] the 5G System and Evolved Packet System (EPS) Architectures including the 3GPP enhancements for multimedia services (including emergency services), IoT, and other market sectors/vertical industries related use-cases. TSG SA5 is currently responsible for [12].

1. Management and Orchestration which covers aspects such as operation, assurance, fulfillment and automation, including management interaction with entities external to the network operator (e.g., service providers and verticals)

2. Charging which covers aspects such as Quota Management and Charging Data Records (CDRs) generation, related to end-user and service-provider

SA6 is currently responsible for application layer specifications, with emphasis on following [13]:

1. Critical communication applications (e.g., Mission Critical services for public safety, railways)

2. Service frameworks (e.g., Common API Framework, Service Enabler Architecture Layer, Edge Application enablement, Messaging enablement)

3. Enablers for vertical applications (e.g., automotive, drones, smart factories)

6. Small Cell Forum

Small Cell Forum (SCF) [13] was founded with the aim of ensuring scale and, explicitly, diversity for the small cell industry. To ensure right conditions for diversity following approaches are considered its priority:

- Diversification of market applications
- Diversification of deployers
- Diversification of supply chains

SCF addresses requirements from service providers, enterprises, and technology providers—all of which will be driving the small cell-enabled 5G future. SCF is helping enterprise, industry and government understand the potential benefits of private networks and providing the technical work that will help to enable them. SCF published a document SCF234 [13] that correlates the interworking between Small Cell Networks and Edge Computing, architectural considerations and service models associated with private networks-based edge computing. Fig. 18 provides distinction of edge locations, while Figure overall functional framework for edge computing correlating Mobile Network Operators (MNO) network functions and on-premise network functions.

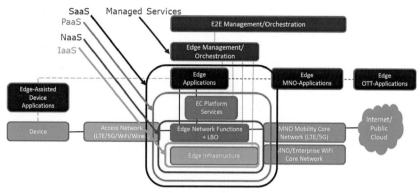

Fig. 18 Functional Framework for Edge Computing.

7. Broadband Forum

BroadBand Forum (BBF) [14] is the communications industry's leading organization focused on accelerating broadband innovation, standards, and ecosystem development. Broadband Forum is an open, non-profit industry organization composed of the industry's leading broadband operators, vendors, and thought leaders who are shaping the future of broadband. Its work to date has been the foundation for broadband's global proliferation and innovation. Broadband Forum's projects span across 5G, Connected Home, Cloud, and Access.

7.1 Connected Home

The IoT enhanced Connected Home [14] presents service providers with unparalleled opportunities and challenges. This project provides the necessary evolution of broadband with the standard—TR-369 or User Service Platform (USP)—has been designed and built by the service providers and vendors of Broadband Forum by leveraging the experience of deployment managed services through complex network environments. High level architecture detailed in TR-369 is provided in Fig. 19 [15].

7.2 5G

5G-Fixed Mobile Convergence (5G-FMC) is one of the areas of high priority to be addressed, to provide:
- Seamless service experience is key for users and drives the need for full FMC

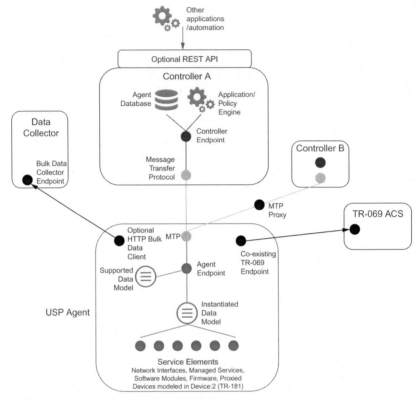

Fig. 19 Architectural Framework of TR-369.

- 5G services are to be deployed in an access-independent context across several access technologies (incl. Wireless and wireline)
- On-demand network services, e.g., different levels of mobility may be required according to the application needs
- Multiple simultaneous attachments will be very common for certain devices and applications

BBF has produced following standards to help evolution of FMC towards 5G:

- 5G Fixed-Mobile Convergence (SD-407)
- End-to-End Network Slicing (SD-406)
- Combined 3GPP and BBF functions (SD-357)
- 5G requirements and enablers (SD-373)

Figure 20 [14] provides high level architecture of correlation between BBF and 3GPP items necessary for FMC.

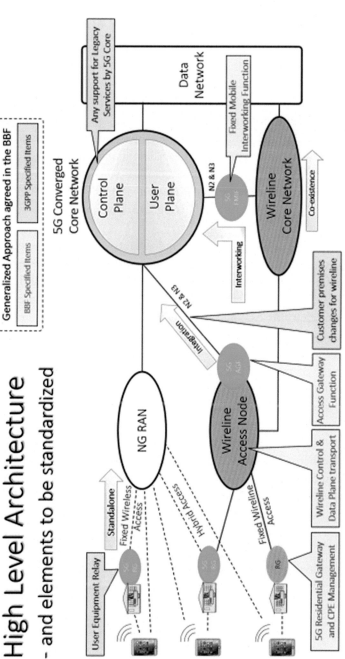

Fig. 20 BBF 5G FMC high level architecture.

8. 5G Alliance for connected industry and automation (5G-ACIA)

To make 5G a success for industrial applications, 5G-ACIA [16] brings together widely varying 5G stakeholders to include organizations of the Operational Technology (OT) (include industrial automation enterprises, machine builders, end users), Information and Communication Technologies (ICT) (includes manufacturers, network infrastructure providers or mobile network operators), academic institutions and more. Together, members discuss and evaluate technical, regulatory, and business aspects with respect to 5G for the industrial domain through working groups addressing these issues. Following working groups are part of 5G-ACIA [16]:

- Working Group 1: Use Cases and Requirements, discusses potential use cases and requirements and defines a common body of terminology.
- Working Group 2: Spectrum and Operating Models, identifies and articulate specific spectrum needs of industrial 5G networks and explores new operator models, for example for operating private or neutral host 5G networks within a plant or factory.
- Working Group 3: Architecture and Technology, Considers the overall architecture of future 5G-enabled industrial connectivity infrastructures. This includes integration concepts and migration paths, the evaluation of key technologies emerging from 5G standardization bodies.
- Working Group 4: Liaison and dissemination, takes care of the interaction with other initiatives and organizations by establishing liaison activities and initiating suitable promotional measures.
- Working Group 5: Validation and tests, deals with the final validation of 5G for industrial applications. This includes the initiation of interoperability tests, larger trials, and potentially dedicated certification procedures.

Various showcases have been created through 5G-ACIA spanning across factory automation, process automation, logistics, maintenance, human machine collaboration. Fig. 21 [16] provides various industrial use case and corresponding service requirements.

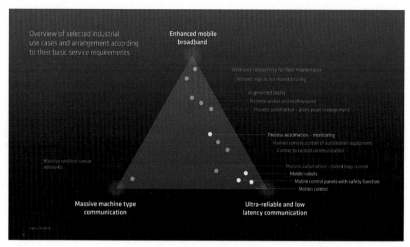

Fig. 21 Industrial use cases according to communication requirements.

9. 5G Automotive Association (5GAA)

The 5G Automotive Association (5GAA) [17] is a global, cross-industry organization of companies from the automotive, technology, and telecommunications industries (ICT), working together to develop end-to-end solutions for future mobility and transportation services. Diverse both in terms of geography and expertise, 5GAA's members are committed to helping define and develop the next generation of connected mobility and automated vehicle solutions.

The transmission modes of shorter-range direct communications, Vehicle to Vehicle (V2V), Vehicle to Infrastructure (V2I), Vehicle to Pedestrian (V2P), and longer-range network-based Vehicle to Network (V2N) communications comprising Cellular-V2X as defined by 3GPP TR 22.885. An example of V2X architecture is described in Fig. 22 [9].

5GAA has produced multiple artifacts that help enable future of automotive industries. Some of the white papers published by 5GAA at the time of writing this book:

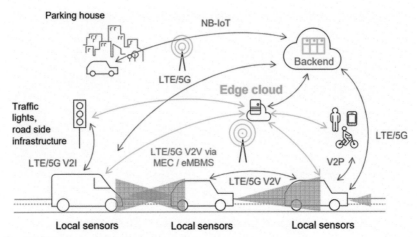

Fig. 22 V2X communications.

- Safety treatment in V2X applications
- Cooperation models enabling deployment and use of 5G infrastructure
- Privacy by design aspects of C-V2X
- C-V2X use cases
- Vulnerable road user protection
- MNO network expansion mechanisms to fulfil connected vehicle requirements
- 5GAA efficient security provisioning
- Making 5G proactive and predictive for automotive industry
- Evaluation of available architectural options
- Edge computing for advanced automotive communications
- Cost-benefit analysis on C-V2X technology and its evolution to 5G-V2X

10. Automotive Edge Computing Consortium (AECC)

Automotive Edge Computing Consortium (AECC) [18] works with leaders across industries to drive the evolution of edge network architectures and computing infrastructures to support high volume data services in a smarter, more efficient connected-vehicle future. The Consortium is creating use cases and requirements on networking and computing for connected services in automobiles. Examples are high-definition map creation and distribution, intelligent driving, and remote diagnostic maintenance. Eventually our work will expand to emerging mobile devices such as drones, robots and other types of vehicles. Few of the consortium activities include:

- Define specific automotive use cases and requirements with a focus on networking and computing for automotive big data.
- Formulate a roadmap strategy from technology development to market introduction, including the network evolution.
- Identify relevant communities for standardization and open-source software development, and support these with inputs for use cases and requirements with measures of success.
- Address efficiency issues in resource utilization such as communication bandwidth, computational power, and storage capacity. Examples of solutions may include in-vehicle systems; edge computing; distributed cloud; process/task migration; network virtualization (SDN/NFV)/containerization (micro-service); network interface/messaging; data center fabric; and multiple-accesses (Wi-Fi/Cellular, etc.).

The AECC provides system architecture that interacts between vehicles, service edge access network and necessary services such as AECC service that is responsible for overall management of AECC system, as described in Fig. 23 [18].

Following publications have been produced through AECC at the time of writing this book [18]:

- Distributed Computing in an AECC System
- Break Down the Barriers to Automotive Edge Adoption
- Connected Cars: On the Edge of A Breakthrough
- Enabling the Connected Vehicle Market to Thrive
- Driving Data to the Edge: The Challenge of Traffic Distribution
- Operational Behavior of a High-Definition Map Application
- General Principle and Vision
- Use-case and Requirement Document

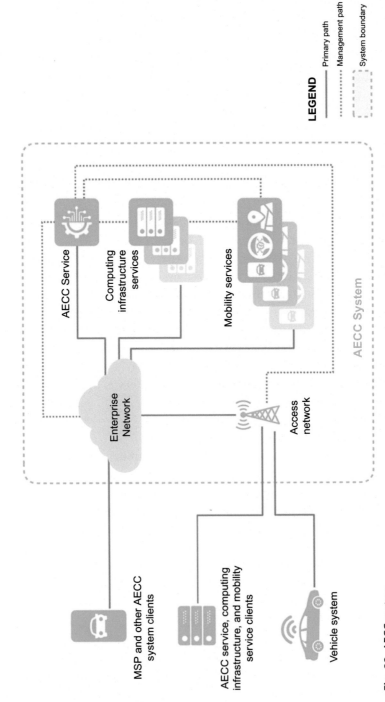

Fig. 23 AECC system.

AECC Service

Computing infrastructure services

Mobility services

AECC System

Enterprise Network

Access network

MSP and other AECC system clients

AECC service, computing infrastructure, and mobility service clients

Vehicle system

LEGEND

Primary path

Management path

System boundary

11. Telecom Infra Project

The Telecom Infra Project (TIP) [19] is a global community of companies and organizations working together to accelerate the development and deployment of open, disaggregated, and standards-based technology solutions that deliver the high-quality connectivity that the world needs. To identify the best market opportunities for connectivity from operators and other connectivity stakeholder, TIP operates as project groups and solution groups each producing set of deliverables and relevant software. Few of the project groups and solution groups that are relevant towards edge computing:

11.1 OpenRAN

OpenRAN's mission [19] is to accelerate innovation and commercialization in RAN domain with multi-vendor interoperable products and solutions that are easy to integrate in the operator's network and are verified for different deployment scenarios. TIP's OpenRAN program supports the development of disaggregated and interoperable 2G/3G/4G/5G NR Radio Access Network (RAN) solutions based on service provider requirements. Fig. 24 provides reference architecture of OpenRAN [19].

Key tenets of OpenRAN from TIP include:
- Disaggregation of RAN HW & SW on vendor neutral, GPP-based platforms

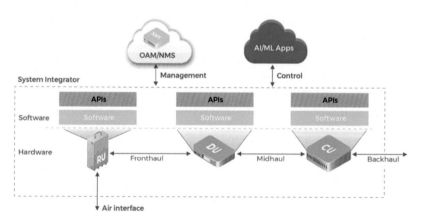

Fig. 24 OpenRAN architecture.

- Open Interfaces—Implementations using open interface specifications between components (e.g., RU/CU/DU/RIC) with vendor neutral hardware and software.
- Multiple Architecture Options, including
- An all-integrated RAN with disaggregation at SW and HW level
- A split RAN with RU, BBU (DU/CU)
- A split RAN with RU, DU and CU
- A split RAN with integrated RU/DU, CU
- Flexibility—Multi vendor solutions enabling a diverse ecosystem for the operators to choose best-of-breed options for their 2G/3G/4G and 5G deployments
- Solutions implemented on either Bare Metal or Virtualized or Containerized Platforms
- Innovation via Adoption of New Technologies (AI/ML, CI/CD…)
- Supply Chain Diversity

11.2 Connected City Infrastructure

The Connected City Infrastructure project group [19] aims to develop a configuration with cost-efficient new and retrofitted street asset solutions and different backhaul technologies. The group will publish a case study to showcase a sustainable business model for street assets, managed by municipalities or public utilities, providing citizen services and generating appropriate revenue streams. Fig. 25 [19] provides reference deployment of a connected city infrastructure.

The group is focused on urban connectivity solutions, specifically:
- The definition and validation of new construction and retrofitted modular street assets with LTE / 5G Small Cells and Public Wi-Fi E2E architectures. The PG members will deploy these street assets, with modular assembly, in a field trial supporting interchangeable backhaul and access technologies.
- The creation of an anonymized business case for provision and operation of operator service based on different backhaul & transmission services (i.e., fiber, mmWave, microwave)—focusing on business driver constructs required to drive a scalable solution. This will also be complemented with deployment and operational guidelines.

11.3 5G Private Networks

The goal of the 5G Private Network Solution Project Group [19] is to make 5G private networks accessible to a broad range of use cases and customers by

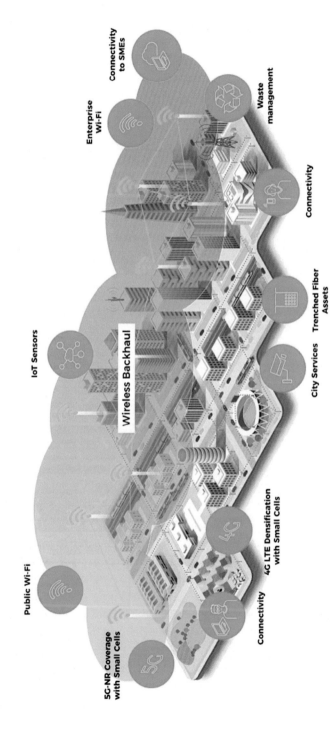

Fig. 25 Connected city reference solution

transforming deployment from a bespoke special project to a standard product with an appropriate cost structure. The group develops requirements for the automated lifecycle management of an on-premises and edge cloud native 5G Private Network solution based on a CI/CD toolset and open, disaggregated hardware components.

This solution group will develop a new approach to the implementation of 5G Private Networks, resulting in:

- Seamless integration of building blocks: 5G connectivity, Cloud Native Edge Computing and use cases
- Lower cost structure from the use of open & cloud native functions for 5G private networks maximizing the use of open-source software
- More efficient and timely installation, maintenance, and upgrade of software through CI/CD automated life-cycle management (LCM) of all the different units composing the architecture: CNP, tools, virtual network functions, cloud native telco functions
- Improved security model and better performance (higher bandwidth, lower latency) through the use of local user plane breakout
- Develop requirements for the automated lifecycle management of an on-premises and edge cloud native 5G Private Network solution based on a CI/CD toolset and open, disaggregated hardware components.

12. IEEE International Network Generations Roadmap Edge Services Platform (ESP)

The purpose of the International Network Generations Roadmap (INGR) [20] is to stimulate an industry-wide dialogue to address the many facets and challenges of the development and deployment of 5G in a well-coordinated and comprehensive manner, while also looking beyond 5G by laying out a technology roadmap with 3-, 5-, and 10-year horizon. Fig. 26 shows the correlation of Edge Service Platform with various aspects of software lifecycle management at the Edge.

The services to run on edge to cloud continuum is proposed as Edge Service Platform Framework (ESPF) [21] which covers the edge service optimizations for low sub millisecond latencies and higher bandwidth and capacities for 5G/IoT/WIFi6 and beyond. Edge Service includes the necessary Platform and Applications that are distributed and delivered to consumers and enterprises. The requirements to support privacy, security, and proximity functionality for location-based delivery of services leads to touch

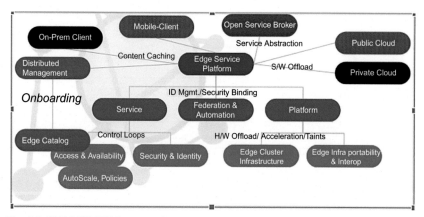

Fig. 26 IEEE INGR ESP framework.

point of edges all over from on-prem, IoT Gateways, light poles, small cells, macro cells, fronthaul, Micro Data Centers, midhaul, central office, provider edge with microservices architecture, loosely coupled composable services from service catalogues across the edge to cloud continuum.

13. KubeEdge

KubeEdge [22] enables cloud-edge synergy, computing at edge, and easy access of a massive number of devices based on Kubernetes container orchestration and scheduling capabilities. KubeEdge is an incubating project under umbrella of Cloud Native Compute Foundation (CNCF).

As shown in Fig. 27 [22] the KubeEdge architecture consists of cloud, edge, and device layers. The control plane is on the cloud. Edge nodes are at the edge. At the cloud layer, the green box on the left represents a Kubernetes master. It is a native Kubernetes control plane without any changes. The light-yellow box on the right represents CloudCore. It contains EdgeController and DeviceController, which process data from the control plane, and Cloud Hub, which sends the data to EdgeHub at the edge. The edge enables application and device management. In the light-yellow box, on the left is Edged for application management. On the right are DeviceTwin and EventBus for device management. The green box on the left represents DataStore. It enables local autonomy. Specifically, when the data of an application or device is distributed from the cloud through EdgeHub, the data is stored in a database before it is sent to

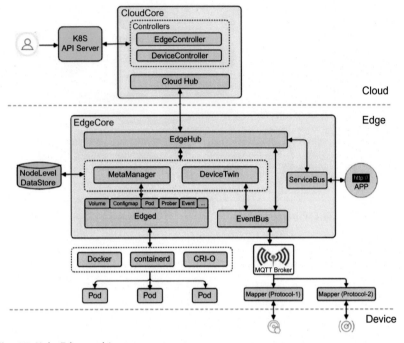

Fig. 27 KubeEdge architecture concept.

Edged or the device. In this way, Edged can retrieve metadata from the database and the service recovers even when the edge is disconnected from the cloud or when the edge node restarts.

14. StarlingX

StarlingX [23] is a complete cloud infrastructure software stack for the edge used by the most demanding applications in industrial IOT, telecom, video delivery and other ultra-low latency use cases. Fig. 28 [23] show high level overview of StarlingX project. With deterministic low latency required by edge applications, and tools that make distributed edge manageable, StarlingX provides a container-based infrastructure for edge implementations in scalable solutions that is ready for production. StarlingX is a project hosted under Open Infrastructure Foundation [24]. The project builds on existing services in the open-source ecosystem by taking components of projects such as Ceph, OpenStack and Kubernetes and complementing them with

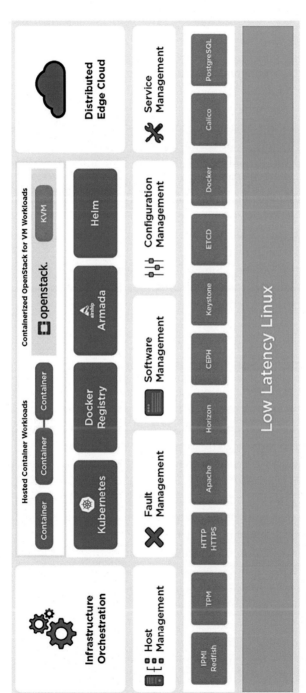

Fig. 28 StarlingX architecture.

new services like configuration and fault management with focus on key requirements as high availability (HA), quality of service (QoS), performance and low latency.

Some of the highlights of StarlingX software:

- Reliability: Fault management, fast secure VM failover and live migration minimizes downtime
- Scalability: Deployable on one to thousands of distributed nodes allowing for a single system to be used from edge to core
- Small footprint: Providing a platform for edge and IoT use cases even for environments with tight resource constraints
- Ultra-low latency: Deterministic, tunable performance optimized for the use case
- Secure: Software security to avoid tampering at the edge, where physical security may be limited
- Lifecycle management: Simplified deployment and operations with full system management through comprehensive orchestration suited for the edge

15. Open Edge Computing Initiative

The Open Edge Computing Initiative [25] is a collective effort by multiple companies, driving the business opportunities and technologies surrounding edge computing. The Open Edge Computing Initiative is shaping the global eco-system around edge computing by:

- Driving the convergence of edge computing platforms and services on a global scale
- Providing attractive edge applications for live edge demonstrations
- Running a real-world edge computing test center (called the Living Edge Lab) for user and technology trials
- Driving the adoption of Open Edge Computing with edge application providers, telecom operators, and cloud service providers
- Tackling key technical challenges of edge computing with its academic partner Carnegie Mellon University

The Living Edge Lab provides an open proving ground for edge computing, as described in Fig. 29 [25].

Few of the publications [25] produced by Open Edge Computing:

- Impact of Delayed Response on Wearable Cognitive Assistance
- The Role of Edge Offload for Hardware-Accelerated Mobile Devices
- OpenRTiST: End-to-End Benchmarking for Edge Computing

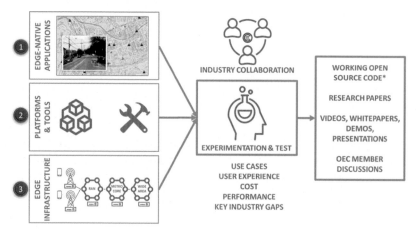

Fig. 29 Working Process and Output from Living Edge Lab.

- Simulating Edge Computing Environments to Optimize Application Experience
- Edge Computing for Legacy Applications
- Towards Scalable Edge-Native Applications
- Seeing Further Down the Visual Cloud Road
- The Seminal Role of Edge-Native Applications
- Towards Drone-sourced Live Video Analytics for the Construction Industry
- Towards a Distraction-free Waze
- EdgeDroid: An Experimental Approach to Benchmarking Human-in-the-Loop Applications
- Edge-based Discovery of Training Data for Machine Learning
- Bandwidth-efficient Live Video Analytics for Drones via Edge Computing
- Experimental Testbed for Edge Computing in Fiber-Wireless Broadband Access Networks
- An Application Platform for Wearable Cognitive Assistance
- You Can Teach Elephants to Dance: Agile VM Handoff for Edge Computing
- Live Synthesis of Vehicle-Sourced Data Over 4G LTE
- Assisting Users in a World Full of Cameras: A Privacy-aware Infrastructure for Computer Vision Applications
- An Empirical Study of Latency in an Emerging Class of Edge Computing Applications for Wearable Cognitive Assistance
- Edge Computing for Situational Awareness

16. Smart Edge Open

Smart Edge Open [26] is a MEC software toolkit that enables highly optimized and performance edge platforms to on-board and manage applications and network functions with cloud-like agility across any type of network. This open-source distribution is designed to foster open collaboration and application innovation at the Network Edge and On-Premise Edge, making it easier for cloud and Internet of Things (IoT) developers to engage with a worldwide ecosystem of hardware, software and solutions integrators to develop solutions for 5G and Edge. A typical Smart Edge Open-based deployment consists of a Kubernetes Control Plane and Edge Node. Fig. 30 [26] shows logical components of Smart Edge Open Developer Experience Kit solution that is built on Kubernetes.

Smart Edge Open offers unique capabilities to accelerate application development at the Edge:

– Abstracts out the network complexity for Cloud and IOT developers making migration of applications from the cloud to the edge easier
– Enables secure on-boarding and management of applications with an intuitive web-based GUI
– Built on a modular, microservices based architecture, it provides the building blocks for various functionalities such as access termination, traffic steering, multi-tenancy for services, service registry, service authentication, telemetry, application frameworks, appliance discovery and control
– It is built on top of consistent and standardized APIs exposed to the developer community

17. Edge Multi Cluster Orchestrator (EMCO)

The Edge Multi-Cluster Orchestrator (EMCO) [27], is a software framework for intent-based deployment of cloud-native applications to a set of Kubernetes clusters, spanning enterprise data centers, multiple cloud service providers and numerous edge locations. It is architected to be flexible, modular, and highly scalable. It is aimed at various verticals, including telecommunication service providers. The mission of the Project is to create a universal control plane that helps organizations to securely connect and deploy workloads across public clouds, private clouds, and edge locations, with end-to-end inter-application communication enabled. Fig. 31 [27] provides logical architecture of EMCO.

Edge Node Component

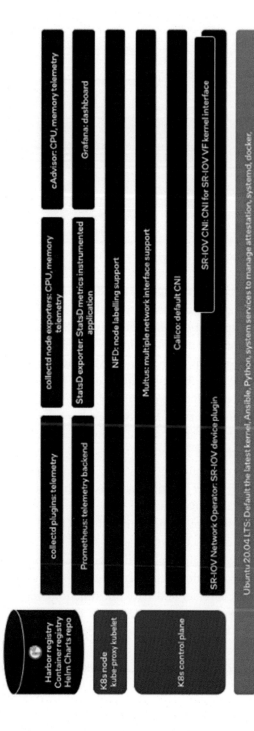

Fig. 30 Smart Edge Open Developer Experience Kit.

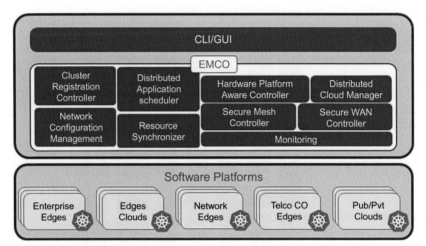

Fig. 31 EMCO architecture.

Number of K8s clusters (Edges or clouds) could be in tens of thousands, number of complex applications that need to be managed could be in hundreds, number of applications in a complex application could be in tens of thousands and number of micro-services in each application of the complex application can be in tens of thousands. Moreover, there can be multiple deployments of the same complex applications for different purposes. To reduce the complexity, all these operations are to be automated. There shall be one-click deployment of the complex applications and one simple dashboard to know the status of the complex application deployment at any time. Hence, there is a need for Multi-Edge and Multi-Cloud distributed application orchestrator.

Compared with other multiple-clusters orchestration, EMCO focuses on the following functionalities:

- Enrolling multiple geographically distributed Smart Edge Open clusters and third-party cloud clusters.
- Orchestrating composite applications (composed of multiple individual applications) across different clusters.
- Deploying edge services and network functions on to different nodes spread across different clusters.
- Monitoring the health of the deployed edge services/network functions across different clusters.
- Orchestrating edge services and network functions with deployment intents based on compute, acceleration, and storage requirements.
- Supporting multiple tenants from different enterprises while ensuring confidentiality and full isolation between the tenants.

18. Global Systems for Mobile Association (GSMA)

The Global Systems for Mobile Association [28], usually referred to as Global Systems for Mobile Communications, or GSMA, represents the interests of mobile operators worldwide, uniting more than 750 operators with almost 400 companies in the broader mobile ecosystem, including handset and device makers, software companies, equipment providers and internet companies, as well as organizations in adjacent industry sectors. The GSMA represents its members via industry programs, working groups and industry advocacy initiatives.

The Operator Platform Group (OPG), within GSMA, has defined the Edge Cloud functionality in the OPG.01 Operator Platform Telco Edge Proposal and the Telco Edge Cloud (TEC) taskforce works on launching a global telco edge computing service that implements it. High level reference architecture of Telco Edge as shared by Operator Platform Telco Edge proposal (OPG.01) is shared in Fig. 32 [28]. The TEC group has produced following artifacts:

– Operator Platform Telco Edge Proposal
– Telco Edge Cloud: Edge Service Description and Commercial Principles
– Operator Platform Telco Edge Requirements
– Edge and Cloud Developments in Europe

The Telco Edge Cloud (TEC) Pre-Commercial Trials Foundry project [28], that is part of GSMA, aims to bring the industry together to collaborate

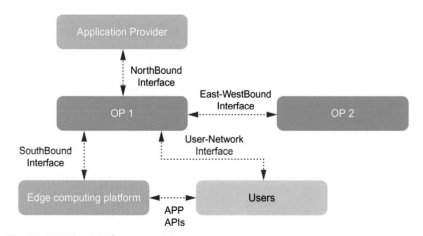

Fig. 32 High-level Reference architecture.

around a number of pre-commercial trials of Telco Edge Cloud capabilities with app developers and service providers sponsored by the industry partners and supported by MNOs. The potential for Independent software vendors (ISVs) is a pilot opportunity with at least two operators in a pre-commercial trial in the second half of 2021. In doing so, generate proof points and promote the benefits of using Telco Edge Cloud infrastructure to stimulate the development of a wider ecosystem.

References

[1] LF Edge, The New Open Edge IOT + Telecom + Cloud + Enterprise + Industrial, LFEdge.org.
[2] Akraino, Welcome to Akraino Wiki, akraino.org.
[3] Edgexfoundary, EdgeX Foundary Project Wiki, edgexfoundry.org.
[4] LFEdge Projects, Project EVE, LFEdge.
[5] Linux Foundation Networking, home page. lfnetworking.org.
[6] ONAP, home page, onap.org/about.
[7] Anuket Artifacts, anuket.io/artifacts.
[8] ORAN Alliance, homepage, o-ran.org.
[9] D.Sabella, et al., "Toward Fully Connected Vehicles: Edge Computing for Advanced Automotive Communications, 5GAA_T-170219.
[10] Open Networking Foundation, homepage, opennetworking.org.
[11] 3GPP, about-us. 3gpp.org/about-3gpp.
[12] 3gpp, 3gpp Specifications Group.
[13] Small Cell Forum. smallcellforum.org.
[14] Broadband Forum, about us, broadbandforum.org/about-bbf.
[15] Qacafe, An Overview of The User Services Platform (USP/TR-369), Resources by Qacafe.
[16] 5G Alliance for Connected Industries and Automation, about-us, 5GACIA.
[17] 5G Automotive Association, about-us, 5GAA.
[18] Automative Edge Computing Consortium, About, AECC.
[19] Telecom Infra Project, homepage, TIP.
[20] IEEE Future networks, International Next Generation Roadmap, futurenetworks.ieee.org/roadmap.
[21] R. Sunku., Edge Automation Platform, International Network Generations Roadmap—2021 Edition.
[22] KubeEdge Maintainers, KubeEdge: Cloud Native Edge Computing, Cloud Native Computing Foundation Blogs.
[23] StarlingX, homepage, starlingx.io.
[24] Open Infrastructure Foundation, homepage, openinfra.dev.
[25] Open Edge Computing Initiative, Homepage, Openedgecomputing.
[26] Smart Edge Open, github.io.
[27] ONAP, EMCO Architecture & Design, wiki.
[28] GSMA, aboutus, GSMA.

About the author

Sunku Ranganath is a Solutions Architect for Edge Compute at Intel. For the last few years, his area of focus has been on enabling solutions for the Telecom domain, including designing, building, integrating, and benchmarking NFV based reference architectures using Kubernetes & OpenStack components. Sunku has been an active contributor to multiple open-source initiatives. He serves as a maintainer for CNCF Service Mesh Performance & CollectD projects and as an elected member on the Technical Steering Committee for OPNFV (now Anuket). He is an invited speaker to many industry events, authored multiple publications, filed eight patents, and contributed to IEEE Future Networks Edge Service Platform & ETSI ENI standards. He is a senior member of the IEEE.

CHAPTER FOUR

IoT-edge analytics for BACON-assisted multivariate health data anomalies

Partha Pratim Ray

Department of Computer Applications, Sikkim University, Gangtok, India

Contents

Abstract

Anomaly detection in Internet of Things (IoT)-enabled systems can significantly improve the quality of the deployed systems. Though existing techniques can detect anomalies from a dataset, more efficient algorithms can be used to minimize the burden of excessive computational overhead on the resource-constrained IoT-edge device pool. In this chapter, we implement the blocked adaptive computationally efficient outlier nominators (BACON) algorithm along with the estimated-expectation/maximization (EEM) method to improve the anomaly nomination for IoT-based health dataset. We deploy the weighted variant of the BACON algorithm package—"wbacon" from the R repository to validate the utilization of anomaly nomination for an IoT-edge-enabled health dataset. The deployed scheme advocates the importance of distance wise Mahalanobis statistical tool for efficient nomination of potential outliers from the IoT dataset.

Advances in Computers, Volume 127
ISSN 0065-2458
https://doi.org/10.1016/bs.adcom.2022.02.002

117

1. Introduction

Anomaly nomination refers to a process to potentially identify a set of observations from a large dataset which can be termed as the outliers [1–3]. The outliers are the data points that form a dataset that behave abnormally or present discrepancies in terms of behavior of homogeneity with other existing observations [4, 5]. Such anomalies can significantly disturb the analytical perspective when not identified during the early stage of data gathering [6–8], especially for the sensor-oriented ubiquitous systems where misinterpretation of potential observation can significantly damage the performance of the ecosystem in later stages [9–11].

In recent years, IoT has emerged as a key enabler of smart applications. The key focus of IoT-edge-enabled systems is to provide minimal delay and maximum throughput of the prospective use cases at the extreme edge (edge) of the existing networking infrastructures [12–14]. Herein, a number of sensors generate data streams and propagate toward micro-controllers which is predefined with a set of algorithms to execute smart tasks. Normally, most of the data is sent to the remote cloud computing platforms for permanent storage where a data analyst can perform given analytical jobs in future [15, 16]. But, this domain of technology is currently facing a serious challenge as follows. Due to excessive use of sensors in IoT-edge-assisted use cases, gradual dumping of huge piles of data is being made to the cloud platforms. Thus, it is regularly consuming the costly network bandwidth and increasing the data traffic. It is also populating cloud storage systems with big sensor data which are often not properly diagnosed. It would have been better if we could imply light-weight edge analytics solutions at the IoT devices that could take instantaneous decisions about an IoT data whether to send it to a remote cloud or not. In doing so, band-width consumption and traffic generation could have been minimized that would in turn make the cloud storage systems more useful for other impor-tant data saving aware tasks [17–19].

This chapter presents the implementation of such a scenario with help of the amalgamated BACON-EEM methods for the IoT-edge ecosystem. We deploy the "wbacon" package [20] from the R repository to identify and nominate the potential outliers from an IoT health dataset [21]. The IoT health dataset contains six variables each with 617 rows. We use human skin conductance (C), saturated oxygen (SO), body core temperature (T), blood pressure low (BL), blood pressure high (BH), and heart rate variability

(HRV) as the variables. Each of these variables contains some intentionally incorporated outliers so as to enable the deployed BACON-EEM algorithm to nominate such observations as the potential outliers. We investigate the efficiency of this algorithm for the IoT-based scenario due to the fact that such provisions are highly resource constrained in nature. We aim to study how the BACON-EEM algorithm combination can be useful to efficiently nominate the introduced outliers from the given IoT health dataset. It is expected that the algorithm can minimize the computational overhead from the IoT devices, thus giving a space for performing other essential computations. We choose the statistical tools instead of conventional machine or deep learning aware algorithms owing to the fact that such schemes are normally computationally intensive thus not always suitable for the IoT-edge deployment.

The major contributions of this chapter can be summarized as follows.
- To implement the BACON-assisted EEM algorithm for an IoT-edge ecosystem
- To analyze the results in terms of robust distances and regression analytics
- To deploy the Mahalanobis distance and distance from median approaches for anomaly nomination
- To assess the basic subset formation and increment for the regression data
- To present regression use case for the HRV against C, SO, T, BL, and BH variables

2. Related works

In Ref. [22], an algorithm is proposed called "FiRE.1" to sketch the linear-time centric global anomaly identification. New outlierness criteria are also proposed herein this paper.

A landmark-based outlier detection approach is presented in Ref. [23]. Such a soft mechanism is based on the candidate and the confirmed outliers. It adopts a sliding window landmark model when a potential candidate is assumed as an anomaly.

An overview of swarm intelligence aware outliers detection methods is discussed in Ref. [24]. A systematic review on security logs for data leak detection is performed in Ref. [25].

A scalable density aware clustering algorithm for anomaly detection is proposed in Ref. [26] where large size datasets can be applied. The algorithm is linear in terms of time complexity.

Study of anomalies in the multivariate dataset for Alzheimer's disease detection is investigated by Ref. [27]. The study uses the parallel independent component analysis, correlation analysis, sparse partial test, genome-wide association, and sparse reduced rank regression out of many such techniques.

Anomaly detection method is investigated in Ref. [28] for identification of intrusions by using a machine learning technique. A parallel tensor factorization based on rational learning is proposed in Ref. [29].

Anomaly detection scheme is also linked with the zero-shot learning for study of signal recognition by using convolution neural network [30]. Privacy for the cyberphysical world is an important factor that must be preserved against intrusions. Thus in Ref. [31], outliers are detected from the intrusion recognizing framework.

Deep neural networks have the potential to detect anomalies in a network. In Ref. [32], an anomaly detection technique is employed to identify faults in an unmanned aerial vehicles network.

Fraud in healthcare-related information spreading can be dangerous for the patients at a large. A sub peer group analysis aware outliers detection method is implied to find fraud in healthcare information sharing [33]. In Ref. [34], outliers detection technique is implemented for an electoral dataset to find the mixed-effects regressions.

Lessons Learned: We learned that none of the works in literature uses the BACON-EEM algorithm for nomination of anomalies for an IoT-edge-enabled health dataset. The novelty of this chapter is as follows: (i) use of BACON-EEM methods for IoT-based ecosystem, (ii) minimization of processing overhead in IoT devices, and (iii) investigation of the robust Mahalanobis distances and distance from medians as key enablers of outliers nomination.

3. System design

The implied system design depends on several factors considering the optimality, breakdown, and equivariance of the observations [35–37]. For example, we can take a case having an IoT data set with n number of data points related to p variables. In this IoT dataset, there exists k which is an unknown value. The unknown value $k < \frac{n}{2}$. In a normal scenario, one can perform a brute-force search to find the outliers having a subset of size $k = 1, 2, 3 \ldots, \frac{n}{2}$. This task is computationally expensive for a resource-constrained IoT-based system due to the large number of subsets, i.e., $\sum_{k=1}^{\frac{n}{2}} \binom{n}{k}$ which require huge amount of processing and memory capacity [38–41]. Moreover, this process may not be successful to find all the subsets nor exclude the anomalies from the IoT dataset.

Nomination of anomalies from an IoT dataset can be done in many ways [42–44], for instance, use of minimum covariance determinant (MCD) and minimum volume ellipsoid (MVE). Thus, we need to first compute the $\binom{n}{k}$, i.e., volume of ellipsoids [45–47]. Second, we can choose a subset that provides minimum determinant or volume where $\binom{n}{k} \ll \sum_{k=1}^{\frac{n}{2}} \binom{n}{k}$ and $h = \frac{n+p+1}{2}$. Again, this method is not feasible for an IoT-based environment due to a large computational complexity.

Use of estimators can improve the optimality condition for both anomaly detection and robust regression environment [48, 49]. An estimator can be expressed as in Eq. 1 to follow an *affine equivariant*, where T is an estimator and A and b are nonsingular matrix and vector, respectively [50, 51].

$$T(XA + b) = T(X)A + b \tag{1}$$

Normally, the brute-force technique is essentially an affine equivariant due to the fact that both covariance matrix and multivariate mean possess similar characteristics. In practice, it is very difficult to find an affine equivariant technique that has a very high breakdown point. In most of the cases, the breakdown point is bounded by $\frac{1}{p+1}$.

In this section, we present two techniques, e.g., (i) multivariant and (ii) regression data that can be applied based on two versions of algorithms. Version 1 is nearly affine equivariant and version 2 is affine equivariant. Version 1 has high breakdown points but version 2 has lower breakdown points.

3.1 BACON algorithm for selection of multivariate outliers nomination

We deploy the BACON algorithm [8] to enable the computation more efficiently for an IoT-based environment. This algorithm can be applied to the large IoT datasets. The methods herein presented have negligible sensitivity to the starting point. Though a robust start can give higher assurance of very high breakdown, robust initial points do not conform to the affine equivariant in its successive iterations.

The algorithm begins with a small subset of data which is presumed to be free of anomaly observations. Later, the subset is increased in size to include all normal data points. The data points that are excluded can be nominated as potential outliers. To save computational chaos in the IoT devices, the basic subset is expected to grow rapidly and validate against a specific criteria. This algorithm removes the need of (i) sorting of large arrays of discrepancies, (ii) minimization of the number of covariance matrices, and (iii) reduction in the number of inverted covariance matrix computations. Algorithm 1 presents the detailed step-by-step methods of this algorithm. An iteration is imposed to increase the size of the basic subset.

ALGORITHM 1 BACON Algorithm for Identification of Multivariate Anomaly Nomination.

1 Input: n — number of observations, m — number of observations for inclusion for the initialization of the basic subset, t — temporary variable

2 Output: Nominate outliers

3 Design initial basic subset where $|subset| \geq m$ based on versions

4 **while** m *does not change* **do**

5   **if** *(Version==V1)* **then**

6   Compute Mahalanobis distances as below

7   **for** $i = 1; \, i \leq n; \, i = 1 + 1$ **do**

8

$$d_i = \sqrt{(x_i - \mu)C^{-1}(x_i - \mu)} \tag{2}$$

9   **if** *(Version==V2)* **then**

10   Compute distance from medians as below and identify smallest of all

11   **for** $i = 1; \, i \leq n; \, i = 1 + 1$ **do**

12

$$t_i = ||x_i - m|| \tag{3}$$

13   Compute correction factor $c_{npr} = c_{np} + c_{hr}$ as below, where p is degrees of freedom, r is size of latest basic subset, $c_{hr} = max\{0, \frac{h-r}{h+r}\}$, $h = \frac{n+p+1}{2}$, setting all observations with discrepancies smaller the $c_{npr}\chi_{p,\frac{\alpha}{n}}$ with $\chi_{p,\alpha}^2 = 1 - \alpha$

14

$$c_{n,p} = 1 + \frac{p+1}{n-p} + \frac{1}{n-h-p} = 1 + \frac{p+1}{n-p} + \frac{2}{n-1-3p} \tag{4}$$

15 Nominate outliers that are excluded by latest basic subset

It also helps to minimize the memberships to the data points against the current basic subset resulting in a state of nonanomalous points. Bigger the basic subset size, more the possibility to yield reliable estimates.

Assume the scenario where there exists a X matrix containing the total nxp number of IoT data points, where columns are variables and rows are observations. The BACON algorithm needs to find the initial basic subset with size m greater than p. The subset can be developed by the BACON algorithm itself or can be supplied as an external input. The initial basic subset must be "clean" where no anomaly can be present. Now, estimation can be done only when the basic subset size is substantially large. Hence, if $m = cp$, the estimation can be based on minimum c observations per parameter of the model.

We can use two versions to create the initial basic subset, e.g., (i) version 1—Mahalanobis distance and (ii) version 2—distance from medians. The Mahalanobis distance-based method is not robust but an affine equivariant, whereas the distance from medians method is robust but not an affine equivariant due to the fact that the coordinate wise median is not considered to be an affine equivariant.

The BACON algorithm helps to nominate the subset of data points as outliers by using either version 1 or 2. Propagating in this way, it later nominates a new subset of central observations and so on. When the initial subset is not closer to the middle of the normal data points, successive progression of iterations tends to drift to the center of the data set. When the basic subset grows more and more, the covariance matrix and mean get more stable.

3.2 Initial basic subset of regression data algorithm

Let us now consider that $y = X\beta + \epsilon$ represents a generic linear model where X is an IoT-based nxp data matrix where p refers to variables having $p < n$. The y represents the n-vector of responses. The β represents the p-vector of several unknown variables and ϵ denotes the random errors having n-vectored size. In this context, mean $E(\epsilon|X) = 0$ and variance $var(\epsilon|X) = \sigma^2 I_n$, where σ^2 and I_n refer to unknown parameter and identity matrix with nth order. The least-square estimate (LSE) of β and σ^2 is given by $\widehat{\beta} = (X^T X)^{-1} X^T y$. The residual mean square (RMS) is given by $\widehat{\sigma^2} = \frac{SSE}{(n-p)}$. The e can be obtained by $e = (I_n - P)y$ that is essentially a vector having generic residuals. The $SSE = e^T e$ represents the residual sum of squares. We can also state that $P = X(X^T X)^{-1} X^T$. We also assume that y_b, X_b are the basic subsets where each of the IoT-based observations is indexed by b, b is a set of indices. The $\widehat{\beta}_b$ can be assumed as the estimated regression coefficients upon fitting the model on b. Herein, we can assume that SSE_b and $\widehat{\sigma}_b^2$ refer to residual sum of squares and residual mean square, respectively.

Algorithm 2 presents the basic subset formation for regression IoT data. At first, the d_i is computed based on the y_m and X_m where the m number of observations are present with smallest possible values. When the X_m is not found to be of full rank, the algorithm increases the basic subset while incorporating the smallest possible values of $d_i(\bar{x}_b, C_b)$. This process continues until it has a full rank. Next, the algorithm computes $t_i(y_m X_m)$ and identifies the $p + 1$ data points where $|t_i) y_m, X_m|$ is considered as the smallest. Later, it declares those as the basic initial subset. The p_{ii} is referred to as the diagonal elements of the projection matrix $P = X(X^T X)^{-1} X^T$. The scaled ordinary least squared residuals are obtained when $x_i \in X_b$ and the scaled prediction errors are generated when $x_i \notin X_b$, $t_i(y_b, X_b)$.

The method starts with $m = \frac{(n + p + 1)}{2}$ by using the distances to define discrepancies. The initial subset is $p + 1$ and it increases until it reaches to the size of $\frac{(n + p + 1)}{2}$ data points. It can increase beyond this limit but gets restricted when the $r + 1$ smallest possible discrepancies aim to exceed $t_{(\frac{\alpha}{2(r+1)}, r-p)}$. Herein, r is the size of the current basic subset for each step and $t_{(\alpha, r-p)}$ refers to $1 - \alpha$ percentiles of t-distribution. The degree of freedom used herein is $r - p$.

This algorithm tries to repeatedly fit the regression model. However, the sorting of discrepancies at each loop step is a computationally intensive process. Algorithm 3 is implied herein to grow the basic subset in blocks. Thus, it can retain the capability for adaptation of the IoT data. It can overlook the unwanted computations within the process, i.e., a robust design is a must.

3.3 BACON robust regression algorithm

The Algorithm 3 aims to nominate the observations which are not included in the final basic subset as the potential anomalies. This algorithm minimizes the regression computations. It also reduces the evaluations of the discrepancies. There is no significance to order or sort the discrepancies rather just simple check against the given constant value. Such checking requires only n number of operations.

ALGORITHM 2 BACON Algorithm for Basic Subset Formation for Regression Data.

1 Input: a vector $nx1$, response variable y, nxp matrix X where X contains covariate data, number of observations m

2 Output: Initial basic subset with minimum m data points, where subset is free from anomalies

3 Apply Algorithm 1, i.e., *BACON algorithm*, on X and find X_m and y_m as the set of m data points where d_i represents the latest distance

4 **if** *(X_m != Full Rank)* **then**

5   **for** $i = 1; i \leq n; i = 1 + 1$ **do**

6   **if** *($X_i \in X_m$)* **then**

7

$$t_i(y_m, X_m) = \frac{y_i - x_i^T \hat{\beta}_m}{\hat{\sigma}_m \sqrt{1 - x^T (X_m^T X_m)^{-1} x_i}} \tag{5}$$

8   **else**

9

$$t_i(y_m, X_m) = \frac{y_i - x_i^T \hat{\beta}_m}{\hat{\sigma}_m \sqrt{1 + x^T (X_m^T X_m)^{-1} x_i}} \tag{6}$$

10 Identify $p + 1$ data points with minimum $|t_i(y_m, X_m)|$ based on the aforementioned equations where $\hat{\beta}_m$ and $\hat{\sigma}_m^2$ represent least-square estimates of β and σ^2, respectively, and declare it as initial basic subset

11 **while** *Until basic subset has m data points* **do**

12   **if** *(X_m != Full Rank)* **then**

13   **for** $i = 1; i \leq n; i = 1 + 1$ **do**

14   **if** *($X_i \in X_m$)* **then**

15

$$t_i(y_b, X_b) = \frac{y_i - x_i^T \hat{\beta}_b}{\hat{\sigma}_m \sqrt{1 - x^T (X_b^T X_b)^{-1} x_i}} \tag{7}$$

16   **else**

17

$$t_i(y_b, X_b) = \frac{y_i - x_i^T \hat{\beta}_b}{\hat{\sigma}_m \sqrt{1 + x^T (X_b^T X_b)^{-1} x_i}} \tag{8}$$

18   Identify $r + 1$ data points with minimum $|t_i(y_b, X_b)|$ based on the aforementioned equations where r is the size of latest basic subset and declare the minimum $t_i(y_b, X_b)$ as new basic subset

ALGORITHM 3 BACON Robust Regression Algorithm.

1 Input: an IoT vector $n x 1$, response variable y, $n x p$ IoT data matrix X where X contains
IoT covariate data

2 Output: A regression model fit on the normal IoT data points, a set of IoT data points with
anomalies, and distances computed by Eqs. (6) and (7)

3 Apply Algorithm 2, i.e., *Basic Subset Formation in Regression Data Algorithm*, to find an
initial basic subset, where the size is $m = cp$ with c as a constant provided externally

4 **while** *Until the size of basic subset does not change* **do**

5   Find abnormalities as per Eqs. (6) and (7)

6   Find new basic subset which has observations with distance $d_i < t_{\frac{\alpha}{2}(r+1),(r-p)}$ with r as
    the size of latest basic subset

7 Nominate all the rejected observations by the final basic subset as the IoT data outliers

3.4 IoT-BACON-EEM algorithm

Algorithm 4 aims to nominate the outliers with a combination of the BACON and estimated expectation/maximization (EEM) method [9]. The EEM acts as the basis of this algorithm where $X = X_o \cup X_m$, where X_o and X_m refer to observed and missing values from an IoT data set X. Here, it is assumed to have the missingness as ignorable and independent of sampling [22, 23]. Herein, the data log-likelihood can be presented as $l(\theta|X) = \eta\theta^T \cdot T(X) + Ng(\theta) + c$, where c is a constant. Such an expression is derived from the factorization of $P(X|\theta) = P(X_o|\theta)P(X_m|X_o, \theta)$ with the normal log-likelihood as $l(\theta|X) = l(\theta|X_o) + log(P(X_m|X_o, \theta)) + c$. The unknown average $P(X_m|X_o,\theta)$ captures the interdependence where over $l(\theta|X)$ and $P(X_m|X_o, \theta^t)$. The θ^t refers to the initial estimate on an unknown parameter, whereas the $\theta^{(t+1)}$ maximizes the next level expectation. The $\eta(\theta) = \eta_1(\theta^T), \eta_2(\theta^T), \eta_3(\theta^T), ..., \eta_k(\theta^T)$ presents the canonical form of θ with additive statistic $T_j(X) = \sum_{i=1}^{N} h_j x_i$. The linearity of $l(\theta|X)$ can be replaced by the $E(T_j(X)|X_o, \theta^t)$ for a multivariate normal distribution $X = X^1, X^2, X^3, ..., X^P$. We can use a vector $T(X) = [T_1(X), T_2(X), T_3(X), ..., T_k(X)]^T$. We can also infer $\sum_{i=1}^{N} x_i^k$ and $\sum_{i=1}^{N} x_i^k x_i^l$ as the sums and sum of products, respectively, with $k, p \geq 1$.

ALGORITHM 4 Iot-BACON-EEM Algorithm.

1 Input: an IoT vector $n x 1$, response variable y, $n x p$ IoT data matrix X where X contains IoT covariate data

2 Output: Nominated outliers

3 Compute the weighted coordinate wise median $med(x)$

4 Determine the Euclidean distance from $med(x)$ as follows while considering the $m = cp$ with minimum a_i data points to form the starting basic subset G

5
$$a_i = ||x_i - med(x)|| \sqrt{\frac{p}{q}} \qquad (9)$$

6 **while** *(G'! = G)* **do**

7 　Compute center as $\hat{m}_G$ and scatter $\hat{C}_G$ by following the EEM method

8 　Update the estimate $\hat{T}_o^G$

9 　Compute squared marginal Mahalanobis distance $MD_G^2(x_i)$ as follows

10 　**for** $i = 1; i < n; i = i + 1$ **do**

11
$$MD_{marg}^2(x_i) = \frac{p}{q}(x_o - m_o)^T C_{oo}^{-1}(x_o - m_o) \qquad (10)$$

12 　The new set G' has all the observations with $MD_G^2(x_i) < c_{\hat{N}\hat{p}\hat{r}} \chi_{p,\alpha}^2$

The sum of products can be analogously presented as $E\sum_{i=1}^{N} x_i^k | X_o$, $\theta^t = \sum_{i=1}^{N} E x_i^k | X_o, \theta^t = \sum_{i=1}^{N} E(x_i^k | x_i^{obs}, \theta^t)$, where $p \geq k \geq 1$. We can also use the Horvitz–Thompson estimators as $T^{k0} = \sum_s w_i E(x_i^k | X_i^{obs}, \theta^t)$ and $T^{kl} = \sum_s w_i E(x_i^k x_i^l | x_i^{obs}, \theta^t)$. The T^{kl} is estimated from the $E(\sum_{i=1}^{N} x_i^k x_i^l | X_o, \theta^t)$. With this the EE-step is concluded. Now, for the M-step we can consider $\theta^{(t+1)} = sop[0] \frac{T^{kl}}{\sum_s w_i}$ where $k \geq 0$, $p \geq l$ and $sop[0]$ refers to the sweep operator for the first column of the IoT matrix.

Algorithm 4 presents the algorithm where BACON and EEM are combined with an IoT-based dataset to roughly estimate the BACON at each iteration. This results in a minimal number of iterations to be involved. Herein, the Mahalanobis distance is assumed as the *marginal* case with a factor $\frac{p}{q}$ with nonmissing number of variables $q = \sum_k r_{ik}$. The partitioning of x is done as $[x_o^T, x_m^T]^T$ and C_{00} refers to the part of the covariance matrix against x_o.

4. Results

In this section, we discuss the results obtained from the study. First, we discuss the robust distance wise analysis and Second, we present robust linear regression analysis for HRV as a case study.

4.1 Robust distance wise analysis

4.1.1 Robust distance

The outliers present in the IoT dataset are clearly shown in Fig. 1. The upper plot shows the robust Mahalanobis distances against the index of IoT-based observations. The robust distance is provided within the range of 5–30. An observation wise plotting is done in this graph. Also, we present the robust distance against the univariate projections in the lower plot. Both plots show 32 potential nominations of observations as outliers shown in red circles. We can also see that there are some black boundary circles near the distance line at 5. It depends on the data analysts to decide whether to consider these circles as potential anomalies.

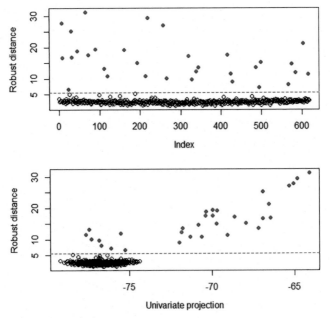

Fig. 1 Robust distance of IoT data points.

4.2 Robust linear regression analysis for HRV

We perform regression between the HRV and other variables of the IoT dataset. We obtain several regression information from the experiment as discussed below.

4.2.1 Residual-fitted

We perform the residual vs fitted values for five different regression studies, e.g., (i) HRV vs C + SO + T + BL + BH, (ii) HRV vs C+ SO + BL + BH, (iii) HRV vs C + BL + BH, (iv) HRV vs BL + BH, and (v) HRV vs BH. Fig. 2 shows all the charts for different regression aspects. Various fitted values are obtained on the x-axis, whereas the residuals are plotted on the y-axis with a range of -3 to 3.

4.2.2 Normal Q–Q

We perform the Q–Q plots for five different regression studies, e.g., (i) HRV vs C + SO + T+ BL + BH, (ii) HRV vs C + SO + BL + BH, (iii) HRV vs C + BL + BH, (iv) HRV vs BL + BH, and (v) HRV vs BH. Fig. 3 shows all the charts for different regression aspects. Various

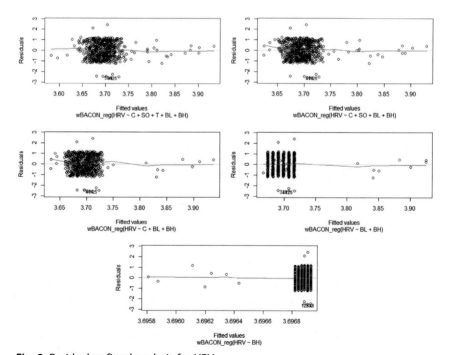

Fig. 2 Residual vs fitted analysis for HRV.

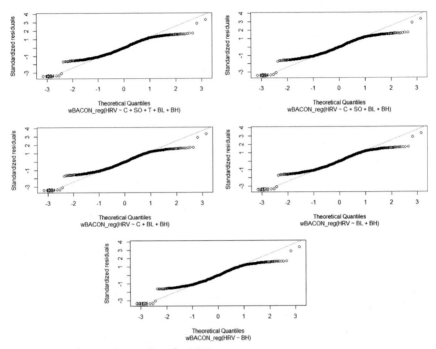

Fig. 3 Normal Q–Q plot analysis for HRV.

theoretical quantile values are obtained on the *x*-axis, whereas the standardized residuals are plotted on the *y*-axis with a range of -3 to 4.

4.2.3 Scale location

We perform the scale locations for five different regression studies, e.g., (i) HRV vs C + SO + T+ BL + BH, (ii) HRV vs C+ SO + BL + BH, (iii) HRV vs C + BL + BH, (iv) HRV vs BL+ BH, and (v) HRV vs BH. Fig. 4 shows all the charts for different regression aspects. Various fitted values are obtained on the *x*-axis, whereas the standardized residuals in square root are plotted on the *y*-axis with a range of 0 to 1.5.

4.2.4 Robust Mahalanobis distance

We perform the robust Mahalanobis distances for five different regression studies, e.g., (i) HRV vs C + SO + T + BL + BH, (ii) HRV vs C + SO + BL + BH, (iii) HRV vs C + BL + BH, (iv) HRV vs BL + BH, and (v) HRV vs BH. Fig. 5 shows all the charts for different regression aspects. Various robust distances are obtained on the *x*-axis, whereas the standardized residuals are plotted on the *y*-axis with a range of -2 to 6.

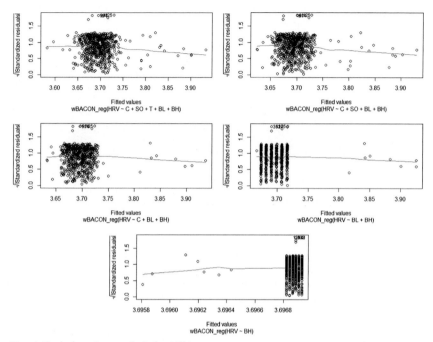

Fig. 4 Scale location analysis for HRV.

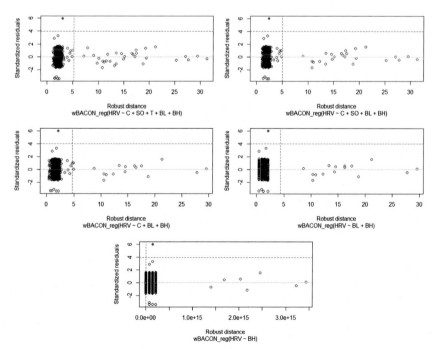

Fig. 5 Robust Mahalanobis distance analysis for HRV.

4.3 Coefficients

Table 1 presents the comparison of regression residuals for five regression studies on HRV. We find almost similar values of Min, 1Q, Median, 3Q, and Max parameters. Median value of all the regression studies lies within -0.01206 to -0.01891. However, Table 2 presents the residual square error (RSE), multiple R^2, adjusted R^2, F-statistic, and p-value. We find minimum RSE for the HRV vs BH regression study. However, the difference of values of the RSE with respect to other regressions is very much similar. Other parameters are also smallest for the HRV vs BH regression study. It has the highest p-value, i.e., 0.9976. It infers that HRV is dependent on all the other variables but mostly on the BH and combination of C + SO + T + BL + BH than others.

We also perform a comparison of parametric values of regression studies as shown in Table 3. Significance is found for all the regressions except HRV vs C + SO + T + BL + BH. Cumulative standard error is higher for this regression than others. In every regression study, the BH has the highest involvement on the dependence factor over HRV.

Table 1 Comparison of regression residuals.

Regression	Min	1Q	Median	3Q	Max
HRV—C+SO+T+BL+BH	−2.48434	−0.58511	−0.01206	0.64300	2.42821
HRV—C+SO+BL+BH	−2.48276	−0.58523	−0.01891	0.63423	2.42865
HRV—C+BL+BH	−2.4782	−0.5817	−0.0134	0.6268	2.4206
HRV—BL+BH	−2.47745	−0.58098	−0.01395	0.63766	2.41625
HRV—BH	−2.46691	−0.58574	−0.01708	0.64618	2.43519

Table 2 Comparison of estimations of regression coefficients.

Regression	RSE	Multiple R^2	Adjusted R^2	F-Static	p-value
HRV—C+SO+T+BL+BH	0.7361	0.002141	−0.006038	0.2618	0.9338
HRV—C+SO+BL+BH	0.7356	0.001893	−0.004642	0.2897	0.8847
HRV—C+BL+BH	0.7352	0.001369	−0.003526	0.2797	0.8401
HRV—BL+BH	0.7346	−0001177	−0.002089	0.3589	0.6986
HRV—BH	0.7345	1.447e−08	−0.001629	8.882e−06	0.9976

Table 3 Comparison of parametric values of regressions.

	Estimate	Std. Error	*t* value	Pr(> |*t*|)
HRV—C + SO + T + BL + BH				
(Intercept)	6.8176680	4.1625086	1.638	0.102
C	0.0167545	0.0479677	0.349	0.727
SO	−0.0118980	0.0218045	−0.546	0.585
T	−0.0137873	0.0353746	−0.390	0.697
BL	−0.0082075	0.0099504	−0.825	0.410
BH	−0.0002322	0.0072381	−0.032	0.974
HRV—C + SO + BL + BH				
(Intercept)	5.5004269	2.4281514	2.265	0.0238*
C	0.0172364	0.0479184	0.360	0.7192
SO	−0.0123191	0.0217626	−0.566	0.5716
BL	−0.0082758	0.0099420	−0.832	0.4055
BH	−0.0001456	0.0072297	−0.020	0.9839
HRV—C + BL + BH				
(Intercept)	4.3106408	1.2151324	3.547	0.000419***
C	0.0167544	0.0478843	0.350	0.726539
BL	−0.0082686	0.0099364	−0.832	0.405649
BH	−0.0002145	0.0072247	−0.030	0.976319
HRV—BL + BH				
(Intercept)	4.3888818	1.1935249	3.677	0.000256***
BL	−0.0084056	0.0099216	−0.847	0.397212
BH	−0.0002168	0.0072195	−0.030	0.976056
HRV—BH				
(Intercept)	3.6994641	0.8728984	4.238	2.6e−05***
BH	−0.0000215	0.0072141	−0.003	0.998

Significance codes: 0 "***" 0.001 "**" 0.01 "*".

5. Conclusion

In this chapter, we implement the BACON-EEM algorithm for the IoT-edge-assisted health dataset. The main focus of this paper is to enable IoT-based devices to nominate the potential outliers present in the given IoT health dataset. We deploy the robust Mahalanobis distance and distance from median schemes to nominate the anomalies from the IoT health dataset. Results show that the BACON-EEM algorithm can minimize the overhead of processing capabilities of the IoT devices.

References

[1] A.S. Hadi, Identifying multiple outliers in multivariate data, J. R. Stat. Soc. B 54 (3) (1992) 761–771.
[2] D. Donoho, Breakdown properties of multivariate location estimators, (Ph.D. qualifying paper), Department of Statistics, Harvard University 1982.
[3] A. Atkinson, Stalactite plots and robust estimation for the detection of multivariate outliers, in: S. Morgenthaler, E. Ronchetti, W. Stahel (Eds.), Data Analysis and Robustness, Birkäuser, 1993.
[4] R.J.A. Little, D.B. Rubin, Statistical Analysis With Missing Data, John Wiley & Sons, Inc., New York, 1987.
[5] R.A. Maronna, Robust M-estimators of multivariate location and scatter, Ann. Stat. 4 (1976) 51–67.
[6] J. Schafer, Analysis of Incomplete Multivariate Data, Volume 72 of Monographs on Statistics and Applied Probability, Chapman & Hall, 2000.
[7] D. Rocke, D. Woodruff, Computation of robust estimates of multivariate location and shape, Stat. Neerlandica 47 (1993) 27–42.
[8] N. Billor, A.S. Hadi, P.F. Vellemann, BACON: blocked adaptive computationally efficient outlier nominators, Comput. Stat. Data Anal. 34 (2000) 279–298.
[9] C. Béguin, B. Hulliger, The BACON-EEM algorithm for multivariate outlier detection in incomplete survey data, Surv. Methodol. 34 (2008) 91–103.
[10] J.F. Magnotti, N. Billor, Finding multivariate outliers in fMRI time-series data, Comput. Biol. Med. 53 (2014) 115–124.
[11] B. Du, L. Zhang, Random-selection-based anomaly detector for hyperspectral imagery, IEEE Trans. Geosci. Remote sens. 49 (5) (2010) 1578–1589.
[12] M. Iqbal, M. Riaz, W. Nasir, Multivariate outlier detection: a comparison among two clustering techniques, Pakistan J. Agric. Sci. 54 (1) (2017) 227–231.
[13] J.A. Jablonski, T.J. Bihl, K.W. Bauer, Principal component reconstruction error for hyperspectral anomaly detection, IEEE Geosci. Remote Sens. Lett. 12 (8) (2015) 1725–1729.
[14] K.M. Sunderland, D. Beaton, J. Fraser, D. Kwan, P.M. McLaughlin, M. Montero-Odasso, A.J. Peltsch, F. Pieruccini-Faria, D.J. Sahlas, R.H. Swartz, S.C. Strother, The utility of multivariate outlier detection techniques for data quality evaluation in large studies: an application within the ONDRI project, BMC Med. Res. Methodol. 19 (1) (2019) 1–16.
[15] A. Cerioli, A. Farcomeni, M. Riani, Strong consistency and robustness of the forward search estimator of multivariate location and scatter, J. Multivar. Anal. 126 (2014) 167–183.

[16] D. Neira-Rodado, C. Nugent, I. Cleland, J. Velasquez, A. Viloria, Evaluating the impact of a two-stage multivariate data cleansing approach to improve to the performance of machine learning classifiers: a case study in human activity recognition, Sensors 20 (7) (2020) 1858.

[17] D. Pelleg, A. Moore, Active learning for anomaly and rare-category detection, Adv. Neural Inf. Process. Syst. 17 (2004) 1073–1080.

[18] J. Gao, F. Liang, W. Fan, C. Wang, Y. Sun, J. Han, On community outliers and their efficient detection in information networks, in: Inproceedings of the 16th ACM SIGKDD International Conference on Knowledge Discovery and Data Mining, 2010, pp. 813–822.

[19] L. Gao, Q. Guo, A.J. Plaza, J. Li, B. Zhang, Probabilistic anomaly detector for remotely sensed hyperspectral data, J. Appl. Remote Sens. 8 (1) (2014) 083538-1–083538-20.

[20] WBACON R Package, Available Online, https://cran.r-project.org/web/packages/wbacon/index.html, 2021. accessed 28.07.21.

[21] IoT Health dataset, Available Online, https://github.com/ParthaPRay/IoTHealth DataSet/blob/main/IoT-multivar-anomaly-with-SO.csv, 2021. accessed 03.08.21.

[22] P. Gupta, A. Jindal, D. Sengupta, Linear time identification of local and global outliers, Neurocomputing 429 (2021) 141–150.

[23] K. Kolomvatsos, C. Anagnostopoulos, Landmark based Outliers Detection in Pervasive Applications, in: 2021 12th International Conference on Information and Communication Systems (ICICS), IEEE, 2021, pp. 201–206.

[24] S. Mishra, R. Sagban, A. Yakoob, N. Gandhi, Swarm intelligence in anomaly detection systems: an overview, Int. J. Comput. Appl. 43 (2) (2021) 109–118.

[25] R. Ávila, R. Khoury, R. Khoury, F. Petrillo, Use of security logs for data leak detection: a systematic literature review, Secur. Commun. Netw. 2021 (2021) 1–29.

[26] S.A.N. Nozad, M.A. Haeri, G. Folino, SDCOR: Scalable density-based clustering for local outlier detection in massive-scale datasets, Knowl.-Based Syst. 228 (2021) 107256.

[27] J. Sheng, L. Wang, H. Cheng, Q. Zhang, R. Zhou, Y. Shi, Strategies for multivariate analyses of imaging genetics study in Alzheimer's disease, Neurosci. Lett. 762 (2021) 136147.

[28] T. Rincy N, R. Gupta, Design and development of an efficient network intrusion detection system using machine learning techniques, Wireless Commun. Mobile Comput. 2021 (2021) 1–35.

[29] F. Al-Obeidat, A. Rocha, M.S. Khan, F. Maqbool, S. Razzaq, Parallel tensor factorization for relational learning, Neural Comput. Appl. (2021) 1–10. Early access.

[30] Y. Dong, X. Jiang, H. Zhou, Y. Lin, Q. Shi, SR2CNN: zero-shot learning for signal recognition, IEEE Trans. Signal Process. 69 (2021) 2316–2329.

[31] I.A. Khan, D. Pi, N. Khan, Z.U. Khan, Y. Hussain, A. Nawaz, F. Ali, A privacy-conserving framework based intrusion detection method for detecting and recognizing malicious behaviours in cyber-physical power networks, Appl. Intell. 51 (2021) 1–16.

[32] A. Alos, Z. Dahrouj, Using multiple deep neural networks platform to detect different types of potential faults in unmanned aerial vehicles, J. Aerosp. Technol. Manag. 13 (2021) e1321.

[33] L. Settipalli, G.R. Gangadharan, Healthcare fraud detection using primitive sub peer group analysis, Concurr. Comput. Pract. Exp. 33 (23) (2021) e6275.

[34] A.M. Di Brisco, S. Migliorati, A spatial mixed-effects regression model for electoral data, Stat. Methods Appl. 30 (2) (2021) 543–571.

[35] Y. Altmann, S. McLaughlin, A. Hero, Robust linear spectral unmixing using anomaly detection, IEEE Trans. Comput. Imaging 1 (2) (2015) 74–85.

[36] A. Arfaoui, A. Kribeche, S.M. Senouci, M. Hamdi, Game-based adaptive anomaly detection in wireless body area networks, Comput. Netw. 163 (2019) 106870.

[37] P. Sawant, N. Billor, H. Shin, Functional outlier detection with robust functional principal component analysis, Comput. Stat. 27 (1) (2012) 83–102.

[38] A.J. Messer, K.W. Bauer Jr, Method of sensitivity analysis in anomaly detection algorithms for hyperspectral images, in: Algorithms and Technologies for Multispectral, Hyperspectral, and Ultraspectral Imagery XXIII, vol. 10198, International Society for Optics and Photonics, 2017, p. 101980U.

[39] S. Sandbhor, N.B. Chaphalkar, Impact of outlier detection on neural networks based property value prediction, in: Information Systems Design and Intelligent Applications, Springer, Singapore, 2019, pp. 481–495.

[40] K.M. Kirtland, Outlier Detection and Multicollinearity in Sequential Variable Selection: A Least Angle Regression-Based Approach, Cornell University, 2017.

[41] S.A. Tomlins, D.R. Rhodes, J. Yu, S. Varambally, R. Mehra, S. Perner, F. Demichelis, B.E. Helgeson, B. Laxman, D.S. Morris, Q. Cao, The role of SPINK1 in ETS rearrangement-negative prostate cancers, Cancer cell 13 (6) (2008) 519–528.

[42] R.J. Johnson, J.P. Williams, K.W. Bauer, AutoGAD: an improved ICA-based hyperspectral anomaly detection algorithm, IEEE Trans. Geosci. Remote Sens. 51 (6) (2012) 3492–3503.

[43] A.S. Hadi, A.R. Imon, M. Werner, Detection of outliers, Wiley Interdiscip. Rev. Comput. Stat. 1 (1) (2009) 57–70.

[44] B. Du, L. Zhang, D. Tao, D. Zhang, Unsupervised transfer learning for target detection from hyperspectral images, Neurocomputing 120 (2013) 72–82.

[45] I.H. Naqvi, Outlier/Event Detection Techniques in Wireless Sensor Networks, 2012.

[46] H. Eldardiry, K. Sricharan, J. Liu, J. Hanley, B. Price, O. Brdiczka, E. Bart, Multi-source fusion for anomaly detection: using across-domain and across-time peer-group consistency checks, J. Wirel. Mob. Netw. Ubiquitous Comput. Dependable Appl. 5 (2) (2014) 39–58.

[47] A. Cerasa, A. Cerioli, Outlier-free merging of homogeneous groups of pre-classified observations under contamination, J. Stat. Comput. Simul. 87 (15) (2017) 2997–3020.

[48] S.G. Beaven, G.G. Hazel, A.D. Stocker, Automated Gaussian spectral clustering of hyperspectral data, in: Algorithms and Technologies for Multispectral, Hyperspectral, and Ultraspectral Imagery VIII, vol. 4725, International Society for Optics and Photonics, 2002, pp. 254–266.

[49] A.J. Messer, On the development of robust anomaly detection algorithms with limited labeled data, AIR FORCE INSTITUTE OF TECHNOLOGY WRIGHT-PATTERSON AFB OH WRIGHT-PATTERSON AFB United States, 2017.

[50] H.H. Pajouh, G. Dastghaibyfard, S. Hashemi, Two-tier network anomaly detection model: a machine learning approach, J. Intell. Inf. Syst. 48 (1) (2017) 61–74.

[51] D. Verma, R. Kumar, A. Kumar, Survey paper on outlier detection using fuzzy logic based method, Int. J. Cybern. Inf. Sci. (IJCI) 6 (1/2) (2017) 29–35.

About the author

Partha Pratim Ray has completed bachelors and masters degree in Computer Science and Engineering and Electronics and Communication Engineering, respectively from West Bengal University of Technology, India. He is a senior member of IEEE. He has published more than 70 research articles in various research avenues. His area of interest is Internet of Things and Next Generation Computing.

CHAPTER FIVE

The edge AI paradigm: Technologies, platforms and use cases

Pethuru Raj[a], J. Akilandeswari[b], and M. Marimuthu[c]
[a]Site Reliability Engineering (SRE) Division, Reliance Jio Platforms Ltd. (JPL), Bangalore, India
[b]Department of Information Technology, Sona College of Technology, Salem, TN, India
[c]Research Scholar, Sona College of Technology, Salem, TN, India

Contents

Abstract

Two of the most interesting topics in the technology world today are edge computing and artificial intelligence (Ai). Individually each of them has done exceedingly well in contributing for the betterment of the society. Now if they converge, we can solidly expect a paradigm shift and the impacts will be mesmerizingly profound and phenomenal. The fusion of these two popular technologies is laying down a spectacular foundation for creating new kinds of experiences and opportunities for people. Newer possibilities can easily come up across business verticals. There will be premium and

Advances in Computers, Volume 127
ISSN 0065-2458
https://doi.org/10.1016/bs.adcom.2022.02.003

people-centric services emerging and evolving for automating and accelerating a number of manual activities in our daily lives. This unique combination can result in fructifying the longstanding demand of people IT. These also smoothen the path and clear the route for the age-old transition from business IT to people IT. This convergence directly influences peoples' lives in a distinguishing and deft manner. This can save time, improve privacy, reduce network traffic, and enable applications or devices to be optimized for specific environments. Precisely speaking, the convergence of these two strategically sound technologies can bring forth noteworthy advancements and accomplishments for the total society. In this chapter, we are to see how this unique linkage is going to be hugely beneficial for institutions, individuals and innovators. The union is being termed as "Edge AI."

1. Introduction

Two of the most interesting topics in the technology world today are edge computing and artificial intelligence (Ai). Individually each of them has done exceedingly well in contributing for the betterment of the society. Now if they converge, we can solidly expect a paradigm shift and the impacts will be mesmerizingly profound and phenomenal. The fusion of these two popular technologies is laying down a spectacular foundation for creating new kinds of experiences and opportunities for people. Newer possibilities can easily come up across business verticals. There will be premium and people-centric services emerging and evolving for automating and accelerating a number of manual activities in our daily lives. This unique combination can result in fructifying the longstanding demand of people IT. These also smoothen the path and clear the route for the age-old transition from business IT to people IT. This convergence directly influences peoples' lives in a distinguishing and deft manner. This can save time, improve privacy, reduce network traffic, and enable applications or devices to be optimized for specific environments. Precisely speaking, the convergence of these two strategically sound technologies can bring forth noteworthy advancements and accomplishments for the total society. In this chapter, we are to see how this unique linkage is going to be hugely beneficial for institutions, individuals and innovators. The union is being termed as "Edge AI."

2. Delineating the two paradigms

In this section, we are to dig deeper and dwell at length to provide detailed information on the two participating technologies: artificial intelligence (AI) and edge computing. Further on, how their union is going to be a game-changer for businesses as well as people in the ensuing digital era. Leading market analysts and researchers are in unison in insisting and illustrating that AI is the future not only for IT but also for all business verticals. In fact, every institution, innovator and individual are to benefit immensely through the distinct developments happening in the AI space. Fresh digital life services and applications can be conceived, concretized and delivered to customers, clients and consumers with the faster maturity and stability of AI technologies and tools.

With the ready availability of big data, the flourish of virtually infinite and affordable cloud storage and computation facilities, the emergence of path-breaking machine and deep learning (ML/DL) algorithms, and highly miniaturized yet performant semiconductors, the AI field is bound to shake up the world in the days to unfurl. Every data-generating domain is to benefit enormously with the breakthrough improvisations happening in the AI space. The process of transitioning data into information and knowledge is being optimized and automated through a host of technologies and tools. Thus, knowledge discovery and dissemination aspects gain prominence. There are a plethora of pioneering personal as well as industrial use cases critically leveraging various AI distinctions. Not only business workloads but also IT systems/services become intelligent in their decisions, deals and deeds with the unprecedented AI power. Digitized entities and connected devices become cognitive in their actions and reactions. There are open source and commercial-grade AI platforms, frameworks, libraries, accelerators and tools in plenty in producing AI model creation, evaluation, and optimization. AI models creation, evaluation, optimization and refinement have become the new normal for worldwide enterprises in order to be ahead of their immediate competitors and to retain the yearned edge in brand value. Precisely speaking, empowering computers, communicators, consumer electronics, handhelds, wearables, portables, implantable devices, gadgets and gizmos with human-like intelligence is the ultimate goal of AI.

The second one is none other than the edge computing paradigm, which is to fulfill the idea of long pending decentralized and distributing computing. That is, computing has to be performed by all the participating devices in a network. In other words, every resource-intensive node in any network has to have the computing and communication capability to team up with one another in the vicinity and with remote ones through networking. The brewing idea is to envisage and ensure state-of-the-art services to businesses as well as commoners. This is quite opposite to the much-trumpeted cloud computing, which is typically consolidated, centralized, automated and shared. Client machines and devices from every nook and corner of the whole world can connect and get the things done through cloud servers, which are increasingly commoditized and cheaper. In the extremely connected world, the computing is done across an entire network instead of centralizing it in cloud environments. That is, not only server machines but also client devices participate and contribute for a bigger purpose.

Toward further clarification, take this renowned example. A security camera put up in a critical junction is capturing and streaming all the video content captured to a nearby or faraway cloud center (public or private) over a network to be analyzed for fulfilling certain requirements such as people and object detection. Now with advanced cameras, that video analysis can be performed locally within the camera itself. This saves a lot. First, it dramatically reduces the amount of data to be transmitted over the network thereby network bandwidth gets preserved. This results in reducing network congestion. Precisely speaking, real-time data capture, storage, processing, decision-making and actuation gets accomplished through edge/fog computing. Data privacy and security are being accomplished through the grandiose participation of networked embedded devices, which also contribute in proximate, timely and trustworthy computation, analytics, knowledge discovery, and action.

There are a number of versatile technological upgrades. We have powerful chips such as multicore CPUs, GPUs, FPGAs, VPUs, TPUs, ASICs and other SoCs. Another interesting development is the open ISA (Instruction Set Architecture) RISC-V. RISC-V provides another option beyond the traditional general-purpose architectures, such as x86 or Arm. With RISC-V, it is possible to customize the instruction set for specialized or extreme application demands, such as artificial intelligence (AI) algorithms. In the case of the surveillance example above, the video processing happens in the camera itself thereby real-time decision and action can be initiated and implemented. Such proximate processing could dramatically reduce the amount of data (video frames) being transmitted to cloud environments over dedicated or the open Internet. By deploying analytics libraries in client devices, which are adorned with specialized and adroit processors, the local and real-time computing gets accomplished. If AI is done right at the device level, then the result is altogether different.

The combination of the versatile AI-specific semiconductor components and the path-breaking AI algorithms in conjunction with the fast, low-cost and local storage offered by flash memory brings forth adequate computing and analytics power. This transition empowers networked edge devices to accomplish proximate data processing in order to emit out instantaneous insights and results. Thus, the new domain of "Edge AI" has blossomed in the recent past. AI pundits, pioneers and proponents are working intently in unearthing people-centric and industrial use cases for edge AI as a way forward to popularize its richness and reach.

3. Tending toward the digital era

There are path-breaking digital technologies and tools emerging to fulfill the digital transformation needs across industry verticals, governments, organizations, and institutions. Every tangible thing in our everyday environments is getting methodically digitized through the application of digitization and edge technologies. And when these digitized entities interact with one another, there is a tremendous amount of digital data gets generated. As widely reported, data has become the new oil or fuel for real and sustainable business transformation. Data is being termed as the strategic asset for any organization to plan and surge ahead. Decision-makers, executives and stakeholders leverage digital intelligence to steer their organizations in the right direction to the destination. The power of digital technologies comes handy in transitioning digital data to information and to knowledge. There are influential technologies and tools for enabling data capture, integration, virtualization, ingestion, wrangling, masking, preprocessing, storage, analytics, and visualization capabilities. The brewing trend and transition is that we are heading toward data-driven insights and insights-driven decisions and deeds. We have listed out the prominent digital technologies below for the benefit of our esteemed readers.

For business transformation, the following activities are being insisted by enterprise architecture (EA) architects, consultants and advisors.

1. Technology Choice
2. Architecture Assimilation
3. Process Excellence
4. Infrastructure Optimization
5. Data Analytics

Digitization Technologies: Let us start with digitization technologies. There are digitization, connectivity and edge technologies. These are collectively called as the Internet of Things (IoT) technologies. These are for transitioning ordinary items into extraordinary artifacts. Edge technologies such as sensors, actuators, stickers, smart dust, specks, tags, barcodes, chips, microcontrollers, beacons, LED lights, etc. are being internally or externally attached on important assets and this nuanced attachment enable edge technologies to acquire the assets' operational and environmental data. This is the way all kinds of physical, mechanical and electrical systems in our personal, social and industrial environments get digitized. Dumb objects become animated entities. Commonly found and concrete objects get methodically digitized. Digital objects can do small-scale and real-time computing and are stuffed with communication modules. Digitized objects can share their health condition, performance, security, operational, and log data. Further on, digitized assets at the ground level gets hooked to cloud-based software applications, services and databases over the Internet. This phenomenon is being termed as cyber physical systems (CPS). Thus, the concepts such as digitization, connectivity, automation and orchestration are acquiring special significance in the digital era. Thus, the Internet of Things (IoT) technologies are destined for establish and sustain the digital living. Digital entities connect, communicate, collaborate, corroborate and correlate with one another. This produces a tremendous amount of multi-structured data. In the subsequent sections, we are to write about how digitalization technologies come handy in squeezing out viable and venerable insights out of such digital data.

The Explosion of Digital Entities: As per the leading market watchers and analysts, there will be trillions of digitized entities, billions of connected devices and millions of microservices in the years to come. With the faster maturity and stability of path-breaking miniaturization technologies, we have state-of-the-art micro- and nanoelectronics products flourishing everywhere. As indicated above, there are competent digitization and edge technologies. The important digitization technologies are disappearing sensors and actuators, which are plentifully and purposefully occupying most of our everyday environments these days. Sensors and actuators are also known as implantable devices. By applying these powerful digitization and edge technologies, all kinds of casual and tangible things in our living and working spaces are entitled to become smart objects. In other words, ordinary items in our daily places become extraordinary. Any tangible/concrete object is bound to become sentient material through the smart application of the above-mentioned edge and digitization technologies.

What is the result of all these empowerments and enablement? All sorts of physical, mechanical, electrical and electronics systems in our buildings, manufacturing floors, entertainment plazas, eating joints, railway stations, air and sea ports, auditoriums, sports stadiums, microgrids, nuclear establishments, shopping malls, etc. are being fruitfully digitized with much care and clarity. These distinct, deeper and decisive digital technologies are intrinsically empowering the whole world to have trillions of digitized entities in the coming years. It is therefore indisputably correct to state that the digitization movement is in full speed. Our shirts become e-shirts, our doors, cots, windows, chairs, tables, wardrobes, kettles, wares, utilities, utensils, etc. will become smart through the formal attachment of digitization and edge technologies. Especially multifaceted sensors and actuators play a very vital role in shaping up the digital world and living. Everything becomes digitized.

Already all our computers (laptops, tablets, desktops, and server machines) are integrated with the Internet communication infrastructure in order to access web content and services. These days our communicators (smartphones) are web-enabled in order to fulfill anytime, anywhere, any network, any device access of web content and services. Additionally, sensors-attached physical, mechanical, electrical and electronics are hooked into the Web in order to be remotely empowered. Such an empowerment makes our everyday devices and machines to be ready for contributing copiously for the realization of digital societies. Thus, physical devices are set to become digital devices, which are intrinsically revitalized to perform edge computing. Digital elements are called edge devices.

The Proliferation of Connected Devices: We talked about ordinary things getting transformed into digital elements through a few pioneering technologies. Now come to the device world. With the fast-growing device ecosystem, we are being bombarded with a variety of slim and sleek, handy and trendy, resource-constrained and intensive, purpose-specific and agnostic, multimedia, multimodal, and multifaceted devices. There is a growing array of handhelds, mobiles, portables, wearables, nomadic, and fixed devices. Further on, we have consumer electronics, medical instruments, home appliances, communication gateways, robots, drones, cameras, game consoles, machineries, equipment, medical instruments, single board computers (SBCs), programmable logic controllers (PLCs), SCADAs, etc. yearning to be digitized and connected to attain all the originally expressed benefits. It is forecast that there will be billions of such higher-end devices soon. With the surging popularity of the Internet of Things (IoT) paradigm, every digitized entity and device is solemnly readied to be Web-enabled. Now with the overwhelming adoption of the cloud idea, every device in and around us is ordained to be cloud-enabled. In other words, every electronics is slated to be connected. The number of connected devices is expected to be in billions soon. Device-to-device (D2D) and device-to-cloud (D2C) integration experimentations and scenarios are gaining momentum due to the steady growth of several implementation technologies.

Precisely speaking, digitized entities and connected devices are stuffing and saturating our everyday environments (personal, social and professional). For enabling context-awareness and multidevice computing applications, the new connected era beckons and dawns upon us. These technologically enabled devices are typically called as edge or fog devices because they are at the edge of the network. They are very near to us. We can see, touch and feel them. For example, the point of sale (PoS) devices in a retail store is an edge device, cameras in our homes, hotels, and hospitals are edge devices, robots in a happening place such as a surgery room inside a hospital is an edge device, the list goes on and on.

Now comes the twist. Edge devices generally generate a lot of data every hour. That is, they collect a lot of useful and usable data about themselves, their environments, owners, users, etc. Edge devices collect operational, log, health condition, performance, and security data. Edge devices are touted as the primarily data collectors about the various temporal, special and behavioral aspects. That is, edge devices plus empowered systems (systems empowered by edge devices) individually and collectively throw a lot of data of their capabilities, capacities, states, change of states, etc. Compared to digital elements, connected devices are computationally powerful. Thus, digital elements are primarily data generator and transmitter. However, connected devices are capable of receiving and processing any amount of digital data in real time.

The Continued Adoption of Cloud Applications, Services and Data Sources: We have discussed about edge devices and their explosion. At one end, we have zillions of digitized entities being made out of physical elements. The second category in the hierarchy is scores of versatile electronics devices capable of capturing, storing, processing, and mining digital data emitted by exponentially growing digitized entities. Now, at the top of the spectrum is none other than cloud assets and applications. Real-time data analytics is being accomplished through electronics devices. Data collection happens through digitized entities at the lower end. Increasingly electronics devices (alternatively termed as edge or fog devices) support in-device AI processing. That is, lightweight AI toolkits are deployed on edge devices to subject digital data to a variety of deeper and real-time investigations in order to extract real-time insights that can be looped back to decision-making systems and people to initiate the process of pondering about the next course of actions with all the clarity and confidence.

In cloud environments (private, public or hybrid), historical and comprehensive data analytics through integrated data analytics platforms and AI frameworks is being done. Thus, the role of cloud infrastructures and platforms in this increasingly data-engulfed world can't be undermined. Especially big data analytics through batch processing can be comfortably done in traditional cloud environments.

The Arrival of Edge Clouds: Considering the need for real-time insights and applications for establishing and sustaining intelligent enterprises, industry houses and cloud service providers are setting up miniaturized cloud environments in their offices and campuses. Telecommunication service providers are using their base stations to have a small-scale cloud centers to ensure real-time customer experience.

Especially with edge devices becoming powerful, edge device clusters/clouds are being formed in a dynamic and ad hoc manner to perform specific tasks. With the forecast of billions of connected devices, there will be a bigger focus on forming and using edge device clouds in order to build and release location-specific, context-aware and real-time applications. Thus, there will be a combination of traditional and modern cloud centers in visualizing and realizing next-generation business and people applications in the years ahead. With devices contributing their unique capabilities, the scope and sophistication of software applications will be deeper, deft, and decisive.

Digitalization Technologies: In this section, we are to see how digital transformation being simplified and streamlined by digital technologies has laid down a stimulating and sparkling foundation for business transformation. Lately, we are being bombarded with multifaceted digitized entities, connected devices and microservices. Because of the accumulation of digitized assets, a massive amount of multi-structured digital data gets produced, gathered, stocked and subjected to a variety of deeper investigations. In short, the faster adoption of digitization technologies results in digitized elements, which generate big data. Now with the grandiose arrival of digitalization technologies, making sense out of digital data gets accelerated. Digital data is being actually gleaned from different and distributed digital entities and devices. That is, the process of transitioning of raw data into information and into knowledge is being automated through a host of revolutionary digitalization technologies such as:

1. Artificial Intelligence (AI) Algorithms
2. Integrated Analytics (Big, Fast and Streaming Data) Platforms

3. Blockchain for Device and Data Security

4. Digital Twins

The above-mentioned technologies are primarily for extracting actionable insights in time out of digital data. There are other supporting technological innovations and disruptions for speeding up the process of developing and delivering highly scalable, available, reliable, and portable digital applications that can run on multiple systems including personal devices, enterprise and cloud servers.

1. Data Virtualization
2. Databases, Data Warehouses and Data Lakes
3. Knowledge Visualization Dashboards
4. Cloud-native Computing
5. Fog/Edge Computing
6. Serverless Computing
7. Microservices Architecture (MSA)
8. Event-driven Architecture (EDA)
9. Message Brokers and Queues
10. Event Meshes
11. Cybersecurity
12. Process Automation
13. Workflow Orchestration

Thus, data to information and knowledge gets facilitated through digitalization technologies. The combination of digitization and digitalization technologies is generally termed as digital technologies.

4. The key connectivity technologies

There are wireline and wireless communication technologies. Lately, 5G is the widely talked about communication technology. This section is allocated to throw more light on the 5G ecosystem. 5G is the new cellular communication standard bringing forth a new kind of network that can connect everyone and everything including digitized entities/smart object/ sentient materials, consumer electronics, medical instruments defense equipment, manufacturing machineries, etc. 5G is being positioned and proclaimed as a new wireless communication technology capable of delivering data transfer of gigabits per second. It facilitates ultra-low latency, higher reliability, and massive network capacity/bandwidth. It has the inherent power to accommodate higher density of devices. Higher performance and heightened efficiency are being seen as the key motivations for envisaging and realization of newer business models and industry use cases.

5G is based on OFDM (Orthogonal frequency-division multiplexing). The OFDM method is famous for efficiently modulating a digital signal across different channels to reduce interference. 5G uses the 5G new radio (NR) air interface and uses wider bandwidth technologies. The 5G NR air interface can enhance OFDM to deliver a much higher degree of flexibility and scalability. This is the key reason for 5G to give access to more people and things for implementing hitherto unheard use cases. 5G can operate in lower bands (e.g., sub-6 GHz) as well as mmWave (e.g., 24 GHz and up). This unique capability will bring in extreme capacity, multi-Gbps throughput and the much-insisted low latency. Mission-critical communications are to be accomplished through the leverage of the 5G standard. With the surging popularity of IoT systems, solutions and services, the role and responsibility of 5G communication acquires special significance. For enabling IoT sensors and devices to talk to one another in the vicinity as well with remotely held cloud-hosted software services and applications, 5G plays a very vital role. The below figure tells all about the distinctions of the 5G communication.

5. The 5G use cases and benefits

Empowering Enhanced Mobile Broadband: Compared to the previous generations, 5G will supply enhanced mobile broadband. The user experiences will be better. For example, virtual and augmented reality (VR and AR) experiences will be more intuitive and immersive.

Enabling New Capabilities: With the faster proliferation of IoT devices and services, 5G contributes immensely in realizing context-sensitive, time-critical and people-centric applications. As indicated elsewhere, as per the reports of leading market watchers and analysts, there will be billions of IoT devices and trillions of IoT sensors in the years ahead. For the forthcoming connected era, 5G guarantee critical connectivity infrastructure. 5G also simplifies and streamlines the aspect of linkage of ground-level IoT systems and cloud IT infrastructures. The Vehicle-to-Everything (V2X) communication will be facilitated. The V2X includes Vehicle-to-Vehicle (V2V), Vehicle-to-Infrastructure (V2I), and smart cars. Telesurgery will become the new normal with the maturity and stability of 5G communication. Telemetry data gets captured and crunched to squeeze out actionable insights in time. With right and relevant insights at hand, a variety of intelligent offerings can be designed, developed and deployed.

Virtual networks (5G slicing) tailored to each use case: 5G will be able to support all communication needs from low power Local Area Network (LAN) – like home networks, for example, to Wide Area Networks (WAN), with the right latency/speed settings. This is addressed today is by aggregating a variety of communication networks (Wi-Fi, Z-Wave, LoRa, 4G, etc.).

The 5G networks under deployment feature lower latency, higher capacity, and increased bandwidth. These improvements will have far-reaching implications on how people across the world live, relax, and work.

Network Speed Increase: 5G networks can give the speed up to 10 Gbps, which is almost 100% increase compared to 4G communication.

Ultra-Low Latency: Latency measures the time it takes for your phone to send a message and get a response. Shorter latency enables quick response interactions. Low latency 5G networks open up lots of new possibilities for services that demand nearly instant response time. That includes telemedicine, self-driving vehicles, remote monitoring, measurement and management of mission-critical assets, etc.

Agriculture, manufacturing, and logistics will benefit from 5G-inspired lower latency. The combination of high speed and minimal lag is also good for virtual reality (VR) and augmented reality (AR) applications, which are likely to be used extensively as connectivity improvements create a more seamless, immersive experience.

Enhanced Network Capacity: 5G networks are poised to deliver up to 1000 times more capacity than the currently pervasive 4G networks. Due to the participation of a large number of networked embedded devices, smartphones, connected machineries, etc. in an industrial environment, 5G is being seen as the powerhouse for the ensuing IoT era. Smart factories, cities, farms, university campuses, warehouses, airports, and other mission-critical places are to benefit immensely with 5G networks. With 5G, one million IoT sensors and devices can be accommodated within a square kilo meter surroundings.

High Reliability: Global Wireless Solutions, Inc. (GWS) has found that AT&T has the most reliable 5G nationwide network, completing 99.5% of the data transfer tasks performed on 5G networks. Verizon Wireless finished second at 98.8% and T-Mobile third at 97.3%.

The next-generation computing needs pioneering communication capacity and capability. The arrival of 5G communication facility is to fairly enthuse pundits and pioneers to ponder about fresh business, technical and user cases. Especially for fulfilling the ideals of digital transformation, the 5G connectivity is indispensable.

6. About edge computing

Edge computing is the computing capability and facility being provided by edge devices. With edge sensors and devices are projected to be in billions in the years to come, the amount of data getting generated by edge devices is humungous. As there is a realization that every type of edge data comprises some useful and usable knowledge, edge data processing gains prominence. It is mandated for extracting actionable insights out of data heaps in time. Data value goes down when time elapses and hence data has to be processed immediately in order to extract time-sensitive business value. Edge computing follows the proven and potential distributed architecture. Cloud computing is quite centralized whereas edge computing is a distributed model. Multiple heterogeneous devices voluntarily cooperate with one another to do edge data collection, storage and processing. The device heterogeneity is being taken care of through containerization. And multiple devices participating in business computing is being fully facilitated through container lifecycle management platforms such as Kubernetes.

Edge devices, instead of sending data to nearby or faraway cloud servers, are being empowered to collect, cleanse and crunch their data instantaneously if they have the requisite data storage and processing power. The brewing trend, as articulated above, edge devices are instrumented with higher processing and memory/storage power. Thereby edge devices innately readied to contribute for real-time computing.

Further on, edge devices can intrinsically form an ad hoc and dynamic clusters/clouds to accomplish bigger and better things. That is, device clusters and clouds are being formed out of edge devices. This technologically enabled transition is seen as a game changer towards establishing and sustaining people IT. Thus far, businesses have been immensely benefiting out of all the noteworthy achievements in the IT space. Now every individual is set to benefit immeasurably through the evolutions and revolutions happening in the IT space. As there are billions of connected devices, the amount of memory, processing power, storage and IOPS capacities are turning out to be simply phenomenal and hence the future certainly belongs to the pioneering edge computing model. There are edge gateways/servers in order to contribute as a master or control server whereas other devices in the network edge may help in data collection, storage and processing. Edge devices may be classified into categories: resource-constrained and intensive. Edge devices such as sensors and actuators are typically blessed with less processing and memory power and there are other edge devices stuffed with more computing capability and storage capacity. The resource-intensive edge devices are normally termed as edge servers as they can participate in data processing captured by edge devices.

In a smart home environment, there may be several single-purpose and multipurpose sensors ranging from gas, temperature, humidity, presence, pressure, and other sensors enabling home-automation activities. These sensors are termed as edge devices. Additionally, there are resource-intensive connected devices such as consumer electronics, Wi-Fi gateways, kitchen utensils, etc. within the home. These powerful devices are being called as edge servers as they can do data processing individually and also, they can form a temporary and purpose-specific device cluster dynamically and swiftly to accomplish bigger and better needs for the home owners and occupants.

Edge computing, in a way, decentralizes processing power to ensure real-time processing without the much-maligned network latency while reducing network bandwidth and storage requirements. Edge computing makes it possible to visualize and realize a number of next-generation people-centric services and applications. Industries also can get a lot of manual things getting automated and accelerated. This newer computing model offers a range of value propositions for producing and running smart IoT applications and use cases across industry verticals. The prime advantages of edge computing are given below:

- **Low latency**: Edge servers/devices are calculatedly instrumented to do the much-demanded proximate data processing in real time as they carry the advantage of being situated near data sources. The computation logic, once centralized, now moves to edge devices to accomplish decentralized and distributed computing. Edge devices are digitized and destined to capture and transmit their own operational state and also their surroundings data to nearby edge servers. Edge servers are close to edge devices and hence data processing happens immediately and decisions are taken quickly.
- **Scalability**: For catering more data, additional devices can be provisioned and configured to accomplish real-time data analytics. This is due to the accumulation of edge devices in any important environment. Device clusters and clouds will become the new normal in conceiving and concretizing multidevice, process-aware and composite applications

- **Security and Privacy**: Devices and transacted data are restricted to the particular environment. There is no need to transmit data over the public and open Internet. Data transmission is within the Intranet and hence data security and privacy are inherently ensured. Additionally, the proven blockchain technology is being utilized in order to bring in an additional layer of security as cyber-attacks and terrorisms are consistently on the rise in the extremely connected world.
- **Bandwidth Efficiency**: The first-level data analytics and filtering happen at the edge devices and hence only useful data gets transmitted to cloud environments for comprehensive analytics. That means the expensive and scarce network bandwidth gets saved through such source-level data analytics.
- **Robust connectivity**: Even if the cloud connectivity is unavailable, edge computing happens without any issue.
- **Real-time Computing**: Data capture and processing happen locally and quickly. Edge computing delivers the real meaning and value for real-time applications and services. Real-time and event-driven enterprises will become the new normal.

Thus, leveraging one or more edge devices to do real-time implementation of advanced and people-centric applications is gaining momentum with the maturity and stability of edge computing technologies and tools.

7. Edge computing architecture

The key information is that edge computing fully complies with the distributed computing patterns. As articulated above, edge devices are all set to become pervasive and persuasive. Every industry vertical is embracing this strategically sound technology in order to be right and relevant to their customers, consumers and employees. Devices are being increasingly stuffed with more processing power and memory capacity and hence edge devices are ceaselessly penetrating into every business domain to bring in deeper and decisive automation. In an edge architecture, there are three types of devices contributing immensely. They are edge sensors and actuators, edge devices and gateways.

- **Edge Devices (Sensors and Actuators)**: As accentuated above, edge sensors are not blessed with high-end memory, processing and storage power. They are for sensing a variety of things and to respond accordingly. We know about gas, heat, pressure, presence, movement, gestures, oscillation, and humidity sensors. Hence, they typically collect data about their surroundings and send it to nearby edge servers. Sensors are miniaturized and hence becoming disappearing. On the other hand, extracted information from aggregated sensor data gets supplied to actuators in time to act with all the confidence and clarity. Sensors and actuators are generally not for data processing. Sensors are sensitive whereas actuators are for real action based on the sensed data.

- **Edge Servers**: These are resource-intensive systems and hence they can run operating systems such as Android, iOS, **Raspbian**, etc. Further on, they can run data analytics platforms/tools/accelerators in order to do real-time data analytics. They can do streaming analytics of edge sensor/actuator data in real time. Time-series data is more prevalent and relevant for edge intelligence. On-device learning is bound to go up with the respective technologies are growing in maturity and stability. Further on, artificial intelligence (machine and deep learning (ML/ DL) algorithms is drawing a greater attention from researchers and professional. The most important contributions of AI are computer vision (CV) and natural language processing (NLP). With the availability of enabling AI toolkits and frameworks, edge devices are being vision-enabled. That is, the concepts of machine vision and intelligence are emerging and evolving fast. Machines are also able to understand human speech and respond insightfully. On-device AI processing could produce predictive, prescriptive and personalized insights in time. These devices are capable of data processing and can talk directly to faraway cloud servers and services. Or they can connect to edge gateway/broker to talk to cloud applications. These devices form quick clusters (called as edge device clouds) to fulfill special needs. Edge clouds are for proximate processing.

- **Edge Gateways** also run any full-fledged operating system. They have unconstrained power supply, CPU power, memory and storage. Therefore, they can act as intermediaries between the cloud and edge devices. These offer the standardized middleware functionalities.

- **Cloud Environments**: We have private, public and hybrid clouds to host business workloads and IT services. These centers involve a large number of commodity server machines, storage appliances and networking solutions in order to host and run web, mobile, and IoT applications and services. Not only applications, databases, platforms, and middleware solutions are also being installed on cloud infrastructures. Simply speaking cloud environments are the highly optimized and organized IT environments to run enterprise-scale, mission-critical, service-oriented, event-driven, operational, analytical and transactional applications. The beauty is that all the cloud resources are optimally used for running multiple applications concurrently in order to achieve resource efficiency. That is, resource utilization has to be on the higher side in order to pass on the monetary benefits to consumers. Further on, all kinds of cloud-inspired services can be supplied to multiple users at the same time. Thus, sharing, automation and orchestration capabilities introduced by the cloud paradigm is still sustaining the interest toward cloudification. All these disruptions bring down the cloud costs drastically. With the leverage of orchestration tools, most of the simple and complex administration activities get automated and accelerated. With the massive scalability, the leverage of cheaper hardware, deeper automation and automated orchestration through the containerization movement, the cloud concept is pitched for affordability, agility, and adaptivity. It is projected that more than the 80% of software applications will reside in cloud environments in the years ahead.

We have discussed the prominent building-blocks of edge architecture. Besides sensors and actuators, devices and device gateways, we need cloud environments in order to stock all kinds of edge data. Further on, in synchronization with current edge data, a kind of comprehensive and historical data analytics can be accomplished in cloud environments. The increased adoption and adaptation of digital twins is being presented as one of the key drivers for the spread of edge computing across industry verticals. Cyber physical system (CPS) is a related discipline gaining significant momentum. All kinds of ground-level physical systems get hooked to cloud-based applications and platforms to envisage smart systems and services. That is, digital systems in our everyday environment are all set to be artistically empowered to exhibit adaptive behavior.

8. Edge cloud infrastructures

Any computing and analytics activity needs a solid IT infrastructure. For certain scenarios, on-device AI capability is needed. That is, a single edge device can suffice. For complicated activities, there is a need for clustering multiple heterogeneous devices together to form a kind of device cloud dynamically. Clustered or cloud environments are mandatory for doing extremely powerful data analytics. As we all know, for certain needs, web-scale companies use thousands of compute nodes to do comprehensive data analytics. For performing edge data analytics, we need distributed edge infrastructures. Thus, edge computing infrastructures include edge devices and their clouds. Some activities are being offloaded to faraway cloud environments. Real-time computing and proximate data processing needs are predominantly being done through edge devices locally.

As indicated above, we will have zillions of digitized entities and connected devices in our earth planet soon. These are subdivided into two major categories: resource-constrained and intensive. Resource constrained digitized entities are focusing on data gathering. Because of the externally or internally embodied communication modules, they can transmit a variety of data to nearby device middleware, which are also called as device bus, IoT gateway and broker, etc. These devices/intermediaries are typically powerful and can comfortably do all kinds of middleware activities. Data fusion, message enrichment, intermediation, security, routing, filtering, tunneling, encapsulation, etc. can be accomplished by IoT gateway software. Further on, they send data to faraway cloud environments in order to be crunched by cloud-hosted data analytics and AI toolkits to extract actionable insights. However, resource-constrained edge devices are not participating in data processing whereas resource-intensive edge devices are capable of not only collecting but also involving themselves in data processing, analytics, mining and learning. Resources-intensive devices can therefore cluster themselves to form ad hoc, dynamic and purpose-specific device clouds in time to accomplish real-time data analytics.

Considering the persistent demands for cloud centers near users, renowned data center service providers are collaborating with telecommunication service providers to set up edge clouds, which are the miniaturized version of traditional clouds (public and private). However, with the ready availability of powerful devices in large number and the maturity of device cloud formation technologies and tools, device clouds can be formed easily and quickly and they are being utilized for a variety of edge computing use cases in a risk-free, affordable, and secure manner.

The prominent reasons for setting up edge cloud centers are as follows. The scalability issues, excessive power consumption, bandwidth wastage, high latency and privacy are some of the critical factors that are driving the demand for edge cloud infrastructure in the form of micro data centers, cloudlets, or edge clouds/clusters. Edge cloud centers support distributed computing architecture.

There are a few important reasons for the surging popularity of edge clouds. Real-time data collection, storage, processing, analytics, decision-making, and action are being made possible with the realization of edge cloud infrastructures. With the accumulation of sensors and actuators, there is a steady stream of sensor and device data. With edge clouds in place, all kinds of sensor data can be captured and subjected to a variety of deeper investigations to uncover hidden patterns in data streams. Real-time customer engagement and fulfillment is being guaranteed through edge cloud centers. Streaming data analytics is becoming one of the most influencing use cases of edge computing.

While Raspberry Pi has long been the gold standard for single-board computing, powering everything from robots to home appliances. This edge computer has a PC-compatible performance, plus the ability to output 4K video at 60 Hz or power dual monitors. Its close competitor, the Intel Movidius Myriad X VPU has a dedicated neural compute engine for hardware acceleration of deep learning (DL) inference at the edge. Google Coral adds to the competition offering a development board to quickly prototype on-device ML products with a removable system-on-module (SoM). There is a transition here from servers-centric cloud environments to devices-centric cloud environments. As there are billions of edge devices, the future belongs to device clouds. Thus, the mass production of single board computers (SBCs) is being touted as the valid reason to rapid adoption of edge computing.

9. Edge analytics

Analytics is being positioned as the key differentiator for the entire human society. With the number of edge devices going up rapidly in and around us, we can expect more device interactions, collaborations and correlations and thereby the amount of device data getting generated is also big. That is, the device world is supplying big and streaming data. The need is to do real-time analytics of device data in order to emit out actionable insights. There are well-known batch and real-time data processing. Similarly, there are big and streaming data analytics methods and platforms. In the recent past, we hear about high-end platforms to complete big data analytics quickly. That is, the real-time analytics of big data is being fulfilled through a host of advancements in the IT space. The unprecedented explosion of IoT devices and the adoption of analytical methods have led to the surging popularity of the IoT analytics domain. Without proper and timely analytics of IoT device data, enabling IoT devices to be intelligent in their operations, outputs and offerings is a tough affair.

With devices individually or collectively contributing for massive amount of multistructured data, data analytics acceleration engines and algorithms are gaining prominence. Customers' needs are constantly changing and meeting their demands mandate for edge cloud and computing. With a surge in IoT sensors and devices in our work spots, living and leisure places, real-time monitoring, measurement, and management turn out to be an important factor for the intended success of IoT devices. By doing real-time analytics on IoT device transaction data, the safety, sagacity and security of IoT sensors and devices are being ensured. Real-time data analytics is essential for creating and delivering real-time applications. Real-time computing is being made possible through edge computing. Newer possibilities and opportunities will emerge with the greater awareness of edge computing and its usefulness. A host of edge-centric services and applications will see the light with the faster maturity of edge-inspired innovations and the penetration and pervasiveness of 5G communication.

The edge analytics paradigm has become more feasible as edge devices are being stuffed with enough firepower. Also, 5G and Wi-Fi 6 communication technologies enable last mile connectivity. In addition, real-time analytics and applications are mandated and hence edge-based data analytics is being taken up with all the sincerity.

Edge analytics optimizes the whole process by handling the bulk of analysis on-site. Edge devices individually and collectively embark on data analytics to uncover hidden patterns and insights in time. The goal is to provide real-time or as close to real-time insights as possible. Additional devices can be easily accommodated to tackle growing data volumes. For example, in healthcare, there is a convincing use case for edge analytics. If we wear a pulse meter and blood sugar monitoring sensor in our body, if there is a sudden spike while walking or exercising, then a notification or alert event is created and an appropriate message gets communicated to our spouse or family doctor immediately. The smartphone in our pocket acts as the powerful intermediary in capturing and conveying the message to the concerned person to enable them to ponder about the counter measures in quick time. This is a streaming analytics phenomenon. A few rules are formed and stocked inside the body-attached device along with a rule engine to take decisions instantaneously.

This is a kind of outlier/anomaly detection in real time. Such a requirement is being visualized across industry verticals. This real-time monitoring, measurement and management of physiological parameters is being facilitated through edge devices (sensors, body area network (BAN) and smartphone). These digitized entities, connected devices, and microservices do a fast calculation based on some rules to articulate what is found. If there is a break-in in the threshold value, immediately the right information is formed as a message and gets delivered to the correct person to initiate the proper actions with all the alacrity and clarity.

Another example is as follows. A device that controls the temperature of a refrigerator at a supermarket detects a dangerous change in internal temperature. That change can cause damage to the products inside in seconds. In the traditional case, that message would have traveled to a central or cloud server and it gets parsed and processed there. The result is then relayed back to the sensor fit inside the fridge. By that time, the refrigerator's goods would have spoiled and fit for nothing. With edge analytics, the same problem could be resolved in a matter of a few seconds with the sensor instantly relaying the problem and implementing a viable solution therein. Similarly, manufacturing plants can use edge analytics to keep better track of machinery health, production output, and be ready to deal with any crisis that arises in seconds instead of minutes.

Similarly, in other industry verticals, such kinds of analytics of streaming data using edge devices or clouds are gaining speed and visibility. Therefore, edge analytics may be an exciting area of great potential, but it should not be viewed as a full replacement for central data analytics. Both can and will supplement each other in delivering data insights and add value to businesses.

Processes are increasingly being automated and accelerated through a host of breakthrough technologies across industry verticals. Processes are being optimized through rationalization and competent technologies contribute in process excellence. Process orchestration is the latest buzzword. Multiple automated process steps are being combined and sequenced to bring in deeper and mission-critical automation. Edge computing is one of the promising phenomena in radically removing process flabs to arrive at highly optimized processes. Edge analytics is seen as an indispensable act for the digital living.

10. The key benefits of edge computing

Businesses can benefit immeasurably through the smart application of edge computing. Here come the prominent and potential advantages.

Real-time Decisions: Due to the participation of a large number of heterogeneous devices in any environment (home, hospital, hotel, etc.), the data creation, capture, storage and processing happen locally and in a distributed manner. Edge devices are the IT infrastructure here. As everything happens near the sources of the data, the transaction time is very less as there is no trip to cloud and back. Mission-critical machine operations are saved from any kind of break or slow down through edge computing. Hazardous incidents are being proactively barred from taking place. Real-time computing is the first and foremost facet of edge computing.

Reliable operations even with intermittent connectivity: Internet connectivity is generally unreliable in remote and rough environments. If the network connectivity is lost, then remotely monitoring and managing critical infrastructures and assets such as oil wells, farm pumps, solar farms, windmills, etc. can be difficult. With proximate processing, there is no data loss or operational failure in the event of any disconnect. Required operations happen locally and fast to fulfill all mandated obligations without flop, falter and fumble.

Security and compliance: Due to local computing, a lot of data transfer between edge devices and cloud environments (private or public) over the porous, public, and open Internet can be avoided. It's possible to filter out sensitive information locally. And only important information gets transmitted to cloud storage for performing historical and comprehensive data analytics.

Renaissance: Edge devices can form clusters/clouds in order to do bigger operations. Local storage, computation, decision-making and actuation eventually reduce operational costs. Innovations will thrive in such an open environment. People-centric services and applications can be visualized and affordably realized.

The Explosion of Edge Services: Microservices architecture (MSA) and event-driven architecture (EDA) are pronounced as the most impactful and extensible architecture patterns and styles for constructing enterprise-class software applications. That is, synchronous and asynchronous microservices are being proclaimed as the most optimal building-block for creating and sustaining software solutions.

Actually, multiplicity and heterogeneity lead to heightened complexity. MSA is for lessening the rising device complexity. That is, heterogeneous devices can be exposed as device services with APIs. Thus, any disparateness and deficiency of devices can be conveniently hidden from the outside world. The much-needed interoperability between current and emerging devices is being facilitated. IoT edge devices can be remotely monitored and managed. Also, device-to-device (D2D) and device-to-cloud (D2C) integration phenomenon has are being simplified and streamlined.

Through service-enablement, different and distributed edge devices can be shrewdly combined to formulate competent and cognitive solutions for complicated and sophisticated applications. Such a smooth and risk-free integration and orchestration of edge devices is seen as an amplifying platform to envisage and produce fresh business opportunities and possibilities. Innovations are bound to thrive in such uncomplicated and open environments. Instrumented and interconnected devices are the initial requirements for empowering devices to be intelligent in their actions. The nuanced service paradigm has penetrated into the edge space toward intelligent, real-time and cloud-native applications.

Improved equipment uptime: Devices' performance and health condition are being constantly monitored and if there is any deviation, it can be proactively and pre-emptively attended. This directly contributes in elongated lifeline for devices, which are contributing for a mission-critical task. Preventive maintenance is enhanced with predictive maintenance with the edge analytics capability enshrined in edge devices. The return on investment (RoI) goes up while the total cost of ownership (TCO) is kept low. The optimal utilization of industrial assets and machineries goes up significantly with edge computing.

Failure prediction: Every module contributing for the process implementation, integration, automation and orchestration is being monitored through automated monitoring and observability tools and all the data thus collected is being subjected to a variety of deeper and decisive investigations in order to understand how each component performs and what is its remaining lifetime? As indicated elsewhere, machine/device data volume is exponentially growing. Edge analytics comes handy in mining data in neatly understanding the health condition of devices and machines. Also, learning from data is another interesting phenomenon gaining the confidence of people through the distinct advancements of machine and deep learning algorithms.

While edge computing has the advantages of local computation and faster decision-making, the traditional cloud gives an illusion of infinite compute power and data storage. Big data analytics can be performed out of cloud environments. The onset of pioneering artificial intelligence (AI) algorithms for emitting out personalized, prognostic, predictive and prescriptive insights out of big data goes hand in hand with cloud environments. The future holds good for the hybrid model. The real-time computing of edge devices in conjunction with massive computing capability offered by cloud environments is definitely the future. Manufacturers are looking for ways and means to enhance the productivity and the responsiveness of their production systems further by leveraging edge computing. Smart manufacturing is the vision of manufacturing industries. Sensing and responding (S & R) capability is being incorporated in assembly lines through edge computing.

Smart manufacturing envisions a future wherein factory equipment can make real-time and autonomous decisions based on what's happening on the factory floor. With multifaceted sensors, actuators, instruments, equipment and machineries joining in the manufacturing process, a number of activities can be measured minutely and linked together to bring in additional astuteness and alacrity. All aspects of manufacturing, supply chain, operations, product design, etc. can be intertwined to bring in flexibility and adaptivity. The solidity of the IoT paradigm, artificial intelligence (AI) advancements, edge/fog computing, 5G communication, blockchain technology, digital twins, software-defined cloud environments, and integrated data analytics supply all that are needed to fulfill the traits and tenets of smart manufacturing. All these improvisations will ultimately lead to the realization of the industry 4.0 vision.

Edge Computing for Smart Retail: As 5G, edge computing, and AI technologies stabilize, we're going to experience a series of hitherto unheard digital innovations and disruptions. Real-time data analytics is the key differentiator of edge computing. Real-time services and applications can be built and supplied to people. Augmented reality (AR) is a newly incorporated phenomenon gaining prominence with the success of edge computing. The fast-evolving 5G is getting empowered through edge computing, which is also enabled through the power of 5G cellular communication. (https://venturebeat.com/2020/01/31/as-retail-evolves-5g-and-edge-computing-keep-you-in-the-express-lane/).

The customer experience in retail stores is bound to go up sharply. Customers' needs are understood beforehand and fulfilled precisely. Inventory and replenishment management tasks are getting automated. Data-driven insights and insights-driven decisions and deals will become the new normal. Computer and machine vision, natural language processing (NLP), human and machine interfacing (HMI), 5G etc. consciously fulfill the widely demanded customer delight.

11. Tending toward edge AI

The unprecedented growth of edge computing will pave the way for a host of sophisticated services and applications at the edge. Edge devices and clouds are capable of enabling edge data analytics. Especially streaming data analytics can be greatly optimized with the smart application of artificial intelligence (AI) algorithms. In Edge AI, the AI algorithms are used locally on a hardware device, without requiring any connection to the outside world. It uses data that is generated from the device and processes it to give real-time insights in less than a few milliseconds.

Your iPhone has the innate ability to register and recognize your face to unlock your phone in a fraction of seconds. Take the example of self-driving cars. Computer vision algorithms play a very vital role in realizing autonomous vehicles. Real-time decision-making capability comes handy for cars to traverse in expressways and city roads without any incident or accident. Vehicle to vehicle (V2V) communication is being simplified through localized computing. The computing happens at the edge wherein data gets generated, captured and crushed. Therefore, edge use cases are growing steadily and sagaciously.

From Google maps alarming you about bad traffic to your smart refrigerator reminding you to buy some missing dairy stuff, AI is pervasive these days. We are quietly moving from AI toolkits deployed in cloud and enterprise servers to AI-enabled edge devices. Smartphones, robots, consumer electronics, drones, cameras, etc. are being activated to be intelligent in their operations through AI algorithms. Handhelds, wearables, portables, and gadgets are being embedded and empowered with AI toolkits. Everyday factors are driving AI processing to edge devices, which are increasingly stuffed with more processing, memory and storage power. Applications like autonomous driving and navigation mandate for sub-millisecond latency requirements that can be made possible only through edge processing. Speech recognition, object detection, and other real-time operations need edge processing.

Computer vision algorithms are being deployed in machines and devices these days in order to enable them to have a clear view and understanding of things around them. This technology empowerment enables devices to understand of personal, professional, location and social needs of people in an unambiguous manner, produce the right and relevant device services and deliver them in an unobtrusive fashion in real time. That is, the new concepts of machine vision and intelligence are being insisted in order to accomplish certain real-time decision-making and action. Deep learning algorithms, a grandiose part of artificial intelligence (AI) domain, are showing a lot of promise in fulfilling the goals of machine vision and intelligence. The deployment model is as follows. Deep learning models are being produced, trained and tested in cloud environments but the inferencing is being performed in edge devices. This way, the much-abhorred network latency is significantly reduced in order to deliver right services to right people at right at right place. Making and refining learning models using cloud infrastructures and platforms are the way forward as cloud centers are being stuffed with a lot of computing, and storage resources. Inference is a relatively simpler task to be performed through edge devices in real time. Also, results can be obtained through edge-based inferencing in time. When new datasets are given to an already curated and corrected model, a new upgraded model gets created and passed on to the edge devices to infer accordingly.

There are formal AI model optimization methods such as compression, pruning, quantization, sparsity, transfer learning, attention, federated learning, knowledge distillation, etc. for facilitating running AI models on edge devices. There are works going on to enable model creation and upgrade directly on edge devices. There are AI-centric processor architectures and implementations for doing on-device AI processing. In short, the edge AI paradigm is all set to flourish with the proper nourishment from semiconductor companies and researchers focusing on efficient, explainable and edge AI models. Resultantly, the prediction, anomaly detection, uncovering of hidden patterns and associations, visualizing fresh possibilities and opportunities can be simplified and speeded up.

12. Artificial intelligence (AI) chips for edge devices

With a series of noteworthy advancements, it is becoming easier to run AI toolkits at the edge today. Edge devices are becoming powerful. AI-enabling chips are being manufactured in large quantities. And the era of in-device AI is all set to drastically transform our everyday life. That is, intelligence is being achieved at the edge. In other words, edge intelligence is becoming the new normal. Devices will become innately intelligent. Intelligent devices will be pervasive and persuasive.

Edge-based AI chipsets and accelerators are being produced and installed in smartphones, connected speakers, head mounted displays (HMDs), automotive systems such as car, vehicles, trucks, etc., laptops and tablets, robots and drones, cameras, game consoles, consumer electronics, kitchen wares and appliances, etc. Depending on the AI application and device category, there are several hardware options for performing AI-inspired processing in the edge. There are multiple options including central processing units (CPUs), graphical processing units (GPUs), tensor processing units

(TPUs), vision processing units (VPUs), application-specific integrated circuits (ASICs), field programmable gate arrays (FPGA) and system-on-a-chip (SoC) accelerators.

Edge AI chips are physically smaller, relatively inexpensive, use much less power, and generate much less heat. These special properties make them suitable and sustainable for both consumer and non-consumer devices. As articulated above, by doing local computation leveraging highly sophisticated chips, a number of benefits can be accrued. Sending large-scale unprocessed data to faraway cloud servers to be processed is fully avoided. Data security and privacy are being fulfilled. Personal, social and professional data get secured through edge computing. Local computations always contribute for time-critical applications. Some scenarios need data processing by powerful servers. Thus, the future belongs to the hybrid model.

Previously for enabling concurrent operations, GPUs have gained immense popularity. These general-purpose chips are more appropriate for cloud and enterprise servers. Machine and deep learning models were run on GPUs without any hitch or hurdle. But running GPUs directly in edge devices needs some refinement. The number of connected devices is growing exponentially and edge devices are gradually empowered to join in the mainstream computing. Edge devices are also hooked to cloud-hosted services, platforms, and databases. There is a need for empowering edge devices to host and run microservices. Hence there is a race to bring forth fresh chip-making approaches and products. For factory automation, smart city applications, next-generation retail experience, etc., companies across the world are working on face recognition, object detection, replenishment management, autonomous vehicles, traffic congestion avoidance, etc.

13. The noteworthy trends toward edge AI

A number of technological innovations and disruptions have laid down a scintillating foundation for visualizing and realizing context-aware and cognitive edge AI services and applications. AI-specific chips are being manufactured in plenty these days and this turns out to be a huge differentiator for the unprecedented success and motivation for edge AI. There are several ways being debated and disserted for pushing AI capability to the edge with the aim of taking edge computing to the next level. Edge devices are becoming powerful and can find, bind and interact with one another in the vicinity as well as remote ones through a network. Also, devices are being hooked into cloud-hosted applications and data sources.

Sometimes one device may not have all the resources in place in order to finish a task. That means, multiple devices (heterogeneous) devices form a dynamic and ad hoc cluster/cloud to accomplish complicated tasks. One or more devices contribute as master/control device whereas other devices play the role of data storage and processing tasks. That is, the formation of device clusters/clouds has laid down a robust and versatile foundation for the surging popularity of the edge AI concept. At the chip level, there are praiseworthy breakthrough processors emerging to accomplish AI operations fast within an edge device. In this section, we are to discuss the key motivations for the intended success of the edge AI paradigm.

- **Distributed computing**: There is a steady movement from centralized computing to distributed computing in order to gain business, technical, and, user advantages. By leveraging the distinct ideals of distributed computing, process and data-intensive AI workloads get divided and distributed across a number of machines to speed up their execution. As we all know, blockchain technology is leaning upon the distributed and decentralized computing models to be efficient and effective for a variety of business verticals. Business problems are dismantled to arrive at competent software solutions, which are being built as a collection of interoperable and portable services. Such disaggregation and aggregation go a long way in agile and risk-free software engineering. That is, enterprise-grade applications are being made through multiple easily manageable services. Each service and its instances and versions/editions are being run on independent runtimes to facilitate service composition. Composite services are process-aware and fulfilling business requirements smoothly. By picking and packing up certain services, it is possible to create different applications. Thus, we are on the road towards next-generation software engineering through configuration, customization and composition methods. In a distributed setup, the communication among participating service components happens via network calls. Thus, applications can be partitioned, data volumes can be subdivided into small pieces, and infrastructure resources are also virtualized and containerized to enhance resource utilization efficiency. Thus, distributed computing is the futurist model. Edge AI is to see a lot of success as it is natively distributed.
- **AI-centric Processors**: The chipset domain is going through a number of innovations and disruptions. As inscribed above, there are generic and specific chipsets emerging and evolving in order to simplify and speed up AI processing at the edge.

- **Advanced Algorithms and Toolsets**: There are several machine and deep learning (ML/DL) algorithms in the industry now in order to bring out personalized, predictive and prescriptive insights in time. A number of state-of-the-art algorithms mimicking human brain function is also being unearthed and experimented. With delectable advancements in the deep learning domain, computer vision (CV) and natural language processing (NLP) requirements are being met comfortably. Now AI algorithms are being taken to edge devices in order to squeeze out real-time insights. There are a litany of enabling frameworks and toolkits to make AI-enabled edge processing simple, smart and swift.
- **AI Model Compression**: AI models are being compressed to reduce model size considerably so that the required computational resources can come down. Also the time being taken to model also comes down sharply. Further on, the power energy consumption goes down thereby the heat dissipation level also reduces. Above all, compressed AI models can comfortably run on edge devices. Thereby, edge clouds formed out of edge devices can be the new environment for AI model creation, evaluation, optimization, etc. With fresh data coming in, edge devices quickly redo the training, testing and deployment.

Implementing AI algorithms closer to where personal, social and professional operations is being termed as the futuristic step. It is all about performing time-critical decisions at the edge and refer to the cloud where intensive computation and historical analysis are needed.

Business behemoths has to address the perpetual AI data management challenges by setting up powerful and highly available edge systems. Generally big AI applications require large and expensive enterprise-scale servers. Or in public cloud centers, commodity servers are being clustered to run AI applications. But in the recent past, we are being bombarded with a plenty of resource-intensive multifaceted devices. By leveraging the proven and potential distributed architecture, AI processing is being done at the edge comfortably. With the growing popularity of connected devices, many industries such as retail, manufacturing, transportation, and energy are generating vast amounts of data at the edge of the network. As indicated above, edge analytics helps to arrive at a cognitive decision.

14. Why edge processing?

The answer is the data privacy and security and the need for real-time analytics of edge data. The real-time feature may be lost if we take edge data to cloud servers to be processed and preserved. That is, proximate processing is mandated for extracting actionable insights in time. Therefore edge processing is getting the attention.

Google's auto suggest application on mobile phones mandates for a 200-ms latency requirement. Similarly, vision-based applications like augmented reality (AR), virtual reality (VR), and mixed reality (MR) need high bandwidth. For these bandwidth-intensive applications, the bandwidth requirement is set to grow from 20 Mbps today to 50 Mbps in the near future. Besides real-time data processing, the cost of doing edge-based data processing turns out to be cheaper than cloud-based data processing. The network bandwidth cost is also on the lower side.

Further on, there are several technological innovations speeding up and facilitating edge side processing. Data scientists produce powerful yet highly miniaturized deep neural network (DNN) models that can run on edge devices without consuming much resources. Efficient implementations of AI algorithms, lightweight AI toolkits, application-specific AI chips, edge cloud formation, etc. enable edge processing. Several frameworks and techniques support model compression, including Google's TensorFlow Lite, Facebook's Caffe2Go, Apple's CoreML, Nervana's Neural Network Distiller, and SqueezeNet.

Edge cloud formation and management platforms are emerging fast. There are lightweight and edge-specific Kubernetes platform implementations. Kubernetes is famous for deploying, managing and maintaining containerized applications across a cluster of nodes. Auto-scaling, auto-healing, declarative APIs, stateful apps through persistent volumes, etc. are the key benefits of container orchestration platforms. Now the widely adopted cloud-native principles are being taken to edge devices.

Thus, there are a plethora of technological revolutions to implement the edge AI vision.

15. Edge-based AI solutions: The advantages

Building and running edge-based AI solutions are not straightforward. As illustrated above, there are limitations on every side. So, there are a few critical decisions to be consciously made. Especially expert decisions ought to be made in data collection and preparation, algorithm selection and training the algorithms continuously, deploying and refining the models, etc. One widespread trend is that training and testing models can be done at cloud environments. Then run the trained and tested model in edge devices to make appropriate inferences in time. The processing/storage capacity at the edge also needs to be taken into consideration while finalizing the edge solution design.

Intelligent enterprises consistently insist for real-time computing. It is indisputably clear that real-time analytics, decision-making, and operations can see the grand reality only through edge computing. Therefore, edge solutions are gaining significance for crafting real-time applications. Further on, to build intelligent devices and applications, AI capabilities are being embedded in edge devices. Edge devices can collectively self-learn, adapt and act cognitively by the sheer power of AI algorithms. And this phenomenal transition is seen as the clear game-changer for bringing forth real business transformations. With cognitive edge devices, a variety of smarter and situation-aware services, systems, and environments can be realized and deployed through the fast-exploding edge AI conundrum. The unique AI competency can be applied on edge system data (log, operational, performance, health condition, security, etc.) in order to bring out actionable insights in time. Experts have pointed out the following advantages of AI-driven edge devices.

1. Edge-based AI is primarily used for constructing sensitive and responsive (S & R) systems. Insights are immediately extracted, delivered and processed in order to fulfill the ideals of real-time operations.
2. Edge-based AI guarantees greater security for edge data as data does not travel outside before taking tactical and time-bound decisions. AI-driven security measures can be applied in order to ensure heightened security.
3. Edge-based AI is highly flexible. Multiple industry domains ranging from smart lighting, energy management, and retail to smart cities are to get immensely benefited out of the advancements happening in the cusp of edge computing and AI.
4. Edge-based AI solutions can be self-contained. With the availability of advanced machine and deep learning algorithms, computer vision and natural language processing applications can be realized. This makes edge AI systems to be autonomous and articulate.
5. Edge-based AI guarantees superior customer experiences. We will fully realize context-aware, people-centric and adaptive services. Owners, operators, and occupants will enjoy the technology-inspired innovations. The customer experience is bound to go up significantly. Trust on technology solutions will see a strong boost.

As we tend toward the deeply connected digital economy, intelligence has to move to the edge. Thus, the powerful combination of AI and the IoT opens up new vistas for organizations to truly sense and respond to events and opportunities around them. Remote and rough locations such as countryside farms, forest areas, oil fields, etc. can benefit immensely from the cutting-edge AI technologies. As IoT moves into more eccentric and disconnected environments, the necessity of edge or fog computing will become more prevalent.

16. Applications that can be performed on edge devices

There are certain types of applications that can be performed on edge devices.

Machine Learning (ML) at the Edge: ML algorithms are extensively used in enterprise and cloud environments in order to make accurate predictions through classifications, clustering, regression, association, etc. algorithms. Now, ML-based prediction capability moves to the edge. Edge devices are being showered with more power and AI-specific chipsets are hitting the market. Therefore, running AI algorithms on everyday machines and devices is gaining momentum. There are several real-world use cases being published by researchers. Intensive care is one area wherein edge-based ML plays an important role. Real-time data capture, processing and decision-making are important for closed-loop systems. There is a need to maintain critical physiological parameters, such as blood glucose level or blood pressure, within specific range of values. Edge AI plays an essential role here. There are several verticals keenly exploring and experimenting with edge AI in order to be ahead of their competitors and to retain the yearned edge of their brand value. In the subsequent sections, we are to discuss some domains wherein the edge AI capability is to bring in distinct advancements.

Machine Learning (ML) Frameworks for Edge Devices: Integrated platforms, in-memory databases, enabling frameworks and accelerators collectively contribute for speeding up AI-inspired data processing at the edge. Like frameworks for accelerated software engineering, ML frameworks are stuffed with pre-trained and tested models for the difficult and repetitive tasks such as speech and face recognition, object detection, natural language processing (NLP), etc. These facilitate the development of custom ML models from the scratch also. We all would have heard Google TensorFlow for AI processing at cloud servers. But with the increased usage of edge devices for real-time computing, Google has released a shrunken version and named it as TensorFlow Lite, which has application programming interfaces [APIs] for many data science-centric programming languages such as Python, Java, etc. This lightweight version is optimized for on-device applications and comes out with an interpreter tuned for on-device machine learning.

Custom models are generally converted in TensorFlow Lite format, and their size is also optimized to increase the much-needed efficiency. ML for Firebase targets mobile platforms and uses TensorFlow Lite, Google Cloud Vision API, and Android Neural Networks API to provide on-device ML features, such as facial detection, bar-code scanning, and object detection, etc. PyTorch Mobile targets the two major mobile platforms and deploys on the mobile devices models that were trained and saved as torchscript models.

It is found that reducing the number of parameters in deep neural network (DNN) models helps decrease the computational resources needed for model inference. Some popular models which have used such techniques with minimum (or no) accuracy degradation are YOLO, MobileNets, Solid-State Drive (SSD), and SqueezeNet.

As articulated above, model compression is a viable technique to take ML models to edge devices Using several compression techniques and caching intermediate results to reuse iteratively, researchers have improved the execution speed of deep neural network models. DeepMon is one such ML framework for continuous computer vision applications at the edge device. More details can be obtained from this page (https://blogs.sap.com/2019/10/16/why-machine-learning-at-the-edge/).

With AI capability is being accomplished through edge devices, there can be multiple openings and opportunities waiting to explode. The IoT device ecosystem will expand further. AI algorithms will be refined to be fit in edge devices. The market for AI-specific chipsets will enlarge ceaselessly. Newer business, technical and user cases will emerge and evolve. The onset of 5G connectivity boosts the deployment of the edge AI paradigm. Forming edge device clouds will be simplified through scores of technological solutions.

Deep Learning at the Edge: One of the most promising domains of integrating deep learning and edge analytics is to achieve computer vision and video analytics. Edge analytics implements distributed structured video processing, and takes each moment of camera recordings and performs analysis in real time. A camera can do this today and when multiple cameras in the vicinity get synchronized, the computational efficiency along with real-time knowledge discovery and dissemination get a strong boost.

The surveillance and security tasks are getting automated through the edge AI concept. Emergency incidents can be tackled efficiently in real time through such kinds of AI-enabled cameras and their clusters. Transmitting videos to faraway cloud storage appliances to initiate and implement video analytics does not solve the real-time requirement. Thus, with the speedy progress of edge AI, real-time computations steadily move to edge environments.

Deep learning neural networks bring in additional automation in order to produce models that substantially improve the prediction accuracy. But the accuracy involves higher computation and memory consumption. A deep learning model generally consists of layers of computations where thousands of parameters are being computed in each layer and passed to the next in an iterative fashion. If the input data has higher dimensionality (that is, if the input data is a high-resolution image), then it is obvious that it needs extra computational capability. GPU farms are being used in cloud environments in order to do this kind of massive computation. Migrating to and performing such a huge computational task is edge devices is definitely not an easy task. However, there is a glimmer of hope with a number of technological innovations simultaneously happen in this space.

Due to the resource-constrained nature of edge devices, the operating model is being tweaked. That is, the deep learning models are trained and tested in powerful on-premises or off-premise cloud servers and then it gets deployed on the edge devices to just make inferences on new data. Inferencing is not a labour-intensive task. The turnaround approach is to design power-efficient ML/DL algorithms, develop specialized hardware, and leverage distributed architecture that predominantly suits for IoT devices. The device distribution and interaction are being supported through the 5G connectivity, which ensures ultra-reliable and low-latency communication. 5G also comes to the rescue in deploying a large number of IoT devices within a crowded environment.

Edge-based Inferencing: This will become a new normal soon. IoT sensors and devices generate a lot of poly-structured data. And powerful ML and DL models are being trained, tested and used in powerful and scalable cloud environments. The refined, curated and refreshed models can, then, be used in edge devices to make accurate inferences on fresh data. AI Edge processing today is focused on moving the inference part of the AI workflow to edge devices. The newly captured data resides in edge devices only and it does not travel to cloud and hence the data privacy and security are fully guaranteed through this futuristic approach. The impact of model compression techniques like Google's Learn2Compress that enables squeezing large AI models into small hardware form factors is also greatly contributing to the rise of AI edge processing. Federated learning and blockchain-based decentralized AI architectures are also part of the shift. With edge device clouds stuffed with edge AI chips, even the complicated training part may move to the edge in the days to unfurl.

Computer Vision (CV) at the Edge: Undoubtedly, machine vision is the next target. Vision-based applications are gaining the popularity. We all know that computer vision (CV) is one of the prime applications of the AI domain. The field of computer vision has grown up solidly and its adoption rate is high with the maturity of deep learning algorithms. Vision-enabling platforms and applications are increasingly deployed and managed in enterprise and cloud servers. Video and vision data analytics have become the new normal for extracting actionable insights out of static and dynamic images. However replicating the same in edge devices is beset with resource challenges and concerns.

The idea is to empower edge devices to view, perceive, understand and act. Due to the resource constraints, there is a need for some unique methods to comfortably run deep learning models in edge devices. Many researchers and experts have come out with breakthrough solutions. One standout way is to use the method of model optimization, which turns out to be the finest technique to make deep learning models to work on low power and cost devices. That is, optimization techniques come handy in fulfilling the need for running deep learning models on edge devices. These techniques include optimizing hyperparameters such as input size, Batch Normalization, Gradient descent, Momentum, etc. The other popular optimization technique is Adam Optimization, which is a method that helps to optimize model performance and loss value during training the model. It computes one's learning rate with different parameters. The Adam optimization algorithm is a combination of gradient descent with momentum and RMSprop algorithms. There is a practical implementation and the details are shared in this page (https://medium.com/datadriveninvestor/edge-ai-computer-vision-on-the-edge-dfa4ad604651).

Natural Language Processing (NLP) at the Edge: Artificial intelligence's resurgence is widely credited to the significant advancements in realizing large computing power. There are and game-changing chipsets such as CPUs, GPUs, TPUs, and VPUs. Then, through compartmentalization (virtualization and containerization), physical machines could get segmented into a number of virtual machines (VMs) and containers. On the other hand, hundreds of distributed VMs and containers can be easily clustered together to realize virtual supercomputers. That is, we have parallel and high-performance computers. The cloudification aspect has made computing as the fourth social utility. These breakthrough improvements have simplified and strengthened the AI processing. Edge devices are being stuffed with powerful processors and hence all the innovations happening in the enterprise and cloud spaces are being replicated in the edge space also.

Edge solutions and systems are being designed with distributed architectures to support data analytics. When analytics becomes the core and central aspect, any edge system can be categorized as cognitive systems. Companies are trying to gain more intimacy and ecstasy with their customers by offering premium services. Right applications are being produced and delivered to right people at right time at right place with all the trustworthiness.

NLP capabilities are being demanded across industry verticals these days. In the healthcare domain, NLP can really help patients in clearly articulating their sentiments and symptoms to care providers. There is no need to send anything to cloud to be processed thereby it is possible to avoid any privacy and security dangers of personal data. The freshly attached human-machine interfaces (HMI) empower edge devices to interact with humans in an intelligent and intimate manner. Edge devices are being enabled to understand human instructions and act upon them with all the alacrity and clarity. If there are any issues with machines, they can express them in an unambiguous manner to professionals. In short, edge devices are also made to have natural interfaces through the incorporation of the NLP capabilities.

One of critical AI components is natural language processing (NLP), which can create intelligent interactions between machines and humans. To fill the gap between people and machines, NLP leverages machine learning, computational linguistics, and even computer science to help understand and even manipulate human languages. This allows for the correct comprehension of the syntax, semantics and structure of various human languages. With this empowerment, it is easy for machines to interact with men in a natural style. The prominent and dominant examples of NLP in action include intelligent assistants like Apple's Siri, Microsoft's Cortana, and the Amazon Echo (Alexa).

17. Edge AI use cases

Machine Learning for Smarter Homes: Smart systems and environments will be realized and sustained with the proliferation of intelligent devices.

For creating sophisticated home applications, multiple devices within a home have to be linked up. Devices need to react for local events. For example, the GPS data from your smartphone or connected car can trigger the smart thermostat at the home to the desired setting when you are approaching your home. Using visual recognition, smart locks in conjunction with smart doorbells can also recognize you, unlock the door, trigger the lights and even start playing music. That is one event or action trigger a whole lot of things in order to facilitate digital living. Such intelligent coordination among home devices is to result in newer experiences for people. That is, machines have to be trained and continuously refined for cognitively reacting to various events. Datasets for training and testing are plentifully available these days. Therefore, empowering machines and devices to learn from datasets is gaining the speed. This empowerment makes devices to be intelligent in their operations, offerings and outputs when new incidents or accidents happen.

Smart Manufacturing: We may have a plenty of multifaceted and differently abled machines in manufacturing floors and product assembly lines. These critical assets have to be minutely monitored, measured and managed. There are observability platforms and tools with real-time analytics capability. Any worthwhile deviation and deficiency can be accurately and proactively identified through such monitoring software. If not identified in time, there can be a huge financial and face loss for the company. When participating machines are inherently enabled to learn their current state, the availability, reliability and versatility of machineries can be succulently increased. Machine learning comes handy in empowering predictive and preventive maintenance of machines. The manufacturing processes can be continuously optimized toward process excellence. The incorporation of edge devices and the power of AI algorithms combinedly contribute for factory automation towards fulfilling the distinct industry 4.0 vision.

Smart Environments: In enterprise and cloud IT environments, the new buzzword of AIOps is gaining momentum in order to optimally run compute machines. Similarly, AIOps capability can be applied on machine and device data (log, operational, performance and security) in order to correctly judge the device throughput and the remaining lifetime. Thus, besides machine intelligence, operational intelligence goes a long way in setting up and sustaining smarter environments across industry verticals. Smart railway stations, airports, warehouses, hospitals, hotels, retail stores, etc. will be easily established and enhanced through edge AI products and services.

Surveillance and Monitoring: This is one of the well-known use cases for the runaway success of edge computing. CCTV cameras are widely used in important junctions and stations for the safety and security of people and properties. Typically images and videos get captured and transmitted to cloud storages in order to be processed to extract any important information. AI toolkits are directly installed and run in cameras. Whenever there is something suspicious and noteworthy, immediately an alarm gets raised and the concerned people can swing into appropriate action to nip any untoward things at the budding stage itself. Real-time processing and notifications go a long way in safeguarding people and properties.

As populations grow, age, and become wealthier, healthcare systems struggle to provide the care needed by people. Digital health is the way forward. The increased deployment and usage of the IoT devices and sensors, cloud infrastructures for healthcare data storage and analytics, path-breaking AI algorithms, edge AI systems, etc. represent the next-generation digital healthcare.

Connected Ambulance: In the healthcare domain, every second and minute is precious. In an emergency situation, the near-zero latency, mobility and proximate data processing capabilities of edge computing can enable faster and more accurate diagnosis and medication by paramedics onsite. Location-based and context-aware services can be made and delivered to various patients and providers. It is possible to do live streaming of processed patient data, such as heart rate from sensors and monitors or live video feeds from first responder body cameras to the hospital with the soaring and roaring 5G power. This gives hospital staff right information on incoming patients' state. This acts as an initial information on vigorously visualizing potential treatment processes. Also, it is possible to do correct analysis of patients' vitals (such as blood pressure, pulse rate, etc.) at the edge for real-time diagnosis and recommendations by first responders.

Further on, augmented reality (AR) glasses (head mounted devices (HMDs) at the edge help to display information about patient history and complex treatment protocols to accordingly empower paramedics. Haptic-enabled diagnostic tools enable remote diagnostics by specialists. More details can be obtained from this page (https://stlpartners.com/digital-health-telecoms/digital-health-at-the-edge/).

In-hospital patient monitoring: As inscribed several times, patients' data privacy and security are paramount. Today healthcare sensors and monitors, medical electronics, instruments, and equipment are generating a lot of useful data but it is getting transmitted to and stocked in cloud environments. Also with time delay, the data value also diminishes. In other words, for emitting out correct insights, data timeliness and trustworthiness are very essential. Transmitting highly confidential data on open and public network (the Internet communication infrastructure) and storing it in third-party cloud storage appliances are definitely not wanted by patients. Therefore, edge devices forming clouds for accomplishing real-time analytics are being demanded these days. An on-premise edge device clusters within the hospital building or campus could process data locally. This setup ensures the unbreakable and impenetrable security for patients' health data. Edge analytics through an embedded streaming analytics platform or a ML/DL algorithm implementation or an AI toolkit enables real quick decision-making and notifications to the concerned of any abnormality.

Remote monitoring and care: With the noteworthy technological advancements, remote diagnoses and monitoring are becoming new normal. Teleconsultation and medication are happening increasingly. In this Covid-19 situation, it is quite difficult to gain personal and physical attention from doctors. Also, chronic and non-communicable diseases are increasingly spreading. With more healthcare issues induced by damaged environments, we have more people who need medical attention often. Considering the growing load, we need more number of care givers and clinicians. Also multispeciality hospitals are being situated in urban areas. Rural people would find it difficult to travel to city areas to have them properly tested. Life expectancy is increasing globally with wealthier people. Elderly people.

Remote patient monitoring is therefore coming handy in alleviating some of this strain. There are competent sensors for minutely and meticulously measuring different body parameters. These sensors can network themselves (body area network (BAN) in order to arrive aggregated value. We have multifaceted wearables, handhelds, smartwatches, etc. Our smartphone can act as an IoT gateway to transmit acquired data to doctors and specialists in time. Cameras can contribute here in enabling monitoring patients remotely. Ambient assisted living (AAL) is an interesting domain. Debilitated and diseased people who are living alone in an isolated situation can be remotely and rewardingly monitored by care givers and family members. Post-surgery care management is also done remotely. We have breakthrough IoT devices and sensors ably supported by 5G connectivity. To ensure local intelligence, IoT devices are empowered with AI chips and toolkits. Thus, data capture and crunching get accomplished locally and fast.

With highly optimized machine learning models incorporated in edge devices, health-related prediction and prescription can be realized in order to ponder about appropriate measures.

Mining, oil, and gas and industrial automation: The business value of edge-based ML becomes obvious in the oil, gas, or mining industry. Employees work in remote and rough environments. AI-enabled robots and machineries capture a variety of decision-enabling data and accurate decisions are being made in time. Such a capability goes a long way in ensuring the safety and security of assets. Critical assets across industry verticals are being technologically monitored so that any kind of deviation and disturbance can be pre-emptively pinpointed and suitable counter measures can be considered and activated.

Ambient Intelligence (AmI): This is quite an old concept. This is a technical domain for achieving computational intelligence everywhere every time. Our everyday environments such as homes, hotels, hospitals, etc. are being stuffed with a wider variety of sensors, actuators, stickers, beacons, micro-controllers, single board computers, LEDs, RFID tags and readers, Wi-Fi gateways, connected consumer electronics, etc. These devices, embedded with AI capabilities, are sensitive, perceptive, and reactive. Devices unambiguously understand people needs, produce appropriate services (context-sensitive) and deliver them to people in time. Not only information and transaction services but also physical services will be supplied to people. Multidevice computing will become the new normal. With the astounding success of containerization and the huge adoption of Kubernetes as the container lifecycle management platform solution, composite services will be realized by leveraging sensors and devices collaboratively, cogently, correlatively and corroboratively.

VR/AR in Retail: Some state-of-the-art stores are starting to leverage cutting-edge technologies and tools in order to lure shoppers. Retail stores offer the ability to try on clothes or makeup virtually by leveraging cameras and augmented reality(AR)-type applications to let potential customers quickly decide the purchase. This capability shows how different color combinations, patterns, and fabrics might look on them via special monitors or smart mirrors. Instead of trying every option physically, these new computer vision-based and machine learning-driven services are enabling consumers to find what they want. Also stores can complete sales quickly than they have in the past.

Fleet management: Logistics service providers generate massive amount of multistructured IoT telematics data, which, when processed immediately and intelligently, results in useful insights. This transition of data to knowledge helps to realize smart fleet management operations. Vehicle-to-vehicle (V2V) communication and vehicle-to-infrastructure (V2I) integration enable drivers to take conscious and cognitive decisions in time in order to take right decisions to steer the vehicle correctly to reach the destination without any problem. Telematics data gets subjected to a variety of deeper investigations by devices inside the vehicle as well as the devices on the road to extract actionable insights in real-time.

Newer yet powerful IoT devices are continuously hitting the market to cater to different personal, social and professional requirements. Miniaturization technologies help to accommodate more inside devices. Also containerized microservices are being installed in IoT devices. In-memory databases are also being deployed in devices' memory module. IoT devices are being hooked into cloud servers in order to be remotely monitored and managed. All kinds of patching and update happen over the air. With an astounding advancements in AI chipsets domain, IoT devices are individually as well as collectively made intelligent in order to design and deliver context-aware services. With the participation of edge devices, the software applications will be profoundly sophisticated. Trend-setting people-centric services will be formulated and provided in time to meet up any spatial, temporal, physical, emotional requirements of people. Enterprise and cloud applications will be considerably strengthened to bring in decisive automation for businesses. There will be deeper integration and complex orchestration in order to visualize integrated and insightful applications. There will be a beneficial integration between edge and traditional clouds in order to envisage next-generation applications that fulfills the digital transformation goals.

The business world is heading toward intelligent and real-time enterprises. The contributions of edge AI technologies, platforms and accelerators are widely applauded.

18. Conclusion

Edge computing is a distributed computing paradigm which brings computation and data storage closer to the location where it is needed. This is to improve response times, data security and save bandwidth. Real-time processing is one of the most robust features of edge AI. It allows users to collate, process, and analyze data then implement solutions in the fastest way possible, making devices highly useful for time-dependent applications. Edge AI has made notable contributions to some industries. Though edge computing addresses connectivity, latency, scalability and security challenges, the computational resource requirements for deep learning models at the edge devices are hard to fulfill in smaller devices.

Mission-critical applications such as factory automation, self-driving cars, facial and image recognition require not only ultra-low latency but also high reliability and fast, on-the-fly decision-making. Any centralized architectures are not able to provide the new performance requirements mostly due to congestion, high latency, low bandwidth and even connection availability. Furthermore, fast decision-making on the edge needs advanced computing capabilities right on the spot, which can be provided only by onboard computers or interconnected edge-computing local nodes working together.

About the authors

Pethuru Raj is working as a chief architect at Reliance Jio Platforms Ltd. (JPL), Bangalore. Previously. worked in IBM global Cloud center of Excellence (CoE), Wipro consulting services (WCS), and Robert Bosch Corporate Research (CR). In total, he has gained more than 20 years of IT industry experience and 8 years of research experience. Finished the CSIR-sponsored PhD degree at Anna University, Chennai and continued with the UGC-sponsored postdoctoral research in the Department of Computer Science and Automation, Indian Institute of Science (IISc), Bangalore. Thereafter, he was granted a couple of international research fellowships (JSPS and JST) to work as a research scientist for 3.5 years in two leading Japanese universities. Focuses on some of the emerging technologies such as the Internet of Things (IoT), Optimization of Artificial Intelligence (AI) Models, Big, fast and streaming Analytics, Blockchain, Digital Twins, Cloud-native computing, Edge and Serverless computing, Reliability engineering, Microservices architecture (MSA), Event-driven architecture (EDA), 5G, etc. His personal web site is at https://sweetypeterdarren.wixsite.com/pethuru-raj-books/my-books.

Dr. J. Akilandeswari is working as Dean—Academics and Professor of Department Information Technology at Sona College of Technology. She has obtained her postgraduate and PhD degree in Computer Science and Engineering from National Institute of Technology, Tiruchirappalli. She is a gold medalist and University first ranker during her undergraduate engineering degree. She has obtained the Fulbright Fellowship during the year 2016. She had been selected as one among 10 administrators in India in Education industry to visit different Universities in US and gain exposure. She also has travelled to Australia as part of CII team and visited more than 15 universities. She has a teaching experience of 23 years and research experience of 16 years. Her research interests include Data Mining, Data Analytics, Web Services, Cloud Computing and Computational Intelligence. She has published more than 60 research papers in both national and international journals and conferences. She has filed three patents. She has membership in professional societies such as IEEE, CSI, IEI and ACM. She has written a book on Web Technology published by Prentice Hall of India. She has executed projects funded by Department of Science and Technology, Government of India. One such project is to take up the training of usage of ICT tools to women farmers.

M. Marimuthu is working as an Assistant Professor in the Department of Computer Science and Engineering at Sona College of Technology, Salem. He is currently doing his PhD at Anna University and his research area of interest is cloud computing and machine learning. He has 13 years of progressive experience in teaching and industry. He has published in more than 15 international journals, attended more than 20 national and international conferences, and participated in more than 50 faculty development training programs (FDP, Workshop, and industry training programs). He has completed 10+ NPTEL courses like Cloud Computing, Deep

Learning, Python programming, etc. and has also completed various online courses like Udemy, Coursera, and IBM. He is a certified Apple Trainer in IOS Application Development and EMC Academic Associate Cloud Infrastructure and Services. He has membership in professional societies such as ISTE, IEI, ISSE, and ISRD. He has received various funds from DST (SEED Division), AICTE (AQIS-PRERANA) and AICTE (ISTE Quality Improvement Scheme) and has currently applied for various funding schemes like AICTE ATAL, DST-NGP & DST—Young Scientist and Technologies.

CHAPTER SIX

Microservices architecture for edge computing environments

Chellammal Surianarayanan
Government Arts and Science College, Srirangam (Affiliated to Bharathidasan University), Tiruchirappalli, Tamilnadu, India

Contents

Abstract

With the advancement in the field of the Internet of Things (IoT), the number of connected devices keeps on increasing and in 2019 itself, it was about 6 billion devices all over the globe. This indicates the generation of enormous volume of data. As the IoT

Advances in Computers, Volume 127
ISSN 0065-2458
https://doi.org/10.1016/bs.adcom.2021.12.001

devices typically have limited computing or processing power, more frequently the data is being transferred to a centralized cloud infrastructure for further analysis. But this poses another issue that the devices generate huge amount of data and if each and every data needs to get transferred into cloud, obviously there would be lot of data traffic between the devices and cloud. Another aspect to be considered is that there are many situations, where the data needs to be analyzed in real time, that too near to the data acquisition point itself. When data is analyzed near to the data acquisition, it enables the implementation of right, time critical tasks at right time. This initiates the evolution of fog computing. Having understood the arrival of fog computing, it should also to be noted here is that though the fog devices are having comparatively richer computing infrastructure, still they are of only moderate devices operated with low power. Hence, it becomes essential to find a right architecture for developing applications that run in fog layer. With this perspective, this chapter tries to show how unique features and key elements of the evolving MicroServices Architecture (MSA) fulfills the requirements of fog computing environment and helps to achieve an optimized computing solution for edge based applications.

1. Introduction

In conventional cloud computing model, data is being gathered by different edge/the Internet of Things (IoT) devices over different edges and transmitted to the centralized large scale data centers for processing and analysis. The conventional cloud computing architecture has been grown out of the fact that at that time, the edge devices had only limited storage and processing capabilities. So, they need to send the data to centralized hyperscale servers, typically to cloud servers. The data is processed and analyzed in the centralized servers. The conventional architecture is shown in Fig. 1. There are some limitations with this architecture, namely, latency, bandwidth, cost, etc. In the conventional architecture, the edge devices send the data to centralized servers for processing and this inherently involves certain latency associated with it. This is latency cannot be tolerated in certain applications. Consider the example of an autonomous car. Assume that a cow is in front of the car. Here, the car has to stop immediately as soon as it detects a cow in front of it while giving appropriate signals to other surround vehicles. Consider another example where the locker area is monitored using IoT based smart security system. Assume that the sensor detects the motion of some intruder during night time. Here also the control action has to be taken without any further latency or delay. Thus, there are applications where processing and taking appropriate control actions in real time become critical. At the time of conventional cloud architecture,

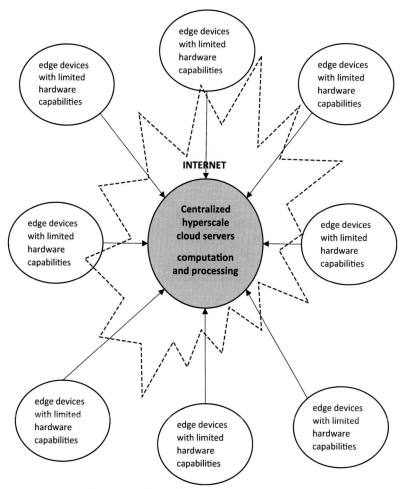

Fig. 1 Conventional cloud computing (edge devices of limited capabilities and centralized processing).

the edge devices were produced with limited capabilities. But nowadays, the edge devices are developed with reasonable and adequate processing and storage capabilities. This makes the processing of data both at the place of data collection and at the time of data collection rather than send it to centralized servers.

As given in Ref. [1], the number of the IoT devices is around 30 billion in 2020 (please refer Fig. 2). The growth of huge number of IoT devices tends to use more bandwidth of the existing network to the cloud. In reality, many enterprises are unable to provide the required bandwidth for the devices.

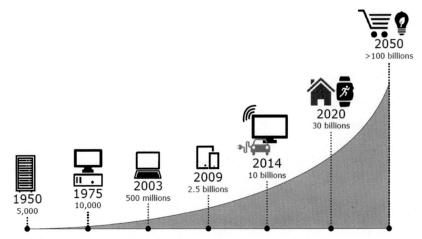

Fig. 2 Growth of IoT devices [1].

In addition, the bandwidth requires additional cost. With this situation on one side, on the other side, nowadays the devices are being developed with adequate storage and processing capabilities. This lays the foundation for edge computing. With the advent of hardware support, the edge devices are being designed with its own computing and decision logic in edge layer itself. The implementation of required processing at edge itself facilitates (i) performing real time analysis on the data without latency (ii) reduces the frequent transfer of data to cloud.

2. Need for edge and fog computing

In order to provide real time analysis and the corresponding control actions in right, the edge devices are being equipped with somewhat reasonable resources including processing power and storage so that an edge becomes capable of performing the required and necessary real time analysis on the acquired data. Also, the data is being sent to its next layer called fog layer as shown in Fig. 3.

The nodes of edge computing layer have limited resources and they tend to carry out very important tasks related to real time analytics. For example, consider an IoT based remote health monitoring system attached to a patient. His vital parameters are being continuously monitored by different sensors such as heart bean sensor, blood glucose sensor, temperature sensor, pressure sensor, etc. It is mandatory that his vital parameters need to be

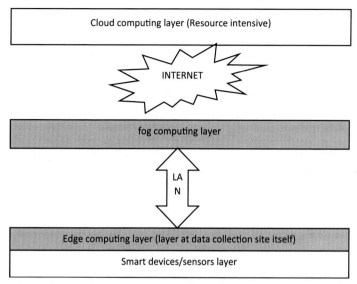

Fig. 3 Different layers of computing in edge and cloud computing environment.

analyzed of their values. If any abnormality is found, immediate it needs to communicate to the concerned physician for timely treatment. Here any kind of tolerance may not be accepted. This kind of immediate analysis is typically incorporated in the hardware itself. The edge nodes are supported with required hardware mechanism to perform the critical actions.

Now the edge nodes communicate the acquired data and analyzed knowledge to fog layer which is almost a centralized one. The two layers are typically established over a Local Area Network (LAN). The fog layer is capable of performing rich analysis when compared to edge layer. The fog layer sits as an intermediate layer. This layer decides which data needs to be sent to cloud for further processing. The computation, storage and processing are being performed along the IoT- to-Cloud path and not simply in cloud [2]. Cloud computing provides resources with high availability but also with higher power consumption whereas fog computing fulfills the needs of edge layer by provide resources with comparatively lower availability and lower power consumption. The key point is the availability of fog resources is very much adequate to meet the computing needs of edges devices. Also, the fog layer is very near to the IoT layer. More important is that the fog layer takes the important decisions in regard to the amount of data that is being transferred to cloud. This greatly reduces the huge transfer of data from edge layer to cloud layer.

3. Nature and requirements in edge and fog computing environment

3.1 Evolving or changing needs of the IoT

In the Internet of Things (IoT) layers, devices and sensors may get added according to need arises. For example, consider a weather monitoring station. In general, certain number of wind speed sensors and temperatures may be used to detect the temperature and speed of wind at different height such as 10 m, 20 m and 50 m from the ground. But the situation may not be the same during all time. Assume that suddenly storm has set and it is required to start additional wind speed sensors, temperature sensors and also rain gauge sensor at many levels. In this scenario, more number of devices is getting added.

As far as monitoring and control actions, more functional modules need to get incorporated. The key point is that there are changes in the functionality of the application according to the needs that can arise with respect to time. In most of applications, the software applications undergo frequent changes in the requirements of the applications. Conventionally, software applications were built for rigid requirements. But modern applications need to incorporate the changing requirements. Here, the applications are dynamic in nature. Also, when the applications grow in size, the deployment of larger sized applications takes longer time. The mandates alternate architecture in contrast to conventional architecture which is typically of three-tier in nature.

3.2 Heterogeneous nature of the IoT layer

The devices present in the IoT layer are of heterogeneous in nature. They operate with different communication protocols such as Bluetooth, Zigbee, etc. Mostly the devices are equipped with short range wireless communication. So, they are connected to the Internet via wireless protocols. Also, the data formats among the devices are may vary. The devices are typically manufactured by different hardware vendors. That is the technologies and protocols used by the devices are of heterogeneous in nature. So, typically the devices interact with one another with the help of the smart IoT gateway which provides the right transformation required which interactions.

3.3 Mobility and low power

The devices are inherently having mobility. Also, the devices in the edge and fog layer are of battery operated.

3.4 Distributed nature of the layer

Various sensors present in the layer are of distributed in nature. Obviously, some of decentralized control is sought for efficient handling of data.

3.5 Low computing/processing power

The nodes present in the IoT to fog layer of are battery operable and though have adequate processing power, it is comparatively lower when compared to that of centralized cloud-based servers which are of unlimited computing power and having very high availability. Consequently cloud servers consume huge power also.

4. Why microservices architecture for edge/fog computing applications?

From the previous section, it is understood that an architectural style of loosely connected independent and lightweight modules of applications could be more conveniently deployed in fog nodes. Microservices Architecture (MSA) is an architectural style for developing individual applications that may undergo frequent changes and frequent deployment. Basically, in MSA, an application is split into several microservices each service is designed based on independent functionality with its own data source so that it can be independently deployed. Modern business applications tend to incorporate the customer needs and demands into the application now and then the demands come. Also, when it is split into independent microservices managed by small team of people, the services may be deployed while without disturbing the integration of overall business process. This facilitates both incremental development and agile software development. In practice, MSA has evolved as a de facto standard for developing enterprise e-commerce applications and cloud native and cloud-centric IoT applications. Microservices inherently support polyglot architecture where a business process in which the microservices can be developed using different languages and different technologies. MSA architecture is language agnostic, hardware agnostic and platform agnostic. This unique feature of MSA makes it as an ideal architecture for edge computing where the IoT devices tend to have heterogeneous communication protocols, hardware and data formats. MSA provides a flexible programming paradigm for edge applications. As far as deployment is concerned, microservices can well be deployed both in virtual and container-based environments. Compared to virtual machines,

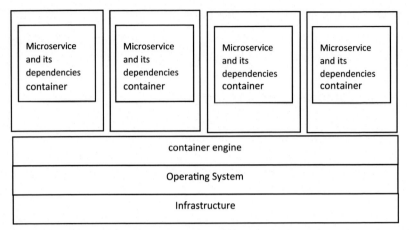

Fig. 4 Containerized environment for microservices deployment.

in containerized environment, the microservices will get deployed in containers and more than one container share the underlying operating system. The key point is that the container does not contain the full operating system. It contains only the microservice and its dependencies as in Fig. 4. This greatly reduces the size and containers become lightweight which is again very appropriate for edge computing where we are supposed to use the computing power only for necessary and crucial processing. Thus, MSA is an appropriate style for developing edge computing.

As mentioned in Ref. [3], the fog computing can benefit from three unique features of microservices, namely, independent deployable, independently scalable and decentralized computing model.

5. How the unique features of MSA fits as a natural choice for edge and fog layers?

MSA promote the development of functionally independent and loosely coupled microservices which are generally designed around the principle of single purpose.

In MSA the application is realized in terms of several number of microservices which can be independently deployable and independently scalable. Not only scalability, any quality of service parameter can be individually set of a microservice according to the need of the application.

For example, in application, consider two microservices say A and B. Imagine that A is interacting with users whereas B is interacting with backend database. Also, assume that the connection to database is done at seldom. Here, the availability for A alone can be analyzed and can be configured individually with required number of CPUs without affecting the rest of the application. This means that, the instances of microservice A alone can be increased whereas the number of instances of B can be kept at minimum to ensure the required connection to database.

Secondly MSA provides inherently a distributed nature of computing model. Microservices are distributed in nature. This means that different portions of an application can be distributed and a decentralized computing model can be developed easily. The microservices can interact with one another using a set of standard lightweight communication protocols such as Representational State Transfer (REST) protocol. With the help of loosely coupled communication among different services, the entire application can be composed using different means of composition namely service orchestration and service choreography. In service orchestration, a centralized coordinator service will be involved with describes the workflow of the services involved in a process whereas in choreography every participating service is aware of the workflow and in this model there is no centralized coordinator services. According to the situations exist, different compositions styles can be realized for the IoT based applications.

Thirdly MSA supports the development of different portions of application with different technologies and different programming languages. I.e. the architecture provides full flexibility to developers that they can use any languages as they wish. They can use different languages such as c, C++, Python, Java, etc. MSA supports polyglot nature of application.

Fourthly, the containerization of microservices and containers management platforms ensures the provisioning of appropriate resources to different microservices and ensures the adequate availability of microservices by the automatic creation and management of required number of containerized instances of relevant modules. More importantly the containers enable the scaling of individual services in a stand-along manner which leads to the optimized usage of resources in fog environment which is one of the critical needs of fog layer.

Fifthly, an IoT system should be resilient where the failure of any single component should not bring the entire system to down. This is facilitated by MSA where the responsibility is decomposed into many services which helps to increase the resilience of the application.

Ultimately, the elements of MSA including Not Only SQL (NoSQL) databases, event-based communication and API gate concepts, service

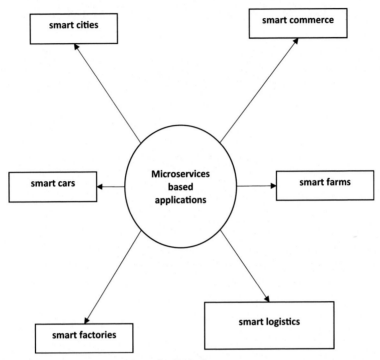

Fig. 5 Different microservices based IoT platforms.

composition can very well implemented in the fog and edge nodes with help of evolving tools and platforms. In addition, the applicability of microservices for IoT can be realized in different domains including smart cities, smart commerce, smart farms, smart cars, smart factories, smart logistics and cross-domain/general purpose applications, as shown in Fig. 5 [4].

6. Overview about elements of microservices

The higher level schematic diagram of microservices application is shown in Fig. 6. From Fig. 6, it is clear that like any other architecture, the MSA architecture also can be viewed as different layers. The bottom most layer is the physical infrastructure. This layer may be virtualized for effective sharing of physical resources. On the top of physical infrastructure, the microservices are deployed in the virtual or containerized environment which is usually managed by the container management tools such as Kubernetes. Each microservice exposes its functionality with well defined Application Programming Interface. The microservices interact with one another with help of standard protocols such as REST API. On the top

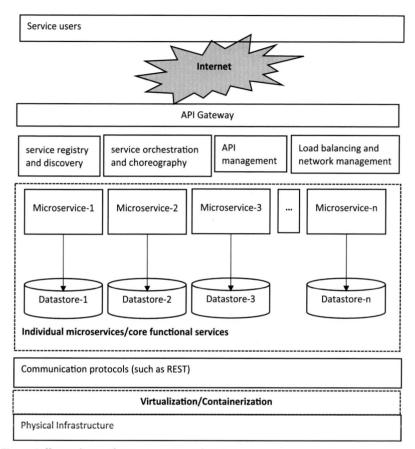

Fig. 6 Different layer of microservices architecture.

of individual services, there is services orchestration layer which is very essential to integrate the individual services according to specific execution pattern so that a particular business process is fulfilled. Also, each and every services needs to register its API in a registry so that other consuming services will come to know about its functionality.

Similarly, once a service is registered in a registry, it needs to be discovered before its actual invocation. Also, when we talk about enterprise level, there may be several services involved in an overall business solutions. So, there is a gateway mechanism which is having the knowledge of all APIs of the individual services and existing load on different service instances. Thus on the top of core services, there is a management layer and on its top there is gateway which will interact with service users. Also, the entire situation is a distributed one, i.e., the services need not to be deployed in same server. The services are distributed and they can interact over the Internet. Service users can consume the services over the Internet.

The key elements of MSA include

- Design Principle of individual services
- Communication protocols of services
- API and microservices
- Service integration techniques
- service registration and discovery mechanisms
- API gateway
- Polyglot support for development
- MSA and transaction support
- MSA and design pattern

6.1 Design principle of microservices

The very basic idea of MSA is to provide as much as independence to each microservice so that each microservice becomes individually deployable and scalable. To fulfill the basic idea of MSA, the microservices are designed around single purpose. That is each microservice is designed to perform only single purpose or single function. To maintain maximum independence, each microservice needs to be designed with its own data store.

The microservices should be cohesive. It means that only the functionality which are really are dependent one another should be kept together to realize a single purpose or responsibility.

The service must be decentralized from other microservices so that a microservice can be individually deployed and can be set for its other quality attributes such as scalability, availability, security, etc. In addition microservices are distributed.

6.2 Communication protocols of microservices

As each microservice is designed with single purpose, individual microservices are required to interact with one another in order to complete any business process. Like in any other districted and decentralized computing paradigm, the microservices can communicate in two different ways. They are (i) synchronous communication and (ii) asynchronous communication. In synchronous communication, the sender and receiver of the message are online with each other. A typical example for synchronous communication is data exchange using HTTP request/response protocol as shown in Fig. 7. The sender sends some message to receiver and waits for response from receiver with a blocked state.

In asynchronous communication, the sender simply sends the message to the receiver and continues to work without any blocked or wait state. The asynchronous communication usually takes place between microservices with the help of a message broker as shown in Fig. 8.

The core components of any message broker include

* message—A message is an information packet
* message producer—is an application that creates and sends message to broker
* message consumer—is an application that consumes the message
* message broker—enables the messages published by a message producer to reach its consumers asynchronously. It has two key components, namely, exchange and queue. Exchange component performs the routing of the message and queue stores the message

6.3 Application programing interface and microservices

Each microservice has to expose its contract or interface to other services so as to enables the other services to interact with that microservice.

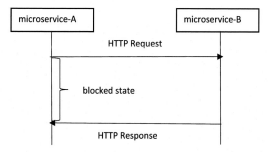

Fig. 7 Synchronous data exchange between microservice-A and microservice-B.

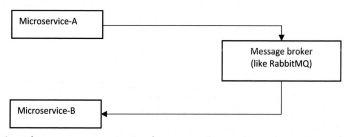

Fig. 8 Asynchronous communication between microservices using message broker.

The interface gives the details of how to access a microservice. Each microservice tends to perform a particular functionality with its required input arguments and gives back the processed results. So, in order to realize the interactions among microservices, each microservice has to invoke another microservice according to the calling semantics which basically includes the data such as (i) which protocol to be used to access the service (ii) IP address or DNS name of services (iii) service and method or function name (iv) arguments and return data type etc.

It is the responsibility of a microservice to expose its functional contract as an Application Programming Interface. A microservice can be accessed only through its API as shown in Fig. 9. The API never exposes the implementation details of the service.

The API can be developed using different specification such as REST API specification, Open API Specification (Swagger), SOAP, Apache Thrift specification, etc. REST API is commonly used by many developers. A brief overview about how REST API specification is given here for reader's reference. According to the specification, the URL of a service contains all the required details for accessing a service and services is considered as a resource. Consider a service, say **"temperature_service"** will return the temperature of a palce whose "pincode" or "zipcode" is given as input argument. The URL of this service is shown in Fig. 10.

In the above example, the arguments are passed as path parameters. The architecture also supports other kinds of parameters such as *Header parameter* which are included in the request header, *Query string parameters* which are specified in as key-value pairs after query string (?), *Request body parameters* which are included in the request body.

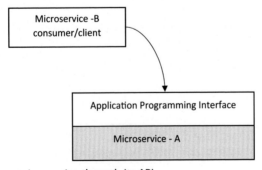

Fig. 9 Accessing a microservice through its API.

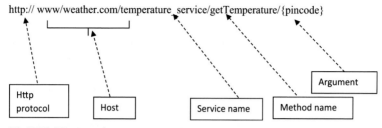

Fig. 10 REST API example.

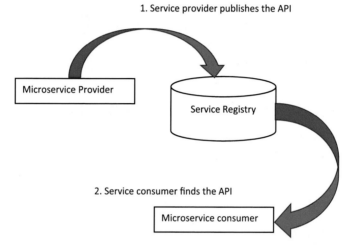

Fig. 11 Service registration.

From the above example, it is clear that HTTP protocol is used to access the service. So, it supports the standard HTTP methods, GET and POST. Now it is clear that the microservice can be very flexibly access using GET as

GET www.weather.com/temperature_service/getTemperature/{620027}

6.4 Microservice registration and discovery

Along with API implementation, a microservice needs to register its API into a registry as shown in Fig. 11, so that other microservices can find the API.

As each microservice tends to provide a single purpose or functionality, as far as a business process is concerned, naturally many services will need to be invoked in a particular order. So, the microservices need to interact with another. As mentioned earlier, a microservice can interact with another microservice only through its API. So, each microservice has to register its API in a registry so that other microservices can look into it and find them. The process where a microservice registers its API in a registry is called service registration and the process where another microservice finds the API from the registry is called service discovery. There are commercially available tools such as Eureka which provides APIs for registering and discovering microservices.

Service discovery is of two types, namely, client-side service discovery and server-side service discovery. In client-side service discovery, the service consumer or client is responsible for determining the location of a microservice. Here, the client makes a request to service registry which returns the available service instances to client. Here determines the matched service and also decides which microservice to use, according to load balancing algorithm. After identifying the right service, the client makes a request to it through its API. The schematic of client-side service discovery is shown in Fig. 12.

In server-side service discovery, client submits its request to API gateway which in turn communicates with service registry to determine the relevant microservice by matching its API against the request made by the client. The schematic of server-side service discovery is shown in Fig. 13.

6.5 API gateway

Typically in microservices based application, since there are several distributed microservices involved, all the APIs of the microservices are required to

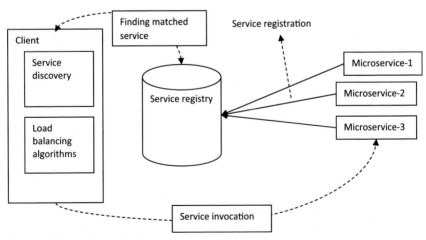

Fig. 12 Client-side service discovery.

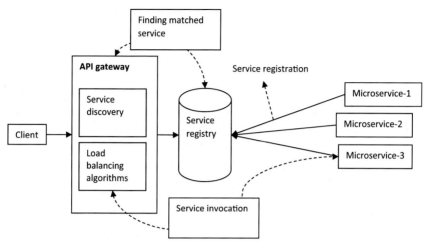

Fig. 13 Server-side service discovery.

be managed effectively. This is done by API gateway. API gateway sits between client and the microservices based application. API gateway is the server-side entry point. Client makes its request to API gateway which in turn needs to identify and locate all the participating services which are capable of fulfilling the client's request and invoke them according the required execution pattern so that the expected business process would be realized. This is shown in Fig. 14.

API gateway provides the following key functions

- It serves as single entry point to application
- It hides the implementation details of individual microservices and provides customized APIs which are specifically designed for different variety of service clients
- It performs service composition
- It performs service discovery and routing
- It hides all the heterogeneity that arises in communication protocols, data formats etc.

6.6 Polyglot support for development

MSA architecture does not provide any mandatory conditions for developers rather the developers can choose their own choice of developing the microservices. This provides flexibility to developers. It enhances the productivity of developers. They need to think about the languages or development environment tools rather they can focus on the core business

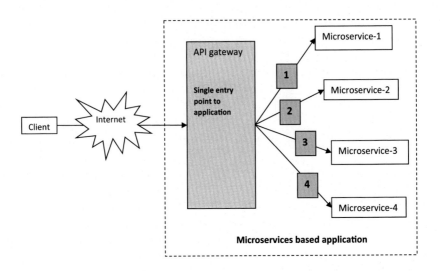

Fig. 14 API gateway as single entry point to the microservices based application.

process and goals. In addition, while designing API also, the architecture does not put any constraint on the specification of API. Developers can choose any kind of API styles such as REST API, Apache Thrift API, Swagger specification for REST API or Open API specification etc. Similarly as far as data formats for communication is concerned, developers can choose any format such as XML, JSON format, YAML etc. Also, the architecture provides support for different varieties of communications namely synchronous REST based communication, message broker based asynchronous communication, event based communication. Here also, various message brokers namely Java Message Broker, Rabbit MQ broker, Apache Kafka broker etc. can be used according to the needs of applications. Thus the polyglot feature of MSA provides greater flexibility of developing applications.

6.7 MSA and transaction support

Any business application involves persistence of data. That is there would be data and information which are considered as assets for any business need to be stored in a durable manner. Obviously there would be databases and datastores which store different data of applications. As microservices are likely to have their own datastores, they databases need to contain logically valid data. Conventionally the requirements of any database transaction needs to fulfill the fundamental Atomicity, Consistency, Integrity and Durability (ACID) properties [5]. Atomicity specifies the need that a transaction should occur in its entirety or it should not happen at all. Consistency specifies that a transaction should follow serialization whenever simultaneous access of same data needs to be accessed. Isolation specifies that each transaction should occur in isolation. Durability refers to the persistence of successful database changes and transactions.

Though conventional applications try to achieve the above ACID properties, modern e-commerce applications have slightly different requirements such as availability. In addition, as mentioned earlier, MSA provides support for polyglot. Here there is no constraint that the data needs to be only in structured format, rather, it can be semi structured, or even unstructured. MSA actually support modern Not only (NOSQL) database servers such as MongoDB, Cassandra, etc. More important aspect is that each microservice is having its own datastore as shown in Fig. 15.

As shown in Fig. 15, in MSA data is associated among different microservices and it is not at one place. This kind of data distribution among different microservices leads difficulty in bringing consistency of data. When one service performs some update in its datastore, then it needs to be updated in all microservices. Updating each and every change in datastore of all involved microservices is a time consuming process. Also, NoSQL databases do not support Two Phase Commit protocol.

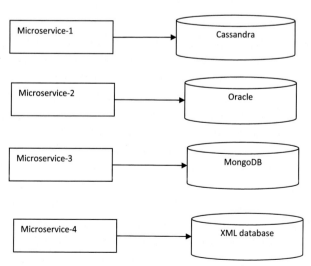

Fig. 15 Distribution of data among different microservices.

In addition to the difficulties in achieving data consistency, modern e-commerce applications are in need of implementing availability rather than consistency at all times. So, modern applications and microservices or cloud native applications are trying to implement what are called Basically Available, Soft state, Eventually consistent (BASE) properties in contrast to ACID properties.

According to BASE properties, an application needs to be always available. It may be in soft state. This means that the state of the application may change over time. The system becomes eventually consistent. Though the state of the application changes over time, it eventually becomes consistent. So, the database becomes logical valid.

Basically, in microservices based applications, they use two kinds of databases, one for read data and the other for writing data. So, the availability is being achieved using read database instance. The write database is containing the updates in datastores with in turn initiate updates in related microservices in order to bring eventual consistency.

6.8 MSA and design patterns

Any edge or fog computing application has its own typical application or domain requirements and Quality of Service (QoS) attributes, namely, availability, scalability, security, resiliency, response time, reliability, etc., MSA applications are being designed and developed using standard design pattern. Design patterns are developed from best practices and provide quality solution for complex, recurring problems.

In order to achieve different architectural and domain requirements for MSA applications, the following design patterns are predominantly used.
- Decomposition patterns—Decomposition pattern such as Domain Driven Design (DDD) and strangler pattern, are used while splitting an MSA application into several, single purpose microservice
- Composition patterns—Composition patterns such as aggregator pattern, proxy pattern, chained pattern, branch microservice pattern, API gateway pattern and client-side UI composition pattern are commonly preferred in MSA applications. Aggregator pattern is used to combine functions delivered by more than one microservices. Proxy pattern is used to prevent the direct exposure or access of a microservice by its clients. Chained pattern provides a variety of composition pattern where functions of different microservices are combined sequentially as a chain. Branch microservice pattern is also a kind of composition pattern where the function of a microservice would be implicitly includes the function

of another microservice. API gateway pattern is used to provide single entry point to an MSA application. Client-side UI composition pattern is used which the GUI of an application is designed using several microservices

- Database patterns—Fundamentally each microservice has to be designed with its own datastore. Commonly used database patterns include database per service pattern, shared database per service pattern, Common Query Responsibility Segregator (CQRS) pattern and saga pattern.
- Observability patterns—To monitor the behavior of an application observability pattern is used. Patterns such as centralized logging service pattern, distributed tracing pattern can be used to monitor the functioning of an application.
- Cross-cutting concern patterns. Whenever changes are performed in the coding portion of microservices, cross-cutting concern patterns can be implemented to avoid or reduce code modification. For example, an application may involve many input parameters which are likely to change. These parameters are brought out the programs and set as properties in configuration files. The configuration files are external to programs. So, even when the values are changed later, it would not require any changes in the code. This pattern is called external store configuration pattern.
- Deployments patterns—A microservice is like to change and it may get redeployed. Whenever changes are performed in microservices, there will be different version of microservices. The deployment of newer versions can be carried out using deployment pattern. For example, the blue green deployment pattern facilitates the deployment of newer version of microservice by maintaining two identical production environment and making one of them live.

6.9 MSA and security

Like any other application, MSA based applications also should be protected against any unauthorized access and security of the application should be ensured. There are various best practices for ensuring microservices security. Some of them include

- API gateway - Implementing API gateway prevents the direct access of microservices by the clients. In addition, the developers can implement overall application level security mechanism in API gateway. They can implement the overall authentication and access control in API gateway. In addition to application level security, security can be implemented for each microservice. For example, OAuth 2.0, provides an industry-standard protocol for authorizing users across distributed application.

- Container Security—When microservices are deployed in cloud environment, containers are used for deployment and in such environments, container security becomes important.
- Regular scanning of source code repository—Typically application are being developed using different third party libraries and dependencies. This makes an application vulnerable to attacks. So, the source code of the application needs to be checked regularly
- Use of HTTPS—Fundamentally HTTPS serves as the basic element in bringing security to any application as it prevents common attacks such as phishing. HTTPS ensures privacy and data integrity by encrypting communication over HTTP.

7. MSA for edge/fog computing

Basically, fog computing devices have limited computing power, storage and processing capabilities. The power requirements are also low. More importantly the devices are of heterogeneous. MSA fits as a natural choice for development and deployment of fog applications due to the following reasons.

1. Services support peer to peer communication.
2. MSA supports polyglot development of applications where different microservices can be developed using different technologies, programming languages, protocols and data formats. This relieves the burden that lies with developers that can freely choose their own development stack.
3. More important aspect is that MSA can run in containerized environment which always optimizes the power requirements. It is one of the desirable key points in fog computing environment where the available infrastructure is minimal.
4. MSA provides a flexible and loose coupled way of developing applications along with individual deployments.

With the above ideas, higher level architecture of MSA for fog computing can be realized as a layer architecture as shown in Fig. 16.

Edge devices—This layer is composed of various IoT devices, connected sensors. These devices are of heterogeneous in nature in terms of communication protocols, data formats, power consumption needs, hardware vendors, processors, computing power etc. Despite the heterogeneity, they are of low powered devices have very limited computing infrastructure.

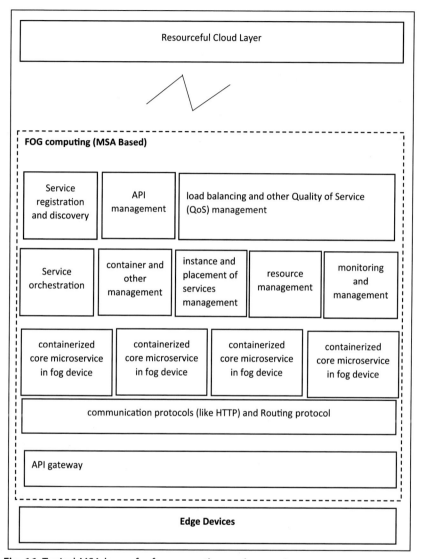

Fig. 16 Typical MSA layers for fog computing environment.

So, they depend on the fog layer which is very near to edge layer and available in the same physical site or same campus geographically.

7.1 Fog computing environment

As far as fog computing layer is concerned, it is having comparatively higher computing infrastructure and thus able to perform the necessary logic and analytics which are mandatorily needs to be done near the sensing point itself. When the required analytics and processing is taken place near to the sensing or edge layer, the unnecessary flow of data to cloud can be prevented which significantly reduces the network traffic and cost.

In the fog layer, the lower most layer refers to API gateway layer. This layer serves as the entry point for the MSA based process deployed in fog computing environment. On the top of this layer, one could find the communication layer. Here, HTTP protocol is shown as it is a lightweight protocol and more suitable for fog environment. The key point here is that HTTP does not require any other additional infrastructure for communication and it is preferred in an optimized environment such as fog.

On the top of communication layer support, we have the layer of core microservices, each of which performs a typical function of the application. These individual microservices are deployed in containerized runtime which can be very well accommodated in fog devices.

On the top of core services, one can find the next higher level layer where service orchestration, container management, management of resources, services instance and monitoring of application. Still on the top of this layer form the other functionalities of service registration, service discovery, API management, load balancing and other QoS management.

All the components of different layers show only the schematic and try to represent that the layers can be deployed in fog devices. At last, from the layer, the processed and analyzed results are communicated to cloud via the Internet. The cloud layer provides rich resourceful computing facility to carry out further analysis and archival of useful knowledge.

8. Challenges

Despite MSA provides many desirable and unique features such as polyglot support, individual and independent deployment, implementation of quality attributes such as scalability at individual service level, loose coupling between services, MSA is inherently associated with the following challenges.

- As the application is split into several services, the operational complexity increased
- Testing of the application becomes complex
- Since microservices need to communicate with one another, there would be latency issues
- Implementation of transaction in databases becomes difficult

9. Conclusion

This chapter discusses the need for edge/fog computing, critical requirements of applications deployed in fog computing layer and how those requirements are naturally fulfilled by the unique features of microservices architecture. For better understanding, an overview of key elements of microservices is presented. Then a higher level conceptual layered architecture is presented to show the applicability and feasibility of MSA for developing fog applications. The challenges associated with microservices application are highlighted.

References

[1] M. Capra, R. Peloso, G. Masera, M. RuoRoch, M. Martina, Edge computing: a survey on the hardware requirements in the internet of things world, Future Internet 11 (4) (2019) 100, https://doi.org/10.3390/fi11040100.

[2] Ashkan Yousefpour, Caleb Fung, Tam Nguyen, Krishna Kadiyala, Fatemeh Jalali, Amirreza Niakanlahiji, Jian Kong, Jason P. Jue, All one needs to know about fog computing and related edge computing paradigms: a complete survey, J. Syst. Archit., Volume 98, September 2019, Pages 289–330, https://doi.org/10.1016/j.sysarc.2019.02.009.

[3] Samodha Pallewatta, Vassilis Kostakos, Rajkumar Buyya, "Microservices-based IoT application placement within heterogeneous and resource constrained fog computing environments", UCC'19: Proceedings of the 12th IEEE/ACM International Conference on Utility and Cloud Computing, Pages 71–81. doi:https://doi.org/10.1145/3344341.3368800.

[4] B. El Khalyly, A. Belangour, M. Banane, A. Erraissi, A comparative study of microservices-based IoT platforms, Int. J. Adv. Comput. Sci. Appl. 11 (8) (2020) 389–398.

[5] https://www.geeksforgeeks.org/acid-properties-in-dbms/.

About the author

Chellammal Surianarayanan is an Assistant Professor of Computer Science in Government Arts and Science College, Tiruchirappalli, Tamil Nadu, India. She earned a doctorate in Computer Science by developing computational optimization models for discovery and selection of semantic services. She published research papers in Springer Service-Oriented Computing and Applications, IEEE Transactions on Services Computing, Springer New Generation Computing, International Journal of

Computational Science, Inderscience, SCIT Journal of Symbiosis Centre for Information Technology, etc. She produced books and book chapters with Springer, IGI Global, CRC Press. She has been a life member in professional bodies such as Computer Society of India, IAENG, etc.

Before coming to Academic service, Chellammal Surianarayanan served as Scientific Officer in Indira Gandhi Centre for Atomic Research, Department of Atomic Energy, Government of India, Kalpakkam, TamilNadu, India. She was involved in the research and development of various need-based development of embedded systems and software applications. Her remarkable contributions include the development of an embedded system for lead shield integrity assessment system, portable automatic air sampling equipment, the embedded system of detection of lymphatic filariasis in its early stage and development of data logging software applications for atmospheric dispersion studies. Totally she has 25 years of academic and industrial experience.

CHAPTER SEVEN

Edge data analytics technologies and tools

N. Jayashree[a] and B. Sathish Babu[b]

[a]Department of Computer Science and Engineering, C Byregowda Institute of Technology, Kolar, Karnataka, India
[b]Department of Computer Science and Engineering, R V College of Engineering, Bengaluru, Karnataka, India

Contents

Advances in Computers, Volume 127
ISSN 0065-2458
https://doi.org/10.1016/bs.adcom.2022.02.004

Abstract

Edge data analytics reduces the volume of data that needs to be sent to the cloud or other available resources for processing. It facilitates with avoidance of an additional processing state via autonomous behaviors of the machine, increased security, and minimized costs of data transmission. Data analytics happened to be of the highest importance over the collected data to draw any meaningful insights (Donitha, 2017) [1]. In this chapter, we discuss the tools and technologies of edge analytics which is the best choice compared to relying on descriptive analytics, diagnostic analytics, predictive analytics, and prescriptive analytics.

Abbreviations

AWS Amazon Web Services
IoT Internet of Things
REST representational state transform
API application programming interface
CQL card query language

1. Introduction to edge data analytics and benefits

The devices generating data in IoT are sensors, ubiquitous computing devices, cameras, etc., used in any IoT application. The streaming data are collected at a centralized body for processing. As IoT technology emerges, the information being monotonously streamed by the IoT devices increases and is rapidly transmitted, and results in an accretion of unmanageable data, most of which is never used. The data produced by various IoT devices need to be stored, processed, and analyzed at the Edge for an effective digital engagement. This is achieved by performing the data analysis at non-centralized data-generating devices in a system. The term "at the edge" represents the data being processed and analyzed nearer to the generating devices.

Edge Data Analytics: A data analysis procedure or tool implemented at the Edge of the network either within or closer to the device that generates data.

The edge data analytics minimizes the amount of data that need to be transmitted to the cloud or other available resources for processing. It allows the application to avoid an additional processing state by enhancing the machine's autonomous behaviors, increasing security, and minimizes data transmission costs. Edge data analytics gains importance with faster decision making irrespective of bandwidth and IoT technology for automation.

Edge data analytics is similar to any other analytics except that it is done *at the Edge* on devices having storage capabilities and processing capabilities, although they have certain communication curbs. The edge data analytics is not a substitution of centralized processing, whereas it can augment with it ensuring the delivery of data discernment. The benefits of edge data analytics are given in Table 1 [1] [2].

1.1 Benefits of edge data analytics

See Table 1.

Table 1 Benefits of edge data analytics.

i. *Improved productivity by real-time analysis with faster and autonomous decision making*: As the analysis is done near the data, it is easy to be done in near real-time and is challenging to be achieved if the data needs to be transmitted elsewhere like the cloud server or any data center.
ii. *Time Efficient*: Edge analytics filters unnecessary information before analysis. This minimizes the processing time and data transmission times. As the data is analyzed at the edge device, it is near real-time analytics.
iii. *Averting latency*: In many applications, it is not advisable to transmit the data to a remote location and wait for the results as they need real-time analytics. Edge analytics reduces latency by real-time analysis and by identifying discernment at the Edge.
iv. *Furnished scalability over time*: As every device analyzes its data, the computation overhead is decentralized, resulting in improved scalability.
v. *Reduced cost of central storage as most processing is done at the Edge and only necessary data is transmitted*: The usual way of data analysis at data centers involves substantial costs. Also, the data storage, processing, and consumption of bandwidth are tied with expenses. Some use IoT device resources to perform the analytics reducing the costs of back-end processing.
vi. *Finer security and privacy by avoiding granular data from storage or transmission*: The analytics applied at the IoT devices avoid entire data across the network. The actual data do not leave the device and hence is secure.
vii. *Minimized bandwidth usage*: Several connected IoT devices will be active worldwide, and the bandwidth required by them is higher as the number of devices increases both to transmit and receive. Edge analytics provides the analytic capabilities to the remote data analysis centers and reduces the bandwidth issue.
viii. *Reduce Bottleneck*: The data generated includes extensive data and causes congestion in the network. The bottleneck is reduced as edge analytics processed the raw data at the Edge rather than forwarding it.

2. Edge data analytics versus server-based data analytics

Edge analytics analyzes data at the device, whereas server-based analytics requires the data to be transmitted to the remote server for analysis. The comparison between the two approaches is given in Table 2 [3].

Table 2 Comparison of the performance factors.

Factors	Edge-based data analytics	Server-based data analytics
Reliability	High reliability with access to the raw uncompressed data. The data generated is treated by the analytics at the device, and the results are more promising.	Less reliable since data are compressed/processed and forwarded to the servers and hence affects the quality of the result.
Downtime	There is no need to transmit data to the cloud or data centers for processing, as the analysis is done at the edge device. Downtime of the network does not affect the analytics.	Require data transmission to the cloud or data centers for analysis, and the process is delayed during downtime.
Bandwidth	Lower-bandwidth requirements as only simple information like alarms are transmitted to the servers and analytics is done at the Edge and hence can be used in a bandwidth-constrained environment.	Requires high bandwidth for the transmission of data to the server and cannot be used in a bandwidth constraint environment.
Latency	Low - Data is analyzed at the device.	High—Data is analyzed at the server.
Time efficiency	Analytics in real-time as the data generated by the device is processed at the Edge.	Analytics has a delay as the data need to be transmitted across the network to the cloud.
Connectivity	It does not require network connectivity to the server.	It requires network connectivity to the server.
Cost	Embedded analytics is set up low cost in the devices.	Focus more on power, reliability than the cost.
Security	There is the least risk of a security attack as the raw data is not transmitted to the server.	There is more risk on data in transit.

3. Architecture and methodology of edge data analytics

Edge analytics involves tools on or nearby IoT devices for collecting, processing, and analyzing the data streamed from the devices instead of transmitting them to the cloud or any other remote servers for analysis. Edge analytics streamlines the data with real-time analysis and ensures the collection of enough data from the device. At the Edge, the devices are designed to have their analyzing capabilities.

The architecture [4] of edge data analytics is given in Fig. 1. The *edge devices* are general-purpose devices with full-fledged operating systems and battery power. The devices can run edge intelligence that runs commands on the data with the help of an *on-premise server* and sends the signals to the actuators. The device is connected to the cloud either directly or through a *Network Switch*. They act between the edge devices and the cloud for location mapping services. The *cloud* sends the signals such as configurations, data queries, or machine learning models. There is a pattern followed by any edge analytics tool:

- The collection of data by any device at the Edge, as the devices generate the data, they are streamed to the edge analytics tool.
- Performing analysis of data at the Edge using in-built analytics features. Data analysis as the data is being streamed decreases the latency in decision-making on the connected devices.
- The results are used to take action, like responding immediately to the received data based on the status and respective measures.
- There will be a transmission of the necessary summarized data from the Edge to the cloud. This reduces the bandwidth requirement in case of transmitting the raw data.

The protocols commonly used to transmit the data from edge device to the cloud include HOOT/HTTPS (Hypertext Transfer Protocol/Secure), MQTT (Message Queuing Telemetry Transport), RTSP (Real-Time Streaming Protocol), WebRTC (A combination of standards, protocols, and JavaScript and HTML5 APIs), and Zigbee (uses packet-based radio protocol).

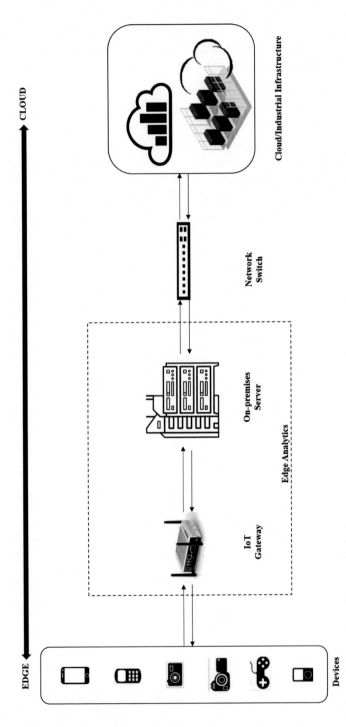

CLOUD

EDGE

Cloud/Industrial Infrastructure

Network
Switch

On-premises
Server

Edge Analytics

IoT
Gateway

Devices

Fig. 1 Edge data analytics architecture.

4. Edge data analytics technologies and solutions

4.1 An extended AWS to the edge devices for function with data generated while using the cloud for necessary resources is named AWS IoT Greengrass [5] (Fig. 2)

It expands cloud capabilities to the local Edge through the accumulation and analysis of the data from the data source, ensuring the autarchic responses to the local events, and communicating securely with the native devices without connecting to a cloud. The customers have the privilege to assemble IoT devices and application logic. It also provides a message buffer for preservation in case of failures and also secures the user data. It consists of software distributions, cloud services, and great features for data analytics.

4.1.1 Benefits

AWS IoT Greengrass helps build intelligent IoT devices faster. Here are the benefits:

i. Pre-built components to add or remove and control device software footprint

ii. Install and govern device software and configuration remotely and at scale without firmware updates.

iii. Bring cloud processing and logic locally to edge devices and operate even with the incomplete connection.

iv. Program devises to transmit only high-value data, making it easy to deliver rich perceptions at a nether cost.

4.1.2 Working of AWS IoT Greengrass

AWS IoT Greengrass (Fig. 3) is an open-source edge runtime and cloud service for building, deploying, and managing device software. The client software enables local processing, messaging, data managing, ML Inference and offers pre-built components to hasten application development. Then the

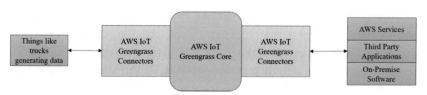

Fig. 2 AWS IoT Greengrass architecture.

Fig. 3 Working of AWS IoT Greengrass.

AWS IoT Greengrass cloud service aids building, deployment, and governing of device software over the squadron of devices.

IoT devices can vary in size. AWS IoT Greengrass Core components, AWS IoT Device SDK-enabled components, and FreeRTOS components are calibrated to interface with one another. If AWS IoT Greengrass Core component loses affinity to the cloud, connected devices can continue to communicate with each other over local network.

The pre-built components facilitate easy extending of edge device functionality without writing code. AWS IoT Greengrass components enable adding features and quickly connecting to AWS services or third-party applications at the Edge.

4.1.3 Scenarios

i. Run at the Edge: AWS IoT Greengrass quickly gets intelligence to edge devices, such as anomaly detection in precision agriculture or powering autonomous devices.

ii. Manage apps: Deploy fresh or heritage apps across squadrons using any language, packaging technology, or runtime.

iii. Control squadrons: Manage and operate device squadrons in the field locally or remotely using MQTT or other protocols.

iv. Process locally: Assemble, amass, refine, and transmit data locally. Manage and control what data goes to the cloud for optimized analytics and storage.

4.2 CSA-A network assessment tool by Cisco named Cisco SmartAdvisor (Cisco discovery service) (Fig. 4)

It is a network assessment tool by Cisco for progressing customer networks mainly used to organize network relocation and superlative network reforms resulting in customer network evolution. It can summarize the data transmissions and manage the reports of the same. The data processing can be applied to assess the network reforms, the services, and the vulnerabilities

Fig. 4 Cisco SmartAdvisor dashboard.

in the network. The client-side tool for CSA as accepted by Cisco is IP Explorer [6] that elevates the network management activity. The objectives of this service are:

- To evaluate the life cycle of the product in spot the outmoded software and hardware components
- To discover issues in security
- To congregate commitment particulars, and
- To gauge the potential of components to launch new services or technologies.

Finally, CSA acts as abet for Cisco associates to realize network concerns and plan optimal network upgrades.

NETvisor's IP Explorer for CSA helps perform Cisco Discovery Service (CDS) to:

- **automatize** network revelation and documentation activity
- **inspect** the network's robustness and end-of-life shape
- **recognize** network-related problems of the customer
- **initiate** Cisco product improvement strategies

4.2.1 Features and benefits

- Contemporary CDS information about components on network revelation
- Handy interface to initiate and mark CDS transaction process

- integrated CDS reports with extended network inventory data
- Requested CDS reports stored in a local database

4.2.2 *Working of CDS* (Fig. 5)

IP explorer collects data from Cisco products through SMTP requests or configuration files and forwards that to the CDS server for comprehensive analysis. IP explorer web GUI has a CDS interface. A new transaction is launched via the CDS interface to initialize a complete examination. Customization is possible through the selection of transaction types which in tracking the status of the activity via the web interface.

The complete analysis results in life cycle reports, security issues, product deficiencies that are not related to security, and contract information. The reports can be accessed through the CSA portal and CDS web interface of IP Explorer and web GUI.

Fig. 5 Working of CSA.

4.2.3 Requirements for the application of CSA

i. User ID and partner level details in Cisco.com

ii. Internet for data transfer between IP Explorer server and CDS servers.

4.3 An analytics software package named Dell Statistica (Fig. 6)

It was developed by StatSoft and acquired by Dell [7]. It facilitates data analytics features combined with machine learning and visualization procedures to identify outliers in the processed data. Statistica has been extended with data preparation and new network analytics capabilities to enhance fraud detection potentiality.

Dell's top objective is to improve operational efficiencies. Dell Statistica has facilitated the reduction of time spent on models by 50%. The development process is much leaner and smoother compared to prior. Dell Statistica was rated an overall leader in customer experience and vendor credibility while scoring substantially above the overall sample based on 33 criteria. It also garnered best-in-class technical support product knowledge and consulting product knowledge. Empowering traditional and citizen data scientists is a key differentiator that likely helped Dell earn a leadership spot.

Reusable Process Templates now empower all types of users to share and distribute analytic workflows. Dell aims to create a foundation of openness and flexibility to elevate Statistica's customer experience. This strengthens Dell's credibility as a trusted, innovative analytics partner and support for heterogeneous environments. This helps customers run any analytics on any data, anywhere, to drive better decisions across their organizations.

Dell Statistica has built a firm basis of business knowledge by endorsing top pharmaceutical, manufacturing, healthcare, financial services, and retail organizations worldwide. For example, deep insights and analytics have helped Sanofi and other customers improve quality control, increase market intelligence, reduce supply chain risk, elevate customer experiences. Dell Statistica leaps out among the different solutions because of its persistent spotlight on making analytics more accessible. The opportunity to modernize analytics remains a guiding force and critical competitive discriminator in its approach, product development, and customer partnerships. The latest version of Dell Statistica includes edge scoring for IoT analytics, native distributed analytics architecture (NDAA), and collective intelligence.

The platform also offers network analytics to visualize entity relationships and graphical associations and combine predictions with human expertise to understand relationships within networks better. Statistica's collective

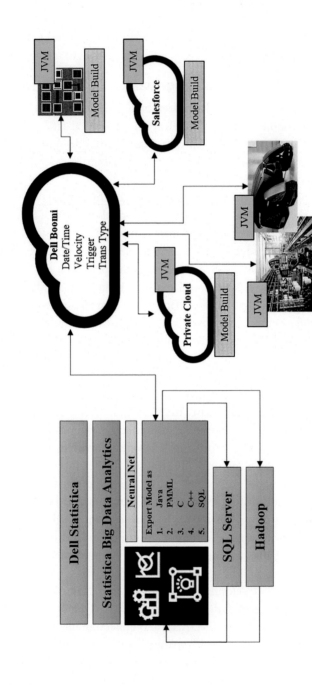

Fig. 6 Dell Statistica architecture.

intelligence feature helps businesses embrace the app marketplace for models. Users can construct models and data or import models written by others. Hence, Statistica has good connectivity to increase its capability based on user requirements. The NDAA serves the purpose of trying to break down data repositories.

4.4 An edge system providing high performance, low latency data processing by Hewlett Packard enterprise named HPE edgeline (Fig. 7)

When used, it is a facility that confederates the process of data collection, control systems, and networks to enable new analytics abilities at the Edge [8,9]. It exhibits high performance and minimizes latency using edge-optimized services and remote system management procedures. It is a converged operational technology (OT) like data acquisition, control systems, industrial networks, and enterprise-class IT in a single, consolidated system that implements data center-level compute and management technology at the Edge.

4.4.1 Significance of converging OT and IT: Connecting workers with data and insights at the edge

• Connecting Edge works with remote experts to diagnose and resolve production issues reduces downtime, making assembly lines more productive.
• Providing line workers with real-time step-by-step guided assembly instructions and other operations-critical information critical through

Fig. 7 HPE edgeline.

the wearable or handheld device and overlaid digitally onto the product and work surface improves product quality and throughput and reduces rework.

- Connecting technical support specialists to customer environments to diagnose and resolve product issues reduces service technicians' number of customer site visits.

As the amount, diversity, and pace of data continue to increase, it's necessary to collect data generated at the Edge, analyze the data in near real-time, and deliver the insight directly to the edge workers and their devices. One of such technology is Augmented Reality with capabilities like:

- Two-way video streaming
- Audio for collaboration
- Content overlay
- Streaming data
- Enterprise system connections
- Quality and compliance standards in enterprise systems
- Access to training and other knowledge management systems

4.4.2 Benefits

i. Gains in productivity and worker efficiency
ii. Reduced impact from a retiring skilled workforce
iii. Improved production quality
iv. Increased asset uptime with guided service and maintenance instructions
v. Reduced time to diagnose and resolve production and product support issues
vi. Reduced time and travel costs associated with service technician site visits

4.4.3 Use cases

- At a process manufacturer of hazardous chemicals, the time needed to identify asset risks in its plant is reduced by 75%. Here, a single spark can result in an explosion. It is possible to digitally retrieve documentation and get 3D rotational views of assets and risk-based inspection graphs. Workers are trained on production processes, using a connected worker solution, on workflows and allow them to experience the environment prior to setting foot on the plant floor.
- Diagnosing and repairing IT devices by nontechnical workers in sizeable retail distribution centers and verifying in-store inventory levels through

mobile devices and wearables to connect to HPE technical support. Here, the number of site visits by HPE technical staff was reduced from 54 to 1 during a five-month pilot.

– A large-format digital printing press manufacturer can diagnose issues twice as fast and reduce resolution times by as much as 70% in its technical support centers. Using HPE Visual Remote Guidance (VRG), the support technician remotely connects with the customer, who dons AR-based wearables to allow the technician to see what the customer sees, diagnose the issue, and visually guide the customer through fixing the problem. This avoids dispatch of support engineer to spot.
– A large enterprise remodeling its location reduces travel time and dollars, according to HPE, using HPE VRG for project status inspections. Construction workers can don the wearables or use a tablet to walk remotely located project managers.

4.5 An edge analytics agent by IBM named IBM Watson IoT edge analytics (Fig. 8)

It is a fully managed, cloud-hosted service with device registration, connectivity, control, rapid visualization, and data storage capabilities [10,11]. It has decisive intelligence to draw out valuable information from the streamed data. It is usually applied to convert the voice signals to textual data for analytics. It uses real-time analytics to monitor current conditions and responds accordingly. It leverages cognitive analytics with structured and unstructured data to understand situations, reason through options, and learn about changed conditions. Only the analytics results are stored at the edge gateways, represented as numerical scores of varying emotions in a conversation.

Operations of Edge analytics agent in IBM Watson include: Filter and reduce data sent to the cloud; Pre-process and transform raw data; Identification of critical conditions to send to the cloud for additional analytics; and Enable local actions. The capabilities of IBM Watson Platform include: Complex analytics; Analytic definition and distribution to Edge; Longer-term trends; and Pattern detection and machine learning;

Below are the use cases:
– Define and maintain cloud and edge analytics in a single view within the IBM Watson IoT platform on the cloud.
– View the results of Cloud and Edge analytics in a single dashboard within the IBM Watson IoT platform on the cloud.

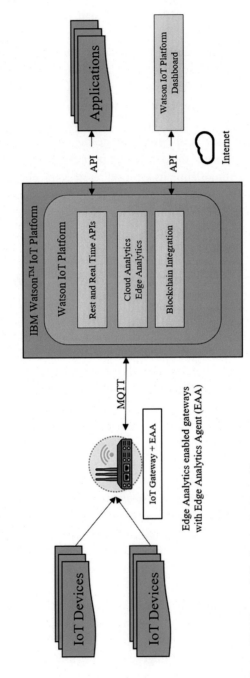

Fig. 8 IBM Watson IoT edge analytics architecture.

- Multiple Edge Analytics Agents support many types of devices and can be connected to a single Cloud instance.
- Edge Analytics capability has expanded from basic value comparisons to Z-Scores, moving average and data smoothing

4.6 An IoT hub to apply analytics at edge devices named microsoft azure IoT edge (Fig. 9)

It transfers data analytics and logic to the edge devices to help the IoT application focus on data discernments rather than data management [12,13]. It is made up of three components: IoT Edge modules, IoT edge runtime, and a cloud-based interface to enhance native data processing capabilities and also manage edge devices.

Azure IoT Edge is a fully managed service built on Azure IoT Hub. On deploying the cloud workloads, it helps them run on Internet of Things (IoT) edge devices through standard containers. Moving few workloads to the Edge of the network enables the device to spend less time communicating with the cloud, respond faster to local changes, and be reliable even in extended offline periods. The feature includes:

i. Certified IoT Edge hardware: Works with either Windows or Linux platforms with container engines
ii. Runtime: It is free and open-source and provides increased command and code resilience.
iii. Modules: Provides Docker-compatible containers from Azure services to the logic at the Edge.

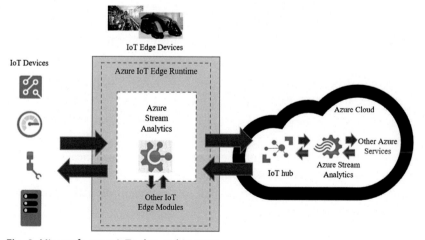

Fig. 9 Microsoft azure IoT edge architecture.

iv. Cloud Interface: Workloads can be managed and deployed remotely from the cloud via the Azure IoT hub.

The security features of Azure IoT Edge include:

i. Ensuring that the devices have the proper software and only authorized edge devices can communicate with one another.

ii. Integrating with Azure Defender for IoT to provide end-to-end threat protection and security posture management

iii. Support any hardware security module to provide strong authenticated connections for trustworthy computing.

4.6.1 Benefits

The locally connected, Edge Compute resources built out with Azure IoT Edge have several benefits for any IoT solution:

- **Lower latency decisions**—With domain logic and Azure services able to run on-premises with an Azure IoT Edge Device, the solution can make quicker decisions and take actions with lower latency.

- **Offline capability**—The IoT Edge Device can operate offline in scenarios during temporary or even longer-term conditions.

- **Data synchronization**—With offline capabilities of Azure IoT Edge, the IoT Edge Runtime will automatically save all IoT remote sensing events on the local device storage and then transmit that data to Azure IoT Hub when connectivity is restored.

- **Lower bandwidth usage**—Low data and IoT device remote sensing must be sent to the cloud by utilizing Edge Compute for further processing. Processing the data locally using Azure IoT Edge allows summarizing the information being communicated to the Azure IoT Edge when all events are not necessary to be sent to or stored in the cloud.

4.7 A solution by Oracle for event processing is named Oracle edge analytics (OEA) (Fig. 10)

It helps to build applications to sieve, tally, and process real-time events by allowing a combination of CQL and Java codes [14]. It can handle fast and streaming data with lower latencies exhibiting high performance, scalability, and security. It includes an explorer facilitating user-friendly web tooling features and a highly scalable, high-performance runtime platform. Some of the features include:

i. Programming model: embedded applications are developed as event processing networks (EPNs)

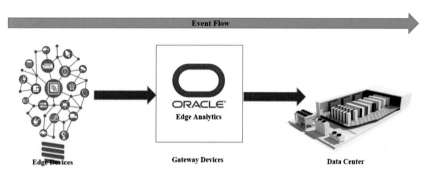

Fig. 10 Oracle edge analytics architecture.

ii. Oracle (CQL): Oracle CQL is supported, except for the functionality provided by the data cartridges like JDBC drivers.

iii. Data source access: Although JDBC drivers are removed from the embedded profile, using data sources to connect to the database through JDBC is supported.

iv. Java data cartridge: The Java data cartridge enhances Oracle CQL to invoke Java code from Oracle CQL code.

v. RESTful web services: implementation of RESTful web services through the Jersey JAX-RS support included with Java Embedded Suite.

vi. Local cache: Local caching service is included

vii. Security: Includes authentication and SSL and utilities such as policygen.

viii. Configuration: Configuration Wizard silent mode to create and configure domains.

ix. Deployer: You can deploy applications from the command line with the Deployer tool.

x. Logging: Oracle Event Processing servers save information to log files for viewing.

xi. Supports high throughput (hundreds of thousands of events per second) and low latency processing.

xii. High-speed real-time data capture and analysis optimized for embedded devices.

4.7.1 Benefits

– Real-time situational awareness, faster decisions, and immediate actions locally ensure customer satisfaction and retention, driving higher revenues.

- Decreased costs and improved compliance with the real-time analysis of data patterns, identifying and immediately responding to critical events and threats
- Cost savings in network bandwidth and processing power. Higher autonomy and resilience in connection loss.
- Improve operational efficiency with on-time insight into the supply chain, integrated systems, and processes, facilitating dynamic resource utilization optimization.
- Provide low cost of ownership and increase productivity with a complete rapid development and deployment platform for event-driven solutions requiring complex event processing.

4.7.2 Use cases

Temperature analysis: Within an Oracle Edge Analytics application embedded in the thermostat devices, Oracle CQL queries aggregate the event data and perform a threshold analysis before sending the events over the Internet. As a result, the events received have already been identified as worth attention.

Server room monitoring: Separate Oracle Event Processing applications provide a two-tiered approach. Sensors at the very]\Edge of the network represent event sources that send data to gateway devices that run Java Standard Edition Embedded and Oracle Edge Analytics. Oracle Edge Analytics applications running on the devices use Oracle Continuous Query Language (Oracle CQL) to query and filter events generated by the sensors. Only event data that meet the filtering criteria is sent to back-end servers in the data center. Oracle Event Processing applications run to send alerts when needed, aggregate and correlate data to identify consistency issues and produce data to be used in reports on patterns.

Grid modernization: Using Oracle Edge Analytics, deployed edge devices establish a node-to-node communication, and a master edge device sends voltage information, power quality, health status to a centralized management system. This solution dramatically reduces operational costs.

4.8 A solution to handle complex analytical processes is named PTC ThingWorx analytics (Fig. 11)

It is enlightened analytics that is user-friendly and easy-to-understand information with visualizations. It exhibits automated complex analytics

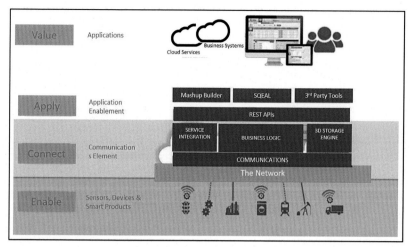

Fig. 11 ThingWorx analytics 8.1 architecture.

procedures to enhance data management, insights, and predictions [15]. It provides tools for facilitating enhanced edge data analytics and arrange the data collected from the devices. It can also represent the current, possible future state of the events and actions to be taken if there is a need to change the outcome. The functionalities include,

i. **Automate analytical processes:** experience in statistical analysis or complex mathematics is not necessary since the AI and Machine Learning technologies used in ThingWorx Analytics automate most of the complex analytical processes of IIoT applications.

ii. **Digital simulations:** It simulates the behavior of physical products in the digital world and incorporates simulation models into the solution. It also utilizes the knowledge when the product is operating in the real world.

iii. **Predictive modeling and scoring:** Using predictive analytic algorithms and machine learning techniques, ThingWorx Analytics analyzes data from connected devices, finds the patterns in the data, and generates a prediction model.

iv. **Real-time anomaly detection:** Helps to observe data from a device, learn about the typical state, and monitor for data points that fall outside the expected range. This helps trigger alerts and allows application users to identify when to take action in IIoT applications.

v. **Descriptive analytics:** Provides services of standard statistical calculations and facilitates statistical monitoring versus historical data.

For example, finding Maximum, Mean, Median, Minimum, Mode, Standard Deviation, Threshold Count, Range Count, and Trend Count.

vi. **Transform microservices:** The Property Transform microserver provides on-demand transformation services to derive value from streaming data entering ThingWorx instead of Descriptive Analytics' historical limitation.

4.8.1 Benefits

- Operationalize insights, predictions, and recommendations across enterprise functions with automated IoT data to enhance decision-making.
- Enable complex analytical capabilities for those who are not data experts with user-friendly interfaces, tools, and applications.
- Enables analysis of historical data and forensic investigation of data after an incident through replay functionality.
- Production-ready deployment enables enterprises to get up and running quickly—at the Edge, on-premise, or in the cloud.

4.9 Components and use cases

- **Analytics server:** Automates complex calculations and generates predictions, simulations, and prescriptions. Using the data from the Things and historical data as a learning source, Analytics Server uses machine learning to build and validate predictive models without assistance. This dramatically reduces or eliminates the need for an expert team in modeling algorithms or technologies.
- **Anomaly detection:** ThingWatcher is a Java API that can be used to build anomaly detection functionality into your IoT applications. ThingWatcher can be used at any location, cloud, or Edge. By deploying ThingWatcher at the Edge, you can monitor a stream of high-speed data that would be difficult or impossible to process in the cloud.
- **Predictive modeling and scoring:** ThingPredictor provides the predictive scoring capabilities of ThingWorx Analytics. It uses prediction models generated by ThingWorx Analytics Server or equivalent Predictive Model Markup Language (PMML) compliant prediction model generation tools to examine a dataset and predict results for each record based on similarities to records analyzed during model training. ThingWorx Analytics predictive scoring can be accessed through a REST API Service, Analytics Builder, or Analytics Manager.
- **Prescriptive scoring and optimization:** ThingOptimizer provides the prescriptive scoring and optimization capabilities of ThingWorx Analytics and allows you to expand your analytical processes beyond

predicting outcomes to seeing how modifications might affect results by automatically identifying influencing factors. ThingOptimizer uses prediction models generated by ThingWorx Analytics Server or equivalent PMML-compliant prediction model generation tool.

- **Analytics builder:** Provides a user interface for simple, intuitive interaction with data. It converts complex data readings into simple-to-understand information and simplifies the advanced analytics process, helping interpret information faster. It also supports data and metadata uploading, predictive model creation (training and scoring), visualization of data analytics (profiles, signals), and data filtering.
- **Analytics manager:** Allows to deploy and execute computational models from external applications within the ThingWorx platform. It leverages product-based analysis models developed using PTC and third-party tools while building solutions on the ThingWorx platform.

4.10 A to-the-edge component to collect, process, and transmit data is named streaming lite by SAP HANA (Fig. 12)

It is a lightweight version of the Smart Data Streaming (SDS) server designed to run on IoT gateways to process data close to the source [16,17]. It is a free-standing server used to situate streaming projects on remote gateway devices. It is a to-the-edge component with the capability to collect, sieve, aggregate, and transmit data. It is a self-contained, independent server that is not a part of the SAP HANA streaming analytics cluster.

4.10.1 Use cases and benefits
- Equipment sensors can stream information about status and events back to a central SAP HANA system. These sensors could be monitoring statistics such as humidity level and temperature or machine status.

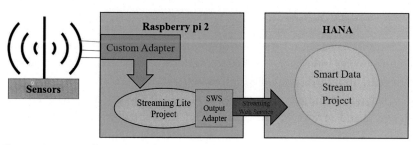

Fig. 12 Streaming lite component architecture.

- The devices can post messages to the SAP HANA system through an adapter or a customized interface such as the Streaming Web Service.
- The messages are consumed by streaming analytics, which applies filters to transform or normalize the data, thus capturing high-value information in the SAP HANA database. For example, suppose you have sensors that track humidity levels and temperature data, as data from these sensors flows in, you can isolate the temperature and humidity values and join them to a table of values from the SAP HANA database.
- The streaming analytics model continuously computes a set of summary information streamed to a live operational dashboard. For example, look at the current data (in the stream) and compare it with time-stamped entries in the SAP HANA table to compare current values to values 5, 10, or 15 min old. You could use this data to identify trends: Is your equipment heating up steadily or too quickly? What impact does humidity play on machine performance?
- The streaming analytics data model also actively monitors the incoming data for conditions that warrant immediate action or attention and generates alerts and notifications (by sending an email or text message) when those scenarios are detected.

5. Working principles and feature comparisons

The application of edge analytics aims to adopt IoT by various industries as a trend in service provision [18]. Its ability to extract tangible and measurable metrics from the IoT devices results in faster data provided with the help of analytics tools and the application of machine learning algorithms for real-time intelligent factors.

Edge analytics also provides the advantage of distributed data aggregation and collects the data across all the industry sectors. They generate critical decentralized data at every instance and require analysis at the Edge with low-bandwidth requirements on large-scale industry data. The edge analytics extends towards a variety of application like logistics, retail services, manufacturing industries, healthcare industries which has a stream of data generated and needs immediate processing.

Analytics at the cloud can store and use all the previous data for the streaming and analyze vast amounts of data from the devices. The industries felt the shortage of cloud analytics capabilities for supporting modern IoT applications. The analytics on the cloud, although is very fast, it is not an instant response. There is a need to avoid network latency, cost inefficiencies, and connectivity.

Transferring most computation work to the Edge resulted in the best communication costs and immediate responses. It was observed that the machines operated more effectively after activating the edge processing mode and edge analytics became a critical factor in the strategy of IoT-based industries. Instead of transferring data to the cloud for processing, it was now considered wise to apply processing tools at the Edge and shift the analytics to the devices that generate the data. Currently, edge analytics is in great demand in various IoT-based applications that depend on instant or real-time responses.

6. Some of the other use cases of edge analytics [19,20]

- **Retail customer behavior analysis:** Retailer data can be collected from a range of sensors, including parking lot sensors, shopping cart tags, and store cameras. By applying analytics to these data, the retailers can provide personalized solutions for everyone with the help of behavioral targeting.
- **Remote monitoring and maintenance for various industries:** Industries may need an immediate response in case of any machine failures and for maintenance purposes. Organizations can identify signs of failure faster when server-based analytics is avoided and take action before any bottleneck can arise within the system. Edge analytics is also used in improving efficiencies by adjusting the operating parameters.
- **Intelligent Surveillance:** It is a real-time intruder detection edge service for their security. With real-time images from the security camera, edge analytics immediately detects and tracks all suspicious activities.
- **Real-time human detection** is a process similar to object detection taking raw images from security cameras and putting them in the camera buffer for processing in the detector & tracker. The identified human figures from detector&trcker are sent to the streamer buffer. Therefore, the whole human detection process can be divided into three threads: camera, detector&tracker, and streamer.
- **Smart cities:** The basic building blocks are the IoT devices embedded into the infrastructure. They help to monitor and collect information

about the behaviors of the components in the infrastructure. As real-time decision-making is necessary for intelligent cities, edge analytics finds application for fast actuators' responses.

- **Healthcare:** Hospitals deploy a wide range of devices that create a heavy load on a cloud for processing and responding faster. Edge analytics is proven to be an optimal method that reduces the overload and is a cost-effective solution considering storage security and minimal connectivity issues.

References

[1] R. Donitha, Edge Data Analytics: What, Why, Who, When, Where, How, Digital Transformation Pro, 2017.

[2] B. Johnson, Analytics at the Edge: Three Benefits of Edge Analytics, Swim Continuum, 2017.

[3] R. Izadi, Video Analytics: Edge vs. Server-Based, DIDARC Trading Co. Ltd Publications, 2017.

[4] E.A. Israel, O.F. Sammuel, C.M. Richard, J.J. Jason, Intelligent energy optimization for advanced IoT analytics edge computing on wireless sensor networks, International Journal of Distributed Sensor Networks (2020).

[5] AWS IoT Greengrass, Developer Guide, Version 2, AWS Architecture Blog, 2021.

[6] Cisco SmartAdvisor using IP Explorer, IP Explorer-Network Discovery, Visualization, NetVisor, 2018.

[7] D. Sweenor, Dell Statistica Stands Out as Analytics Market Heats Up, Dell Technologies, 2016.

[8] HPE Edgeline Converged Edge Systems and OT Link, Hewlette Packard Enterprise Products, 2021.

[9] J. Lang, Connecting Workers with Data and Insights at the Edge, IDC Technology Spotlight, 2020.

[10] An Introduction to IBM Watson IoT Edge Analytics, IBM Internet of Things, 2017.

[11] Watson IoT Platform: A Fully Managed, Cloud-Hosted Service With Capabilities for Device Registration, Connectivity, Control, Rapid Visualization and Data Storage, IBM Technologies Products, 2017.

[12] Azure IoT Edge: Cloud Intelligence Deployed Locally on Edge Devices, Microsoft Azure Services, 2021.

[13] Azure IoT, Quickly Turn Your Vision Into Reality With Secure, Scalable, and Open Edge-To-Cloud Internet of things solutions, Microsoft Azure Services, 2021.

[14] Oracle Edge Analytics, Oracle Data Sheets, Java Embedded Products, 2015.

[15] Thingworx Analytics, Delivering Powerful, Operationalized Analytics to Solutions Built on the Thingworx Platform, ThingWorx Analytics Product Brief, 2017.

[16] R. Waywell, Learning How to Use Streaming Lite With HANA Smart Data Streaming, SAP HANA Community Blogs, 2016.

[17] SAP HANA Streaming Analytics Use Cases: Master Guide, SAP HANA Community Blogs, 2016.

[18] B. Posey, Edge Analytics: Integrated Analytics Gives Users Operational Intelligence Edge, 2020.

[19] D.E. Sweenor, Six IoT Use Cases for Edge Analytics: How to Acquire Manufacturing Superpowers in the Internet of Things, TIBC Software Inc, 2017.

[20] Data Analytics at the Edge, Pathfinder Report, SAS Dell Technologies, 2019.

About the authors

Jayashree N received her bachelors and master's degree in Computer Science and Engineering from Visvesvaraya Technological University. She had over 10 years of teaching experience, and presently working as Assistant Professor in the Department of Information Science and Engineering at Cambridge Institute of Technology, Bangalore. Her research interests include Information and Network Security, Artificial Intelligence and Machine Learning, and privacy protection in selected application domains. She has Co-Authored in research papers based on security, privacy and queuing models in IoT and cloud computing.

Dr. B. Sathish Babu received his bachelor's and master's degree in Computer Science and Engineering from Bangalore University. He obtained his Ph.D. in Protocol Engineering Technology Unit, Department of ECE at Indian Institute of Science, Bangalore, in 2009. He has over 30 years of teaching experience and over 15 years of research experience. He is presently working as a Professor and HOD of the Department of Artificial Intelligence and Machine learning at RVCE Bengaluru. His research interests include; Information and Network Security, Cognitive computing applications for Network controls, Soft computing solutions for Cloud Computing Scheduling and Virtualization, Data Science, Opportunistic Networks routing and security, Building Machine learning and Deep learning models in selected application domains, Quantum Computing, and others. He has published over 100 international journal/conference papers in his area of research, with many publications featured in Scopus-Quartile journals and Web of science journals. Dr. Babu co-authored books on Mobile and Wireless Network Security published by Tata McGrawHill, and Communication Protocol Engineering published by PHI-India. These books are widely used as text and references in more than 15 Indian universities. He also co-authored many chapters in the edited books published by the CRC Press and IGI Global International. He has worked as a co-investigator in the joint research projects with the Indian Institute of Science, Imperial college-London, and Florida International University-USA.

CHAPTER EIGHT

Edge platforms, frameworks and applications

Kavita Saini[a] and Pethuru Raj[b,c]
[a]School of Computing Science and Engineering (SCSE), Galgotias University, Delhi, Uttar Pradesh, India
[b]Site Reliability Engineering (SRE) Division, Reliance Jio Platforms Ltd. (JPL), Bangalore, India
[c]Reliance Jio Cloud Services (JCS), Bangalore, India

Contents

Abstract

Edge computing, a rapidly growing technology, satisfies the business needs of most of the smart business cultures today. Smart agriculture, smart cities, smart manufacturing or any other smart business can be benefited from Edge computing. The chapter discuss how to about the cloud computing to edge computing. Various edge computing technologies and applications are also discussed in detail.

Advances in Computers, Volume 127
ISSN 0065-2458
https://doi.org/10.1016/bs.adcom.2022.02.005

5G Communications, Remote Monitoring, Healthcare and other such applications require tremendous accuracy, high latency I less cost. Augmented and Virtual reality applications or gaming are just few applications. There are many more to discuss about. The chapter discuss how edge computing is helpful for all such applications.

Unlike cloud computing, edge computing enables data analysis, its processing, and transfer at the edge of the network. Basically data is analyzed locally where it is stored. The analysis is done in real time without latency and allows for quicker data processing and content delivery.

1. Introduction to cloud computing

Cloud computing by which remote servers hosted on the Internet store and process data, rather than local servers or personal computers [1]. It is ready to move to the next level, i.e., *"Edge Computing."* Icloud, onedrive, Google are the few examples of cloud computing [2].

As cloud computing is the "On-Demand" availability of the computer system resources, especially data storage and computing power, without direct active management by the user [3].

To Leverage 5G wireless technology and artificial Intelligence to enable faster response times, lower latency (ability to process very high volumes of data with minimal delay), and simplified maintenance in computing. Cloud provider are responsible to manage, restore and backup. In short data centers are being managed by cloud service providers [4,5]. They are also providing high-end computing power, software and so on.

This is where Edge Computing comes in, which many see as an extension to the cloud computing, but which is in fact, different in several basic ways [6].

2. Cloud computing to edge computing

The basic *difference between cloud computing and edge computing lies in the place where the data processing takes place.* At the moment, the existing Internet of Things (IoT) system performs all of their computations in the cloud using data centers. Experts believe the true potential of edge computing will become apparent when 5G networking go mainstream in a year from now. Edge offers an extra added scalability which will be needed for locally applicable responsibilities in the environment with a huge quantity of data producers and customers [7,8]. Multifaceted, extensive and data determined responsibilities which aren't time threating would be benefited greatly from the richness of ascendable sources in clouds [9,10]. Edge can make the communications more consistent, in the manner that it could provide an option if the network link to cloud breaks. It can be an interesting especially for crisis management scene where edge provides optional substructure to preserve critical responsibilities thriving. Edge sources could generally be accessible within one hop from the wireless gateways which operators are linked with. Preferably, edge system can identify and back handler flexibility, e.g., by moving its information and computing to the subsequent nearby positions. User will be able to enjoy consistent connectivity without even realizing it [7].

3. Edge computing: A brief overview

Edge Computing enables data to be analyzed, processed and transferred at the edge of a network. The idea is to analyze data locally, closer to where it is stored, in real-time without latency, rather that send it far away to a centralized data center [4,10]. Time-sensitive data is processed using edge computing, while data that is not time-sensitive is processed using cloud computing which is the main advantage of edge computing over cloud computing. The prime aid edge offers consist lower latency, higher bandwidth [11], device processing and data unburden as well as reliable computations and storing. Concludingly, 5G requires edge to drive need for its facilities [5].

In edge computing some content will be offload before using which will help in accessing data without any delay. In short in Edge computing there is a use of 5G wireless technology and AI. Edge could be considers as an additional layer between the CC and users, applications and devices (such IoT devices).

With the rapid advancement in IoT and Edge Computing, the traditional cloud computing was facing communication latency and network bandwidth as a biggest challenge. A new driving technology coming into role has now moved the functionality of centralized cloud computing to the network edge node for addressing the difficulties in traditional system. Several edge computing systems have emerged from various backgrounds in order to reduce latency, increase computational capabilities and handle huge machine connectivity [12]. This study provides an in-depth look at three common edge computing technologies: mobile edge computing (MEC), cloudlets, and fog computing [6].

4. Essential of edge computing

Pushes the intelligence, processing power, and Communication Capabilities of an edge gateway or applicant directly into devices [13]. The idea is to analyze data locally, closer to where it is stored, in real-time without latency, rather than send it far away to a centralized data center. So whether you are streaming a video on Netflix or accessing a library of video games in the cloud, edge computing allows for quicker data processing and content delivery [12].

As of now all the applications using IoT devices store and compute the data on cloud data center. With edge computing it is possible to perform all the analysis and computation at the edge on cloud and saves lots of time. Not just saving time edge helps in computation in real time. Playing online games, live video streaming are the examples where computation should be finished in fraction on second. This is possible with edge computing [14,15]. This is how edge computing different from cloud computing.

Only important data is sent over the network instead of whole data. In this way edge computing is reduces the amount of data traversal over the network. Transferring data from cloud to devices and vice versa take time, also known as Internet latency. It can be achieved using AI, 5G wireless tech and IoT devices. It will help in reducing amount of data stored over cloud as some data some data could be stored only on edge [8].

Edge Computing simplifier this Communication chain and reduces Potential Print of Failure. In edge Computing, physical assets like pumps, motor, and generators are again physically wired into a control system, but this system in controlled by an Edge Programmable Industrial Controller, or EPIC. Edge Computing saves time and money by Streamline IoT Communication, reducing system and Network Architecture Complexity and decreasing the number of potential failure in an IoT application Reducing system Architecture Complexity is key to the Success of IoT applications [7,16].

5. Advantages of edge computing

Edge computing has many important features unnoticed in preceding network generations. These consist of huge data generation. Edge directs the computing information, application, and facilities aside from the Cloud server to the edge of networks [10]. Content suppliers and app creators can utilize an edge computational system by providing handlers the facilities nearer to themselves. Edge computation is categorized as employing higher bandwidths, lower latencies, and real time admittance to the system data which could be utilized by numerous appliances [13,17,18]. Edge is favored to supply the wireless communication needs of next-gen technologies, like virtual reality & augmented reality, that are collaborating in behavior [14]. Edge technology enables the computing to be executed at the networking edges.

5.1 Latency reduction

Cloud computing can't adequately support the volume of data being processed every second. Having spoken about latency within the cloud computing world, there is a lot that cloud computing does not provide to cloud-based applications [5]. Given the amount of stored data within the cloud, there are two problems that transpire during the processing stage—latency in processing and high number of wasted resources. These issues exist especially in decentralized data centers, mobile edge nodes, and cloudlets [17].

By reducing latency, edge computing enhances network performance. The information does not go travel as far as it would in a traditional cloud architecture rather devices process data natively or at a local edge center. It has been noticed earlier if source is sending any mail at destination at same workplace also, there could be some delay in standard network. This delay does not exist if the procedure occurs at the edge and the company's router handling office emails [19].

5.2 Safer data processing

DDoS (Distributed Denial of Service) assaults and power outages are common in cloud environments. Systems are less prone to interruption and unavailability because edge computing spreads processing and storage. There is no single point of failure in the setup [15].

Furthermore, because many procedures take place locally, cyber attackers are unable to prevent data from being sent. Even if a data breach occurs on one machine, the attacker can only access location data [20].

5.3 Inexpensive scalability

Edge computing empowers a company to increase its capacity with a combination of IoT devices and peripheral servers. Adding additional resources does not require investing in a more expensive private data center to build, maintain, and expand [5]. Instead, the company can set up regional edge servers to expand the network quickly and inexpensively.

Edge computing also reduces growth costs as each new device does not add additional bandwidth requirements across the network.

5.4 Simple expansions to new markets

The company can partner with the local edge data center to quickly expand and explore new markets. The expansion does not require expensive new infrastructure. Instead, the company only sets up end-to-end devices and starts serving customers without delay. If the market seems unwanted, the extraction process is quick and inexpensive.

This benefit is important for industries that need rapid expansion in areas with limited connectivity.

5.5 Consistent user experience

As the edge servers work closer to end users, the problem of a remote network is less likely to affect customers. Even if the local facility is disconnected, the peripheral devices may continue to operate due to their ability to handle important traditional tasks. The system can also extract the data route in other ways to ensure that users retain access to the services [15].

5.6 Speed

Speed is absolutely vital to any company's core business. Take the financial sector's reliance upon high-frequency trading algorithms, for example. A slowdown of mere milliseconds in their trading algorithms can result in expensive consequences. In the healthcare industry, where the stakes are much higher, losing a fraction of a second can be a matter of life or death [1,20].

5.7 Edge computing technologies

There has been a substantial growth in connected smart devices and IoT nodes, which resulted into increased data generation at these nodes. Handling this massive amount of raw data is a crucial challenge because of limited computational and energy resources [2,4,9,19]. Due to the requirement of large processing and storage capacity, the existing cloud computing platform can easily handle the tremendous heaps of data generated by IoT devices. But this is not conceptual for dissipated IoT systems or the On-time operation of deliciated latency IoT applications as they require centralized manner of operation and are concerned with the associated delay also. Edge computing can minimize end-to-end latency, save bandwidth in backlog links, and mitigate the computational pressures on cloud-servers, by providing cloud-like computing environment, storage, and communication facilities at the network edge [21].

The presence of "Edge devices" reduces the compute burden at data centers by handling some of the requests directed to the cloud locally, without the need for cloud involvement. As a result, the delay in resolving requests is reduced, and a subset of requests can be handled in real time. Because of their widespread availability and geographical distribution, edge devices also aid mobility. There are various technologies that can be used to develop edge computing, which is based on the idea that it can expand the settings of IoT usage by complementing the cloud [22]. Let's take a close look at some IoT based edge computing technologies, including cloudlets, mobile edge computing (MEC), fog computing, and a novel idea called the Cloud of Things [11].

These technologies are also predicted to be important in the development of edge computing platforms.

5.8 Cloudlets: An overview

A cloudlet is basically a regional cloud that can bring far-flung cloud services nearer to the user. Cloudlets are small-scale, mobility-enhanced cloud data centers that sit at the network's edge. The cloudlet's primary goal is to support furious resource and interactive mobile applications by delivering strong computing resources to mobile devices with reduced latency. A wireless local area network with single hop at comparatively higher speed, allows User Equipments (UEs) to connect to the computing resources in the neighboring cloudlet [15]. To ensure crisp reaction time, cloudlets constitute the intermediate tier in a 3-tier hierarchical architecture containing Edge device layer, cloudlet layer, and cloud layer.

For security reasons, cloudlet is bounded in a tamper resistant box for safeguarding safety in unregulated regions [9]. On the inside, cloudlets consist of a group of source rich multicore computer with high-speed internet connection and higher bandwidth wireless LAN for the use by closer mobile gadgets (Fig. 1).

The Cloudlet is an architecture model that enables cloud computing at the mobile network's edge. Low latency and high bandwidth characterize this environment, forming a fresh ecosystem in which network service provider can open their network edge for third party user, allowing them to quickly and flexibly install creative and innovative services.

6. Significance of cloudlets

The purpose of a cloudlet is to improve the response time of mobile apps by employing low-latency, high-bandwidth wireless communication

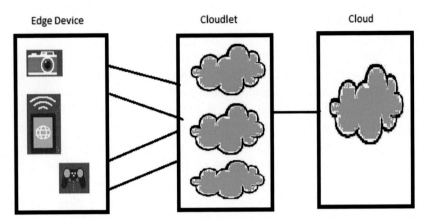

Fig. 1 Three tier view of cloud data centers at edge network.

and physically bringing cloud computing resources, such as virtual computers, closer to the mobile devices that access them [23].

Cloudlet vs Cloud: Cloudlets differ from Cloud in many cases:

1. Clouds are normally monitored by their service providers whereas cloudlet is self-managed.
2. Provider locate the cloud in purpose specific areas at his premises, on the other hand cloudlet are located in business premises in the form of a data center in a box.
3. Cloud uses internet bandwidth/latency whereas cloudlet uses the latency/ bandwidth of a local area network.
4. Since cloud has a centralized ownership, cloudlet is owned by the local business/organization.
5. Cloud can accommodate hundreds and thousands of users at a time but cloudlet can accommodate only few users.

Benefits: Cloudlet resides as a middle layer in its three tier view, offering various benefits in the edge infrastructure

1. *Simple to set up:* The fact that cloudlet servers are exiled makes maintenance easier; adding or replacing of a cloudlet just takes few minutes for setup and simple steps of configuration.
2. *Enhancement of security:* The cloudlet's proximity to mobile nodes makes the architecture more resistant to DoS attacks (secondary variants). It can also help to avoid data leakage from traffic analysis by limiting the range of end-to-end connection, which prohibits snoopers from accessing traffic data from afar.
3. *Resilience:* Even with shaky connectivity to a remote cloud provider, a cloudlet collection can provide dependable cloud computing services.

Mobile Edge Computing (MEC): Multi-access Edge Computing, also known as Mobile Edge Computing (MEC), is a network design that allows computational and storage resources to be placed within the Radio Access Network (RAN). The MEC aids in improving network efficiency and content delivery to end-users. This device can do this by adapting over the load available on the radio link, resulting into increase in network efficiency and reducing the requirement for long-distance backlogs. MEC offers mobile and cloud computational abilities among the access system, and goals to unite the telco and IT at the mobile edge network [19]. As in near locations to users, MEC can provide a service environment with ultra-low latencies, high bandwidth, and direct entree to real time system data [23,24]. Mobile Edge Computing is the principal technology among the next gen system technologies.

As per European Telecommunications Standards Institute (ETSI), mobile edge computing is defined as "Mobile Edge Computing offers an IT service environment and cloud computing capabilities at the mobile network's edge, within the Radio Access Network (RAN), and close to mobile customers." Listed below are few characteristics of MEC:

On-Premises	Mobile edge computing works in silos, which improves performance in a machine-to-machine setting. MEC's ability to isolate itself from other networks makes it more secured.
Proximity	Because mobile edge computing is placed at a nearby area, it has an advantage in analysing and materialising huge data. It's also useful for gadgets that require a lot of processing power, such as AR (augmented reality) and video analytics.
Reduced latency	Mobile edge computing services provides user devices in close vicinity, separating network data transfer from the core network. As a result, the user experience is considered to be of excellent quality, with extremely lower latency and higher bandwidth.
Location awareness	For information sharing, edge dispersed devices use low-level communication. MEC gets data from edge devices in the local access network in order to locate devices.
Network context information	By incorporating MEC into their business model, applications that providing network information and services of On-time network data can definitely benefit enterprises and events. These apps can estimate the congestion of the radio nodes and network bandwidth typically based on RAN having real-time information, which will enable them make smart decisions for improved customer delivery in the future.

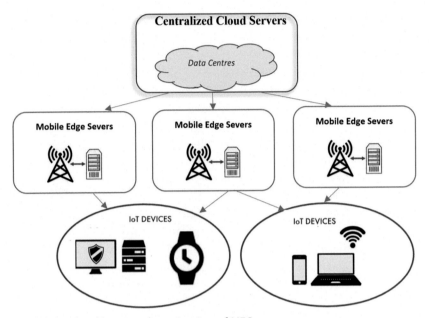

Fig. 2 General architecture: three tier view of MEC.

MEC is a layer that sits between mobile devices and the cloud. As a result, the infrastructure is organized into three layers: cloud layer, MEC layer, and mobile device layer. Mobile edge computing, for the most part, works in tandem with cloud computing for supporting and improving the performance of end devices (Fig. 2).

6.1 MEC benefits

Listed below are some advantages of Mobile Edge Computing (MEC) that are proving to be beneficial to both Mobile Network Operators (MNOs) and application service providers.

- Mobile network operators might provide third-party suppliers with real-time access, allowing them to deploy their applications and services in a more flexible and agile manner. These services could make money by charging for things like storage, bandwidth, and other IT resources.
- MEC-enabled infrastructure-as-a-service (IaaS) platforms at the network edge node could benefit application service providers by allowing them to scale their services while maintaining higher bandwidth and reduced latency. ASPs may also gain real-time access to radio pursuit that is likely to develop.

- End users may perform faster computation by offloading the MES servers.
- Driving business model evolution—The capacity to collect or process data from any facility's base leads to the development of a wide range of business models. Evolution of the Industry 4.0 is a best example for this. In this case, mobile edge computing can act as an enabler for developing a proactive maintenance business model to extend the life of any resources.
- A new way to run your business more efficiently—Mobile edge computing will allow SMEs to grow their marketing efforts, reach more customers, and improve their services at considerably lower costs than the public cloud.

6.2 FOG computing

Fog computing is a framework hierarchical architecture defined by the OpenFog Consortium as a horizontal architecture that divides processing, storage, control, and connectivity services and goods anywhere along the spectrum from the fog to things [5,19]. Fog computing differs from edge computing in that it includes tools for across networks and between edge devices for spreading, coordinating, controlling, and protecting resources and services [1].

It is also known as Fog networking or fogging. It is a decentralized infrastructure where application resides between data stores and the cloud. It is an architecture which uses edge device for computation, storage and communication [19]. This edge device is the device that controls the data flow at the boundaries of any two networks, example: router, switch, IAD, gateways, hub, multiplexer or bridge. It is actually a mediator between hardware and remote server. Fog is a distributed network environment and is very close to cloud computing and IoT device. It has a high security network. Instead of sending selected data to the cloud for processing, fog actually process the data which saves the network bandwidth and reduces latency requirement which in turn helps in fast decision making capability (Fig. 3).

Definition: Some researchers have defined Fog computing as:

"Fog computing is a highly virtualized platform that provides compute, storage, and networking services between IoT devices and traditional cloud computing data centers, typically, but not exclusively located at the edge of network."

"Fog computing is a scenario where a huge number of heterogeneous (wireless and sometimes autonomous) ubiquitous and decentralised devices communicate and potentially cooperate among them and with the network to perform storage

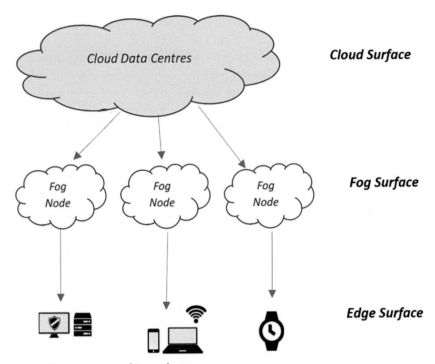

Fig. 3 Three tier view of Fog infrastructure.

and processing tasks without the intervention of third parties. These tasks can be for supporting basic network functions or new services and applications that run in a sandboxed environment. Users leasing part of their devices to host these services get incentives for doing so."

"The term Fog computing or Edge Computing means that rather than hosting and working from a centralized cloud, Fog systems operate on network ends. It is a term for placing some processes and resources at the edge of the cloud, instead of establishing channels for cloud storage and utilization."

Considering these definitions we can define Fog computing as: "A distributed computing platform in which end or edge devices perform the majority of the work. By existing in between users and the cloud, it is also associated with the cloud for non–latency-aware processing and long-term storage of important data."

6.3 Benefits

6.3.1 Confidentiality

Fog computing can be used to keep the quantity of data shared to a minimum. Instead of sending sensitive personal information to a centralized cloud platform, any confidential material can be reviewed regionally. In this way, the IT staff will be able to monitor and control the device. Any portion of data that has to be analyzed can also be sent to the cloud [6].

6.3.2 Efficiency

Clients can employ fog procedures to make the machine function effectively they want it to. Publishers may easily design these fog applications with the right combination of tools. They can use it whenever they want once the job gets completed [13].

6.3.3 Safety and security

Fog computing supports multiple devices to be connected to the same network. As a result, rather from being consolidated, processes in a complex decentralized system take place at multiple end terminals. This makes it easier to identify potential threats before they have a large-scale impact on the network [5].

6.3.4 Bandwidth

The cost of bandwidth required for data transmission depends on the availability of resources and it can be costly [7]. The throughput requirements are greatly decreased because the chosen processing is done locally instead of being transferred to the cloud. This bandwidth reduction will be particularly significant as the number of Internet—connected devices grows. When the number of IoT devices grows, this bandwidth savings will be especially useful. Fog computing allows multiple devices to share a common network. As a result, instead of being centralized, processes in a complex distributed system occur at multiple end points. This makes it easier to identify potential threats before they spread throughout the network.

6.3.5 Latency

Another advantage of processing data locally is the reduction in latency. The data can be analyzed or processed at the data source that is expected to be geographically nearer to the user. This can result in instantaneous responses, which is highly valuable for services that require quick reactions [4].

Several edge computing systems have emerged from various backgrounds in order to reduce latency, increase computational capabilities and handle huge machine connectivity. This study provides an in-depth look at three common edge computing technologies: mobile edge computing (MEC), cloudlets, and fog computing.

6.4 Edge computing applications

There are various applications of Edge computing, some are listed below (Fig. 4):

6.4.1 Smart systems

It comprises of IoT devices used as home essentials like smart TVs, smart phones, smart lights, CCTV cameras etc. Smart systems also include devices used for monitoring air quality index, weather conditions, traffic management, smart gardening and others which will help in making smart city. All those systems used in healthcare like fitness tracker, apps for pandemic diseases (Arogya setu) are also a part of smart systems [25].

6.4.2 Video streaming

Various reports have highlighted the statistics that video streaming on internet will be capturing almost 83% of all internet traffic by 2022. This video streaming requires good bandwidth resources and cache requirement which

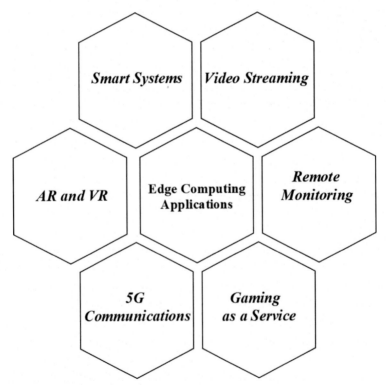

Fig. 4 Edge computing applications.

directly affect the cost parameter and the video quality. Edge computing is undoubtedly providing a reasonable mechanism to cache the local video. These devices are embedded with filtering capability, to figure out useful data [16].

6.4.3 Remote monitoring and predictive analysis

With the help of edge computing enabled IoT devices manufactures are now able to remotely monitor their assets more carefully and at early stage before they create any disaster. Bringing the processing capabilities closer to the device, has helped in predicting the real-time health status of the machines [20]. This may beneficial in analyzing and detecting the changes required in production line before any failure occurs.

6.4.4 Gaming-as-a-service

It is a kind of online video game that runs on a cloud server and directly stream on the player's device [26]. Xbox, a Microsoft soft product is a cloud gaming service, where game itself is hosted and processed in cloud data centers but feed directly to the gamers device through real time streaming. Edge and 5G connectivity integrations provides the bandwidths required to aid high quality, multi-player gaming experiences. Edge is not only restricted to the gaming but it is also the future of mobile applications. It significantly contributes into the future of industries adopting hybrid multi-clouds & edge structures as it plays a crucial role in the digital structures. Mobile experience will be more real-time, more interactive, and rich in handling/ operating just because it will be because of edge computing [23].

6.4.5 5G communications

For lower latency and higher throughput 5G communication is required. 5G is considered as a next gen. Cellular network. With edge computing it is possible to bring the cloud computing capabilities closer to the end user or to the edge of network. Cloud computing where there is high latency, low throughput and less security, edge computing combat with all mentioned problem and improves the user experience.

5G and edge computing technologies are capable to improve the application's performance significantly. Specially where data is to be processed in real time, 5G increases speeds 10 times in comparison to 4G and edge computing reduces the latency by bringing the computational capabilities to the edge of the network. This deadly combination of 5G and edge computing improves the users real time application's experience.

6.4.6 5G smart health care

5G smart health care, on the other hand, has far more grave safety and seclusion apprehension than traditional healthcare services. Traditional medical services are altered by 5G medical applications, which extend them from the healthcare center to an online examine mode that involves many users, data systems, and medical devices, is embedded by posing serious security, massive medical data transmission, and privacy challenges. The quality of medical services and the routine operations of medical facilities will be severely harmed by security weaknesses in terminals, networks, and systems.

Science and technology in healthcare area is a considerable study part for countless scholars. Similar to various manufacturing, health care subdivision can also be assisted from edge, for example heart patients suffering from heart attack [14]. Health care application are generally regarded time sensitive applications in Internet of Things. At first, cloud computing was utilized for health care implementations but was not great success due to latencies problems. Introduction of edge solved these problems & made cloud accurate for health care IoT uses. Into the smart clinics edge is currently used in numerous ways, like in wireless health regulation the data received from the affected ones directed to the physicians which eases them to tackle the emergencies [1]. Meanwhile, there are no 5G security standards or 5G medical engineering safety standards in place, and it's unclear to concern security protection for 5G health check appliances.

6.4.7 Security monitoring

Various IoT devices like Intrusion detection systems (IDS) are deployed on either ends of communicators for an efficient network monitoring. It helps in security and privacy related issues while transmitting data either from one network to another or from data centers to edge device [19].

6.4.8 AR and VR

Augmented Reality (AR) and Virtual Reality (VR) is one of the rapidly growing industry. Reason is both helps in reducing the cost and time to perform any task, for example remotely managed any operation [2]. By removing the geographical barriers any operation and be performed by the experts sitting on same or different places.

But presently the biggest problem with deployment of such applications is comprehensive deployment of mobile AR is that current devices. Still there is a need to improve the computational power, data communication and graphical performance.

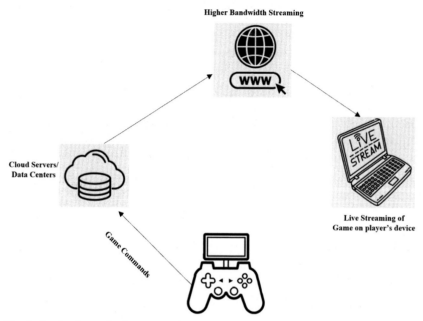

Fig. 5 Applications of edge computing.

With the edge computing all computational task could be taken care by the edge of network and will help in improved performance in all aspects. Challenges that can now be overcome by 5G and edge computing by performing computation at the network edge and not need to send and receive data frequently to/from cloud and will support more and more real-life applications [19] (Fig. 5).

6.5 Edge computing and future

Edge computing is currently in its early stages and has a great capability to pave the pathway for more effective dispersed computations [19]. The prime goals of edge computing are to give real time interactions, local processing, higher data rates, and higher availabilities. Edge advances network actions to assist and locations varied scenes, like distant surgeries [17]. The flaws and lacunas of cloud and fog are completed by the edge computing to a greater amount which has consequence in low response time, lower latency, lower bandwidth price, lesser energy usage and superior and higher data privacy. The significance and need of edge computing is acceptable with the assistance of few of the upcoming sectors and appliances that are extensively utilized in recent times like smart city, smart houses, online marketing, etc.

So, to offer an advanced and well-organized facility to IoT smart applications, the idea of edge computing is able to execute all the functions by going beyond the cloud capabilities. Edge visions to get facilities and efficacies of Clouds near to the handler for safeguarding fast processing of data-concentrated apps. Video analytics, online shopping, smart cities, smart homes, digital health care, mutual edge and high privacy apps are few of the extensively utilized famous areas & applications that uses the idea of edge.

Transferring data processes through the edge networks will be helpful to companies so that they can take benefit of the increasing quantities of IoT gadgets, advance network speeds, and increase customer experiences. Scalable behavior of the edge also makes it an appropriate option for the rapidly developing, sprightly industries, particularly to those who uses data centers and utilizes cloud substructures [1]. Edge gives an unprecedented benefit of flexibility and reliability which will increase of each product combined or integrated with it resulting into customer satisfaction. Edge computing provides varied benefits over conventional type of network structures and will definitely play a significant part in near future. With the advancement in internet associated gadgets coming in markets everyday advanced administrations have just scratched the uppermost part of what's possible with Edge.

6.6 Shortcomings of edge computing

So far the chapter talks about edge computing and benefits of its use over the cloud computing. Every technology be it cloud or edge computing have some drawbacks and give scope of improvements in the technologies. It also help in discovering the scope and hope for new technologies [4].

Like other technologies, Edge computing also has some Shortcomings. This section will elaborate the same. With Edge devices, (used for edge computing) open the door for the various attacks [10]. With different attacks, an attacker can inject malicious activity. Edge devices can be infect with these attacks which may lead to A software or even entire network may be infected. These attacks are difficult to manage adequately due to distributed environment.

Another problem is cost. Though with the edge computing we can reduce the computing but at the same time its maintenance cost will increase. The main reason for higher maintenance cost is typically used of various edge devices, maintenance team should be knowledgeable enough to manage this technology.

7. Conclusion

This chapter briefly introduced cloud computing and given a more comprehensive definition of edge computing. The chapter talk about various technologies around the edge computing. After a deep discussion about the technology various benefits out of use of the same explained widely. The chapter also discuss the journey from cloud computing and edge computing in detail. Various applications after detailed explanation also major part of the chapter. Though, there are noteworthy security threats within IoT gadgets, which means edge security is most vital than ever. Along with the threat resolvase techniques, edge also gives a chance to improve the security by researching the related areas. The chapter concludes that the edge computing has tremendous benefits over the cloud computing and helpful to overcome the limitations of traditional and cloud computing. With 5G and edge technology there is very bright future of many real time applications.

References

[1] K. Saini, V. Agarwal, A. Varshney, A. Gupta, E2EE for data security for hybrid cloud services: a novel approach, in: IEEE International Conference on Advances in Computing, Communication Control and Networking (IEEE ICACCCN 2018) Organized by Galgotias College of Engineering & Technology Greater Noida, 12–13 October, 2018, 2018, https://doi.org/10.1109/ICACCCN.2018.8748782.

[2] I. Goodfellow, et al., Generative adversarial nets, Proc. Adv. Neural Inf. Process. Syst. (2014) 2672–2680.

[3] H.T. Dinh, C. Lee, D. Niyato, P. Wang, A survey of mobile cloud computing: architecture, applications, and approaches, Wirel. Commun. Mob. Comput. (2013).

[4] P.J. Werbos, Backpropagation through time: what it does and how to do it, Proc. IEEE 78 (10) (1990) 1550–1560.

[5] S.R. Jena, R. Shanmugam, R. Dhanaraj, K. Saini, Recent advances and future research directions in edge cloud framework, Int. J. Eng. Adv. Technol. 2249-8958, 9 (2) (2019), https://doi.org/10.35940/ijeat.B3090.129219.

[6] V. Mnih, et al., Human-level control through deep reinforcement learning, Nature 518 (7540) (2015) 529.

[7] S.R. Jena, R. Shanmugam, K. Saini, S. Kumar, Cloud computing tools: inside views and analysis, in: International Conference on Smart Sustainable Intelligent Computing and Applications under ICITETM2020, Elsevier, 2020, pp. 382–391.

[8] F. Bonomi, R. Milito, J. Zhu, S. Addepalli, Fog computing and its role in the internet of things, in: Workshop on Mobile Cloud Computing, ACM, 2012.

[9] C. Szegedy, et al., Going deeper with convolutions, in: Proceedings of the IEEE Conference on Computer Vision and Pattern Recognition, 2015, pp. 1–9.

[10] K. Saini, P. Raj, Handbook of Research on Smarter and Secure Industrial Applications Using AI, IoT, and Blockchain Technology, IGI Global, 2021. ISBN13: 9781799 883678, ISBN10: 1799883671, EISBN13: 9781799883685.

[11] Z. Liu, Z. Dai, P. Yu, Q. Jin, H. Du, Z. Chu, D. Wu, Intelligent station area recognition technology based on NB-IoT and SVM, in: Proceedings of the IEEE 28th International Symposium on Industrial Electronics, 2019, pp. 1827–1832.

[12] J. Marescaux, J. Leroy, M. Gagner, et al., Transatlantic robot-assisted telesurgery, Nature 413 (2001) 379–380.

[13] I. Stojmenovic, S. Wen, The fog computing paradigm: scenarios and security issues, in: Federated Conference on Computer Science and Information Systems (FedCSIS), IEEE, 2014.

[14] M. Satyanarayanan, P. Bahl, R. Caceres, N. Davies, The case for VM-based cloudlets in mobile computing, IEEE Pervasive Comput. 4 (2009) 14–23.

[15] C.-C. Hung, et al., VideoEdge: processing camera streams using hierarchical clusters, in: Proceedings of the IEEE/ACM Symposium on Edge Computing (SEC), 2018, pp. 115–131.

[16] K. Hong, D. Lillethun, U. Ramachandran, B. Ottenwalder, B. Kold-ehofe, Opportunistic spatio-temporal event processing for mobile situation awareness, in: Proceedings of the ACM International Conference on Distributed Event-Based Systems, 2013.

[17] L.M. Vaquero, L. Rodero-Merino, Finding your way in the fog: towards a comprehensive definition of fog computing, in: ACM SIGCOMM CCR, 2014.

[18] I. Stojmenovic, Fog computing: a cloud to the ground support for smart things and machine-to-machine networks, in: Telecommunication Networks and Applications Conference (ATNAC), IEEE, 2014.

[19] S. Hochreiter, J. Schmidhuber, Long short-term memory, Neural Comput. 9 (8) (1997) 1735–1780.

[20] J. Zhu, et al., Improving web sites performance using edge servers in fog computing architecture, in: SOSE, IEEE, 2013. [12] H. Madsen, G. Albeanu, B. Burtschy, and F. Popentiu-Vladicescu, "Reliability in the utility computing era: Towards reliable fog computing," in IEEE International Conference on Systems, Signals and Image Processing (IWSSIP), 2013.

[21] S. Yi, Z. Qin, Q. Li, Security and privacy issues of fog computing: a survey, in: International Conference on Wireless Algorithms, Systems and Applications (WASA), 2015.

[22] D.N. Le, R. Kumar, B.K. Mishra, J.M. Chatterjee, M. Khari, (Eds.),, Cyber Security in Parallel and Distributed Computing: Concepts, Techniques, Applications and Case Studies, John Wiley & Sons, 2019.

[23] B. Chen, S. Qiao, J. Zhao, D. Liu, X. Shi, M. Lyu, H. Chen, H. Lu, Y. Zhai, A security awareness and protection system for 5G smart healthcare based on zero-trust architecture, IEEE Internet Things J. (2020), https://doi.org/10.1109/JIOT.2020.3041042.

[24] B. Ottenwalder, B. Koldehofe, K. Rothermel, U. Ramachandran, Migcep: operator migration for mobility driven distributed complex event processing, in: Proceedings of the ACM International Conference on Distributed Event-Based Systems, 2013.

[25] K. Hong, D. Lillethun, U. Ramachandran, B. Ottenwalder, B. Kold-ehofe, Mobile fog: a programming model for large-scale applications on the internet of things, in: ACM SIGCOMM Workshop on Mobile Cloud Computing, 2013.

[26] S. Yi, C. Li, Q. Li, A survey of fog computing: concepts, applications and issues, in: Proceedings of the 2015 Workshop on Mobile Big Data, ACM, 2015.

About the authors

Kavita Saini, is presently working as professor, School of Computing Science and Engineering, Galgotias University, Delhi NCR, India. She received her PhD degree from Banasthali Vidyapeeth, Banasthali. She has 18 years of teaching and research experience supervising Masters and PhD scholars in emerging technologies.

She has published more than 40 research papers in national and international journals and conferences. She has published 17 authored books for UG and PG courses for a number of universities including MD University, Rothak, and Punjab Technical University, Jallandhar with National Publishers. Kavita Saini has edited many books with International Publishers including IGI Global, CRC Press, IET Publisher Elsevier and published 15 book chapters with International Publishers. Under her guidance many MTech and PhD scholars are carrying out research work.

She has also published various patents. She has also delivered technical talks on Blockchain: An Emerging Technology, Web to Deep Web, and other emerging areas and handled many special sessions in International Conferences and Special Issues in International Journals. Her research interests include Web-Based Instructional Systems (WBIS), Blockchain Technology, Industry 4.O, and Cloud Computing.

Pethuru Raj working as a chief architect at Reliance Jio Platforms Ltd. (JPL) Bangalore. Previously. worked in IBM global Cloud center of Excellence (CoE), Wipro consulting services (WCS), and Robert Bosch Corporate Research (CR). In total, I have gained more than 20 years of IT industry experience and 8 years of research experience. Finished the CSIR-sponsored Ph.D. degree at Anna University, Chennai and continued with the UGC-sponsored post-doctoral research in the Department of Computer Science and Automation, Indian Institute of Science (IISc), Bangalore. Thereafter, I was granted a couple of international research fellowships (JSPS and JST) to work as a research scientist for 3.5 years in two leading Japanese universities. Focuses on some of the emerging technologies such as the Internet of Things (IoT), Optimization of Artificial Intelligence (AI) Models, Big, fast and streaming Analytics, Blockchain, Digital Twins, Cloud-native computing, Edge and Serverless computing, Reliability engineering, Microservices architecture (MSA), Event-driven architecture (EDA), 5G, etc. My personal web site is at https://sweetypeterdarren. wixsite.com/pethuru-raj-books/my-books https://scholar.google.co.in/ citations?user=yaDflpYAAAAJ&hl=en.

CHAPTER NINE

Edge computing challenges and concerns

Kavita Saini[a], Uttama Pandey[a], and Pethuru Raj[b]

[a]School of Computing Science and Engineering (SCSE), Galgotias University, Delhi, Uttar Pradesh, India
[b]Site Reliability Engineering (SRE) Division, Reliance Jio Platforms Ltd. (JPL), Bangalore, India

Contents

Advances in Computers, Volume 127
ISSN 0065-2458
https://doi.org/10.1016/bs.adcom.2022.02.006

259

Abstract

In the current trends of Internet of things and with the excessive use of end devices like laptops, smartphones, sensors that are continuously generating large amount of data is encouraging the furtherance of Edge Computing. Since there is a demand for lower latency network, the edge computing has always supported the light weight end devices to work in an efficient way toward complicated problems and provide desired services to the end user. The major issue in providing a better service to the customer lies in the delay and the network traffic which indirectly includes our concern for the issues like network bandwidth, cost efficiency, energy saving, data security on edge device. This chapter discusses about the several critical challenges that are faced in different areas, also the trends in the area of edge computing.

1. Introduction

These days the amount of data generated, captured, consumed and created is significantly increasing. Now the worry is to store and manage the data to hard drive on your PC or any other device, will definitely require the wide storage space which is highly cost effective and unmanageable [1,2]. So instead of saving data to the nearby hard drive, you can store it on a cloud based system which allow data to be collected from various IoT devices and make it available anywhere in the world. It is expected that more than 15 billion IoT devices will connect to the enterprise infrastructure by 2029.

The amount of data generated by these IoT devices are almost doubling every year and it's going to be approximately 79ZB by 2025. This large and excessive amount of data produced at an IoT device is sometime unable to be transferred from a remote network to another site [3,4]. The reason being, that the various devices transmits the data at the same time and sending this over sufficient quantity of data to the cloud or to the central repository for computation may create the latency and bandwidth issues. This results in shifting to more efficient computing alternative called Edge Computing. The need is to process and segregate the needful data generated by the nodes in the network, close to the edge node. For this the edge device should be made more capable to compute and pre-process the data on its own. Instead of only generating the data and transferring it as it is to the central repository for computation and storage, the device will be provided the figuring capabilities to extract the meaningful data. Since irrelevant data is no more transmitted to the data centers, this may lower the latency and save the network cost [5].

In coming few years, industries are going to completely adopt cloud edge technology along with the IoT applications for controlling and managing their industrial services in more efficient and energy saving way. Various authors have tried to define Edge computing in most natural and suitable way [6].

2. Cloud, fog and edge computing

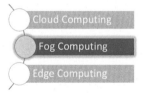

2.1 Cloud computing

Cloud computing works as an Internet store which provides a remote server where you can process your data. It does not require any direct active management by the user, means you do not have to worry about the storage management. Instead of saving data to the nearby hard drive on your PC or any other device, you can store it on a cloud based system which allow data to be collected from various IoT devices and make it available anywhere in the world [7].

IoT Device: These are the connected devices that continuously generate large amount of data. If all the data is send to the cloud the cost will definitely increase [8,9]. So, instead of sending the data directly to the cloud it is send to the devices near to the network to analyze and compute.

2.2 Fog computing

It is also known as Fog networking or fogging. It is a decentralized infrastructure and application resides between data stores and the cloud. It is an architecture which uses edge device for computation, storage and communication [10,11]. This edge device is the device that controls the data flow at the boundaries of any two networks, example: router, switch, IAD, gateways, hub, multiplexer or bridge. It is actually a mediator between hardware and remote server [12].

Fog is a distributed network environment and is very close to cloud computing and IoT device [13]. It has a high security network. Instead of sending selected data to the cloud for processing, fog actually process the data which saves the network bandwidth and reduces latency requirement which in turn helps in fast decision making capability. Fog nodes can be planted on very crucial places like under the sea, railway track, etc.

2.3 Edge computing

Edge computing and Fog computing, both are the extensions of cloud network. Edge computing technology saves time and resources in maintenance of operations by collecting and analyzing data in real time [14]. As in cloud computing all the data generated by IoT device is stored on a cloud based system and the maintenance and processing of data is done there, in edge computing data is not transferred anywhere [15,16]. The maintenance and processing of data is done on the device that initially created it. The computational power is given to the edges on the network. Edge computing is more secure and it reduces the unnecessary traffic to the central repository, as only the selected data after computation from edge is sent to the cloud.

"Edge computing is a new paradigm in which the resources of an edge server are placed at the edge of the Internet, in close proximity to mobile devices, sensors, end users, and the emerging IoT." [1].

"Edge computing refers to the enabling technologies allowing computation to be performed at the edge of the network, on downstream data on behalf of cloud services and upstream data on behalf of IoT services" [2].

Table 1 Comparison of cloud, edge and fog computing.

Parameter	Cloud	Edge	Fog
Location of Data Processing	Data is processed on the cloud server	Data is processed on the edge itself, may be IoT sensors	Data is processed on the edge device connected to the LAN hardware, say gateways
Capacity	Does not provide any reduction in data while sending or transforming data	Reduces the amount of data sent to cloud	
Purpose	Suitable for the long-term, in-depth analysis of data and storage	Both are more suitable for the quick analysis which is required for real-time response	
Latency	High	Low	Very Low
Security	Less security compared to Edge and Fog	High security	

"Edge computing is part of a distributed computing topology where information processing is located close to the edge, where things and people produce or consume that information" [7].

Edge computing can also be defined as a technology where the computation and control for optimized data is given to the node (an IoT device) which itself is generating the data, in order to improve the performance of a distributed network [7] (Table 1).

3. Implications and challenges in adopting edge computing

In vehicle automation, video surveillance, and other areas, edge computing has the ability to provide superior intelligent services with a faster reaction on programs that run in real time. But still, it suffers from various challenges. Few are listed below:

3.1 Accessibility

Edge applications frequently create logistical challenges when it comes to deploying human IT resources to administer them, and they don't allow for high operator expenses. Companies cannot afford to hire a professional administrator to oversee and maintain each and every Edge location. For example, for every sensors located at the oil wells for keeping a check on the status of wells or devices deployed at the agriculture farm land to continuously monitoring the health of soil and plant, these operator restrictions—either due to distance, device volume, geographic accessibility, or other cost factors—need Edge applications that are not only small in terms of computing footprint but also in terms of technical overhead, from installation to ongoing operations.

3.2 Control and management

For making the edge infrastructure more reliable Differentiation, Elasticity, Segregation (Isolation) and reliability should be considered as an important factor. With the rapid rise of IoT deployment, we expect different services, such as Smart Home, to be provided at the network's edge. These services will be prioritized differently. For example IoT devices used in disaster management must be processed earlier than other device.

Elasticity could be one of the major challenges at the network's edge because, the edge device must be extremely dynamic. Is it possible for the owner to add a new item to the current service without any difficulty? Or, if anything wears out and needs to be replaced, can the prior service quickly adopt a new node? These issues should be addressed with a flexible and extensible approach.

Isolating the edge device from the network or from operating system and device provides further protection against malware and other assaults. In case of distributed networks topology, if a device or an application crash, the whole system is required to e rebooted. Also, we have an option of token management to solve the issues of locks. But in case of edge device computing where the data is processed closer to the device, the data may be shared and accessed as a resource, among different applications. Say for example, in smart surveillance systems with multiple CCTV cameras in a shop, if one device crashes or fails, the owner would still be able to control and monitor mishaps without the failure of entire OS. The introduction of a deployment mechanism could potentially fix this problem. If the OS detects a conflict before an application is installed, the user will be notified and the potential access issue will be avoided.

From a reliability standpoint, it's critical for the EdgeOS to keep track of the entire system's network topology, and each component must be able to communicate status/diagnosis data to the EdgeOS. It is sometime very difficult to find out the potential reason of a service of a device failure at the host end. An IoT device sometimes fails to report the data constancy due to battery outage or bad connection condition under some unpredictable situations. Some communication protocols must be used which provides dynamic connection with sensor nodes. Data sensing and data communication between sensor and the application is also an important issue.

3.3 Scalability

IoT device collects data and transmit it to more powerful computing nodes, where all of the original data is further processed and analyzed. Nonetheless, individual edge node computational power is restricted, making scalability of computational capacity for edge computing a difficult task.

Though the number of edge devices are increasing with the increased demand of cloud based services, the application on edge are expected to work more consistently despite of increased load on the edge. In order to comprehend the status of the network, Edge computing necessitates a large number of monitors, as well as servers and network devices. Finally, viewing the status of the entire edge environment becomes challenging.

3.4 Privacy and security

Edge computing devices are more vulnerable to attack because if you are putting your data on to the edge instead of cloud, it is exposed to the real world. So we have to make it secure either by physical security which may or may not be possible, or by securing the machines by secure boot which make sure that when the data is coming out, it must be encrypted. Encryption will definitely require strong authentication to login to the device for a direct access remotely. Potential connectivity is also a challenge, as there is no guarantee that your device is always connected. It may shutdown due to power failure or battery discharge. Edge devices that are very small in size may suffer from physical isolation.

3.5 Data storage

There are large numbers of data generators and sensors in an IoT network which are deployed to sense the data and report to the network gateway. For example in smart security surveillance system, the security cameras keep on sending the videos to the gateway where the data is just stored in the database and nobody is consuming the data. After certain period the video is flushed with the latest videos. The edge node must have less human involvement, and must be capable enough to process the data at the edge level resulting in the optimized data after event detection and privacy protection. Choosing the appropriate level of data abstraction might be tricky at times. Some applications or services may not be able to learn enough knowledge if too much raw data is filtered out. However, there are several limitations if we want to store a significant amount of raw data. Data storage would be a challenge.

3.6 Latency

Latency is defined as any lag in communication between a network and its devices. Edge computing, which is based on distributed network, can alleviate latency issues by ensuring that there is no disconnect in real-time information processing and providing a more stable network. Application latency and decision-making delay are lowered by placing computation capability closer to the data at the edge. Faster request and response is expected from back-and-forth movement from the edge to the core. Application data travels the network in directions, sharing data and dealing with access permissions, with computing placed at both the core and the edge.

3.7 Performance

IoT devices are now used as building blocks in smart cities. These devices are integrated into the city's infrastructure, allowing for the monitoring of device operation and the collection of periodic data on these assets. Edge computing sends critical technical data such as traffic, flood, safety, and infrastructure monitoring. It allows for more on-device computation for real-time decision-making. Monitoring the performance of these edge devices from edge to cloud will be a challenge for the consumer. Technology which will provide end to end monitoring must be deployed to keep an eye on edge to cloud transactions.

4. Concerns with edge computing

4.1 Cost

Edge computing devices with processing capabilities are expensive (Fig. 1). Increased requirement of equipment at the site, requires deployment of new devices along with the modifications to older versions that lack such processing capabilities, resulting in additional expenditures. An edge computing framework's configuration, deployment, and maintenance are all costly endeavors.

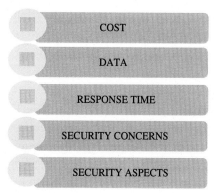

Fig. 1 Concerns with edge computing.

4.2 Data

Since the edge computing provides on-device computation capability to the edge node for extracting the meaningful data out of the raw data generated by the IoT device. This subset of useful data is used for further processing and lot of raw data is wasted. There could be a possibility of loss of some important information which could provide additional insights in real time decision, but is filtered now. Elimination of the critical data, for improved efficiency is an important concern in different business applications running over cloud.

4.3 Response time requirement

Any lag in communication between edge node and cloud should be reduced in order to make an effective communication. The request and response time between user and cloud service provider termed as latency, is required to be as low as if user is working on a system with database on his device. For example as mobile shopping becomes increasingly popular, it is critical to improve the customer experience, particularly in terms of latency. The latency will be considerably decreased if the shopping cart updating is offloaded from cloud servers to edge nodes in this situation.

4.4 Security concerns

Edge computing aims at protecting the processed data traveling from edge to centralized data centers. Since edge computing is a decentralized system it is important for the users to implement proactive threat detection technologies for early detection of known and unknown vulnerabilities. There is a need to secure the edge application beyond the network layer by firewalls, encryption policies and access control mechanisms.

4.5 Security aspects

Due to the increased data generation at the edge node, Edge Computing will face new and unanticipated security and privacy concerns. Confidentiality, Integrity, and Availability are the three primary requirements in attaining security and privacy in a shared environment like Edge Computing. IoT functionality necessitates service migration between local and global scales, making the network more vulnerable to malicious activity. Furthermore, because users' personally identifiable information would be exchanged and/or kept on edge servers, security and privacy become critical concerns in such a distributed framework. As a result, EC-assisted IoT networks are more exposed to cyber threats and attacks. Malicious attacks can occur during any of the three core functions of Edge servers: communication, computation, and storage.

5. Security and privacy attacks on edge computing enabled devices

5.1 Physical attacks

It is basically the tampering of the valuable assets. If attackers have physical access to the edge nodes/devices, they can launch this assault. In this situation, valuable and sensitive cryptographic data can be recovered, the data can be tampered, and the software/operating systems can be tweaked or changed.

5.2 Sniffing

It is sometimes called as eaves dropping. Over communication channels, adversaries listen in on private discussions such as usernames, passwords, and other sensitive information. Attackers can get vital information about the network if sniffed packets contain access or control information of the Edge nodes, such as node configuration, node identification, and the shared network password.

5.3 Unauthorized access

To access or share the information, neighboring Edge nodes communicate with one another. However, if an attacker gains access to one of the unsecured nodes, they can take control of the entire neighborhood and can violate the network's stated security policy.

5.4 Routing table attack

DoS, Packet mistreating, RTP (Routing Table Positioning), Hit and Run are some of the common attacks to router. At the communication level, attackers can modify routing information by redirecting or discarding data packets. The malicious edge nodes could: (1) Drain all network packets, (2) Drain selective packets, (3) Record packets at one network location before migrating them to another, or (4) Broadcasts "Hello" to all network nodes claiming to be their neighbor.

5.5 Distributed denial of service attack (DDoS)

DDoS assaults are carried out via networks of nodes that are linked to the Internet. These networks are made up of malware-infected IoT devices that can be manipulated remotely by an attacker. Individual devices are known as bots (or zombies) and a botnet is a collection of bots. The attacker can direct an attack once a botnet has been built by delivering remote instructions to each bot. When a botnet targets a victim's server or network, each bot sends requests to the target's IP address, potentially overloading the server or network and causing a denial-of-service to normal traffic. Outage, Sleep deprivation and battery draining are considered to be famous DDos attack.

5.6 Malicious hardware and software injections

An unauthorized software and hardware components is injected into the edge network by the attacker, resulting into a devastating impact on the efficiency of existing edge servers and devices and even exploiting service provider, in which entities that provide the software and hardware solutions that enable edge computing unwittingly begin executing hacking processes on the attacker's behalf. Camouflaging and Node Replications are some common frightening practices by attacker.

5.7 Integrity attack

In this attack, the attacker attempts to encrypt or corrupt the sensitive data and may demand for some ransom to restore the data. In edge enabled computing infrastructure the attacker changes the teaching process of the learning model by injecting wrong datasets or may misuse the vulnerabilities of the learning model without performing any changes to it.

5.8 Privacy leakage

The functionality of Edge nodes may require the extraction of personal information from data generated by user devices. Data owners must own all information regardless of its importance like Personal activities, preferences, and health state may be sensitive; and others, such as the air pollution index, public information, and social events, may not be. Unfortunately, without the authorization of the information owners, they can be shared with other users or network organizations, making them vulnerable to attackers during data transfer and sharing.

5.9 Logging attacks

This type of attack might cause damage to IoT systems in your Edge infrastructure, if log files are not secured. Infrastructure developers must keep track of events like application faults and failed/successful authorization/ authentication attempts, and log these events timely.

5.10 Data storage and protection

Data gathered and processed at the edge does not have the same level of physical protection as data stored in more centralized locations. Vital information can be compromised and leaked easily by removing a disc drive from an edge resource or copying data from a simple memory stick. It can also be more challenging to provide reliable data backup due to restricted local resources.

6. Countermeasures to security and privacy attacks in edge infrastructure

6.1 Solution to physical attack

Organizations should consider innovative approaches to improve the physical security of any edge nodes. Additional robust techniques may be used during manufacturing of devices, or locking mechanisms and other physical protections may be implemented in the field.

6.2 Solution to sniffing

Connect to a trusted network as the attacker as Public networks are usually put up and not monitored for any incursions or bugs. Attackers can either sniff that network or construct their own network with similar names to deceive users into joining it. An attacker sitting in a coffee shop can set up a Wi-Fi network called "Free coffee_Wi-Fi," and adjacent users can join to it, transmitting all data through the sniffer node of the attackers. Ensure that all the data transmitted through the edge node must be encrypted, so that even if your data is sniffed the attacker will not be able to sense the traffic. Network administrators must check the network for any attempts at infiltration or rogue devices set up in span mode to record traffic.

6.3 Solution to unauthorized access

Unauthorized access can be defended by authorization and authentication. Entities must mutually authenticate one another across different trusted domains in the Edge computing environment. This comprises single-domain and cross-domain authentication, as well as handover authentication. Authorization prevents attackers or malicious Edge nodes from receiving answers. It examines if a service provider or an Edge node/device or a router has access to, control over, alter, or share data.

6.4 Solution to routing table attack

Effective countermeasures against routing information assaults include developing trustworthy routing protocols and installing a high-quality intrusion detection system (IDS) that monitors for hostile traffic and detects policy breaches. Nodes that use reliable routing protocols can build a table of trusted nodes to share sensitive data.

6.5 Solution to distributed denial of service attack (DDoS)

Organize a DDos Response plan, if a DDoS attack is successful, prepare a checklist to ensure that your assets have advanced threat detection, as well as any technical competencies and experience that would be required. Establish an incident response team and inform the stakeholders. Prevent DDoS attacks by securing your infrastructure and implementing prevention management systems that combine firewalls, VPN, anti-spam, content filtering and other security layers to monitor activities. Basically we need to ensure that the network's normal regulations are not breached. In a nutshell, regulate the behavior of devices in a network.

6.6 Solution to malicious hardware and software injections

To defend malicious hardware and software injections researchers have proposed three measures. First, side-channel signal studies, which use timing, power, and spatial temperature analysis to detect hardware trojans. This approach detects malicious firmware or software placed on edge nodes by looking for anomalous system behaviors like increased execution time and power consumption. Second, to detect and model harmful attacks, Trojan activation methods compare Trojan-afflicted integrated circuits with non-Trojan-afflicted integrated circuits. Lastly, Circuit modification or replacement to provide circuit-level protection and even allowing the node to self-destruct in the case of an assault.

6.7 Solution to integrity attack

Encrypting your data is the most effective way to ensure its integrity. This applies to data transmission as well as data in storage. Unauthorized parties cannot access encrypted data, thus even if your data silo is breached, your data will be totally protected. We now live in a time where password authentication is no longer sufficient. Credential stuffing assaults are constantly bombarding network security systems, and users continue to click embedded URLs and email attachments that install key loggers. These constant attacks, along with our mobile environment, where remote logins are the standard, create genuine dangers for any organization [5]. Emails with enclosed data attachments should not be forwarded, and what applications interacting with data files should minimized. Microsoft's Windows Information Protection and Azure Rights Management are two examples that are providing data protection.

6.8 Solution to privacy leakage

Identifying the critical data and using Data Loss Prevention (DLP) software to protect the sensitive information may help in defending against privacy leakage. Organizations should implement a data protection plan, focusing on sensitive documents and their treatment, because DLP is strongly reliant on correct classification of information. Closely monitor traffic on all networks. In order to detect unauthorized actions Data Activity Monitoring (DAM) solution can provide another layer of protection for this. Data Encryption is also another way to keep your data secure [10].

6.9 Solution to logging attacks

Attackers frequently change the logs of the victim machine to escape detection by system, network, and security administrators. The attacker will try to delete specific events from the logs related to gaining access, escalating privileges, and installing RootKits and back doors. The first step to ensuring the integrity and utility of your log files is straightforward: Logging should be enabled on all of your sensitive systems. A policy or standard must be created in the organization that states that logging is required. Setting adequate permissions on log files is another common-sense defense for preserving essential system logging and accounting [8]. Another step could be to put up a separate logging server. We can raise the bar for attackers by deploying a separate logging machine. Encrypting log files is another effective method of log protection. Without the encryption key, attackers will be unable to make any serious changes to the data.

6.10 Solution to data storage and protection

Because of the rapid advancements in IoT devices, some artificial intelligence (AI) functions can now be moved from the centralized cloud to Edge devices/nodes. Security, privacy, and latency will all be improved due to this [10]. In order to eliminate data redundancy and maximize bandwidth in IoT networks, the replicate copies of data on intermediate nodes must be removed. Unfortunately, intruders will have access to important information as a result of this. Secure data deduplication is used to counteract this problem, allowing intermediaries to access replicated data without obtaining any knowledge of it [5].

6.11 Embedding blockchain on edge infrastructure

Blockchain is a new technique that aims to create a trusted, dependable, and secure foundation for data regulation and information exchanges among various operating network edge entities. It establishes rules that let decentralized systems to make decisions regarding the execution of specific transactions in concert, based on voting and consensus algorithms. This will: (1) provide safe audit-level tracking of EC-assisted IoT data transactions; and (2) remove the need for a central trusted intermediary between interacting IoT edge devices.

7. Future of edge computing

1. Edge computing's future will undoubtedly be open. Edge computing will converge with the utilization of data to transform knowledge into actions that benefit businesses and their consumers, thanks to artificial

intelligence and machine learning. It will eventually be regarded like any other area where applications can be installed in a consistent and uncompromised manner.

2. We will see boundaries between the edge, data centers—if they even exist—and the cloud because there will be so much computing and storage available. Multiple ecosystems will start providing data from the edge directly, posing challenges for data consolidation across the three realms. At the edge, AI and machine learning will have progressed to the point where more sophisticated autonomous use cases will be available. Data management will continue to be important, but the problems will be new.

3. Edge computing's future will improve in tandem with sophisticated networks such as 5G and satellite mesh, as well as artificial intelligence. It has suddenly opened up the globe to some potentially futuristic possibilities by having greater capacity and power, better access to fast and widespread networks (5G, satellite), and smarter machines within computers (AI).

4. Shifting data processing to the network's edge can help businesses capitalize on the expanding number of IoT edge devices, boost network speeds, and improve customer experiences. Edge computing's scalability makes it a great alternative for fast-growing, agile businesses, particularly if they already use colocation data centers and cloud infrastructure.

5. Companies can optimize their networks by leveraging the power of edge computing to provide flexible and reliable service that strengthens their brand and keeps customers pleased. Edge computing has various advantages over traditional network design and will undoubtedly continue to play a significant role in the future for businesses. Innovative enterprises have likely only scratched the surface of what's possible with edge computing as more internet-connected products hit the market.

6. Edge computing addresses a growing demand for decreased latency, processing of increasing amounts of data at the edge, and network resilience.

7. "Edge computing" is broad enough to accommodate a variety of sub-markets, but it will grow from thousands of custom patterns to a few dozens, with cloud providers playing a key role all the way to the edge, or complementing edge solutions. Enterprises must emphasize a distributed cloud-based solution as the default and future-proof edge solutions by relying on partnerships and ecosystems rather than a single-vendor strategy, according to McArthur.

8. Conclusion

Cloud and IoT are the two advancements in networking technologies which has made Edge computing possible. It provided the storage capacity and processing power near to the user's infrastructure, resulting into reduced latency and higher bandwidth. With the rapid proliferation of IoT devices, as well as the resulting massive data traffic generated at the network's edge, placed additional strains on the current state-of-the-art. Because of the bandwidth and computational power available, a centralized cloud computing model has emerged to overcome with the shortage of resources. Since Edge Computing technology has unique features and provided extended QoS (Quality of Service), it involves huge risk in data security and privacy. This paper provides the Security and privacy aspects of edge computing. It also discusses the future aspects of this emerging trend.

References

[1] W. Shi, G. Pallis, X. Zhiwei, Edge computing, Proc. IEEE 107 (8) (2019) 1474–1481.
[2] W. Shi, et al., Edge computing: vision and challenges, IEEE Internet Things J. 3 (2016) 637–646.
[3] S.R. Jena, R. Shanmugam, R. Dhanaraj, K. Saini, Recent advances and future research directions in edge cloud framework, IJEAT 9 (2) (2019), https://doi.org/10.35940/ijeat.B3090.129219. ISSN: 2249 – 8958.
[4] K. Saini, V. Agarwal, A. Varshney, A. Gupta, E2EE for data security for hybrid cloud services: a novel approach, in: IEEE International Conference on Advances in Computing, Communication Control and Networking (IEEE ICACCCN 2018) organized by Galgotias College of Engineering & Technology Greater Noida, 12–13 October, 2018, https://doi.org/10.1109/ICACCCN.2018.8748782.
[5] M. Simsek, A. Aijaz, M. Dohler, J. Sachs, G. Fettweis, 5G-enabled tactile internet, IEEE J. Sel. Areas Commun. 34 (3) (2016) 460–473.
[6] N. Hassan, S. Gillani, E. Ahmed, I. Yaqoob, M. Imran, The role of edge computing in internet of things, IEEE Commun. Mag. 56 (11) (2018) 110–115.
[7] Y. He, F.R. Yu, N. Zhao, H. Yin, Secure social networks in 5G systems with mobile edge computing, caching, and device-to-device communications, IEEE Wirel. Commun. 25 (3) (2018) 103–109.
[8] Y. He, J. Ren, G. Yu, Y. Cai, D2D communications meet Mobile edge computing for enhanced computation capacity in cellular networks, IEEE Trans. Wirel. Commun. 18 (3) (2019) 1750–1763.
[9] S. Choy, B. Wong, G. Simon, C. Rosenberg, The brewing storm in cloud gaming: a measurement study on cloud to end-user latency, in: 2012 11th Annual Workshop on Network and Systems Support for Games (NetGames), IEEE, 2012.
[10] S. R. Jena, R. Shanmugam, K. Saini, S. Kumar, "Cloud computing tools: inside views and analysis", International Conference on Smart Sustainable Intelligent Computing and Applications under ICITETM2020 (ELSEVIER), pp. 382–391.
[11] I. Sarrigiannis, E. Kartsakli, K. Ramantas, A. Antonopoulos, C. Verikoukis, Application and network VNF migration in a MEC-enabled 5G architecture, in: 23rd IEEE

International Workshop on Computer Aided Modeling and Design of Communication Links and Networks (CAMAD), 2018, pp. 1–6.

[12] E. Ahmed, A. Ahmed, I. Yaqoob, J. Shuja, A. Gani, M. Imran, M. Shoaib, Bringing computation closer toward the user network: is edge computing the solution? IEEE Commun. Mag. 55 (11) (2017) 138–144.

[13] Z. Zhao, K. Hwang, J. Villeta, Game cloud design with virtualized CPU/GPU servers and initial performance results, in: Proceedings of the 3rd workshop on Scientific Cloud Computing Date, ScienceCloud ACM Press, 2012.

[14] W. Hu, Y. Gao, K. Ha, J. Wang, B. Amos, Z. Chen, P. Pillai, M. Satyanarayanan, Quantifying the impact of edge computing on mobile applications, in: Proceedings of the 7th ACM SIGOPS Asia-Pacific Workshop on Systems - APSys '16, ACM Press, 2016.

[15] K. Saini, P. Raj, Handbook of Research on Smarter and Secure Industrial Applications Using AI, IoT, and Blockchain Technology, IGI Global, 2021 (ISBN13: 9781799883678 | ISBN10: 1799883671 | EISBN13: 9781799883685).

[16] J. Gonzalez, G. Nencioni, A. Kamisinski, B.E. Helvik, P.E. Heegaard, Dependability of the NFV orchestrator: state of the art and research challenges, IEEE Commun. Surv. Tutor. 20 (4) (2018) 3307–3329.

Further reading

[17] J.A. Silva, R. Monteiro, H. Paulino, J.M. Lourenco, Ephemeral data storage for networks of hand-held devices, in: 2016 IEEE Trustcom/BigDataSE/ISPA, IEEE, 2016.

[18] Y. Li, J. Xue, W.Z. Wang, T. Li, Edge-oriented computing paradigms, ACM Comput. Surv. 51 (2) (2018) 1–34.

[19] C. Xu, K. Wang, P. Li, S. Guo, J. Luo, B. Ye, M. Guo, Making big data open in edges: a resource-efficient blockchain-based approach, IEEE Trans. Parallel Distrib. Syst. 30 (4) (2019) 870–882.

[20] K. Ahmad, A. Kamal, K.A.B. Ahmad, M. Khari, R.G. Crespo, Fast hybrid-MixNet for security and privacy using NTRU algorithm, J. Inf. Secur. Appl. 60 (2021), 102872.

[21] D.N. Le, R. Kumar, B.K. Mishra, J.M. Chatterjee, M. Khari, (Eds.), Cyber Security in Parallel and Distributed Computing: Concepts, Techniques, Applications and Case Studies, John Wiley & Sons, 2019.

[22] B. Chen, S. Qiao, J. Zhao, D. Liu, X. Shi, M. Lyu, H. Chen, H. Lu, Y. Zhai, A security awareness and protection system for 5G smart healthcare based on zero-trust architecture, IEEE Internet Things J. (2020), https://doi.org/10.1109/JIOT.2020.3041042. early access, Nov. 30.

[23] S.S. Vedaei, A. Fotovvat, M.R. Mohebbian, G.M.E. Rahman, K.A. Wahid, P. Babyn, H.R. Marateb, M. Mansourian, R. Sami, COVID-SAFE: an IoT-based system for automated health monitoring and surveillance in post-pandemic life, IEEE Access. 8 (2020) 188538–188551.

[24] Z. Liu, Z. Dai, P. Yu, Q. Jin, H. Du, Z. Chu, D. Wu, Intelligent station area recognition technology based on NB-IoT and SVM, in: Proc. IEEE 28th Int. Symp. Ind. Electron. (ISIE), 2019, pp. 1827–1832.

[25] J. Marescaux, J. Leroy, M. Gagner, et al., Transatlantic robot-assisted telesurgery, Nature 413 (2001) 379–380.

[26] https://iotbusinessnews.com/2020/08/10.

About the authors

Kavita Saini is presently working as Professor, School of Computing Science and Engineering, Galgotias University, Delhi NCR, India. She received her Ph.D. degree from Banasthali Vidyapeeth, Banasthali. She has 18 years of teaching and research experience supervising Masters and Ph.D. scholars in emerging technologies.

She has published more than 40 research papers in national and international journals and conferences. She has published 17 authored books for UG and PG courses for a number of universities including MD University, Rothak, and Punjab Technical University, Jallandhar with National Publishers. Kavita Saini has edited many books with International Publishers including IGI Global, CRC Press, IET Publisher Elsevier and published 15 book chapters with International publishers. Under her guidance many M.Tech and Ph.D. scholars are carrying out research work.

She has also published various patents. Kavita Saini has also delivered technical talks on Blockchain: An Emerging Technology, Web to Deep Web and other emerging Areas and Handled many Special Sessions in International Conferences and Special Issues in International Journals. Her research interests include Web-Based Instructional Systems (WBIS), Blockchain Technology, Industry 4.O, and Cloud Computing.

Ms. Uttama Pandey born on 13 Oct 1984 at Faridabad, Haryana, she is currently working as an Assistant Professor in the Department of Computer Science, DAV Centenary College, Faridabad, Haryana. She had completed her B.Sc. from Delhi University, MCA from MD University and M.Tech (CSE) from MD University. Pursuing Ph.D. in Computer Applications from Galgotias University, Greater Noida, Uttar Pradesh. She has published papers and attended various national and international conferences. She has participated in faculty development programmes and national level awareness programmes on "Research Methodology",

"Angular and Business Intelligence", "Python 3.4.3" and various others. Also participated in International E-Panel Discussion Programme organized by ISM Patna.

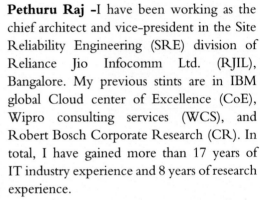

Pethuru Raj -I have been working as the chief architect and vice-president in the Site Reliability Engineering (SRE) division of Reliance Jio Infocomm Ltd. (RJIL), Bangalore. My previous stints are in IBM global Cloud center of Excellence (CoE), Wipro consulting services (WCS), and Robert Bosch Corporate Research (CR). In total, I have gained more than 17 years of IT industry experience and 8 years of research experience.

Finished the CSIR-sponsored Ph.D. degree at Anna University, Chennai and continued with the UGC-sponsored postdoctoral research in the Department of Computer Science and Automation, Indian Institute of Science (IISc), Bangalore. Thereafter, I was granted a couple of international research fellowships (JSPS and JST) to work as a research scientist for 3.5 years in two leading Japanese universities. Published more than 30 research papers in peer-reviewed journals such as IEEE, ACM, Springer-Verlag, Inderscience, etc. Have authored and edited 16 books thus far and contributed 35 book chapters thus far for various technology books edited by highly acclaimed and accomplished professors and professionals.

Focusing on some of the emerging technologies such as IoT, data science, Blockchain, Digital Twin, Containerized Clouds, machine and deep learning algorithms, Microservices Architecture, fog/edge computing, etc.

CHAPTER TEN

A smart framework through the Internet of Things and machine learning for precision agriculture

Veeramuthu Venkatesh[a], Pethuru Raj[b], and R. Anushia Devi[a]
[a]School of Computing, SASTRA Deemed University, Thanjavur, India
[b]Reliance Jio Cloud Services (JCS), Bangalore, India

Contents

Advances in Computers, Volume 127
ISSN 0065-2458
https://doi.org/10.1016/bs.adcom.2022.02.007
279

Abstract

Today, development in IoT, Information and Communication Technology (ICT) and Wireless Sensor Networks (WSNs) have the potentiality to turn out few economic, environmental, and technical challenges along with opportunities in the ecosphere. It generates a large amount of data with numerous modalities, temporal and spatial variations. It is mandatory to develop a higher knowledge level based on an intelligent system to analyze this big data for accurate decision making, estimating, and dependable sensor management. There are several applications based on IoT with machine learning techniques that are globally available. This chapter focuses on advanced techniques used in smart agriculture systems based on IoT and machine learning algorithms. Several types of research are completed on this system to offer smart services for real-time monitoring of any agricultural environment. The IoT based smart agriculture systems are the most gifted approach to enhance the productivity of food items by reducing power and water consumption.

Abbreviations

IoT   Internet of Things
ML   machine learning
SA   smart agriculture
WSNs   wireless sensor networks (WSNs)

1. Introduction

The IoT paradigm has progressed to be one of the largest advancements in the technology of the new age science in the past few years [1]. It combines the advantages of several pre-existing technologies such as the Wireless Sensor network, cloud computing, RF identification, middleware components and end-user applications. IoT enabled devices like tablets, computers, and mobile phones can access information on the surroundings and similar objects without any intervention by humans. Among the key features of IoT are RFID and WSN. A Wireless Sensor Network is meant to monitor the environment it is deployed in and records the physical conditions with relatively cheaper data acquisition techniques and is composed of spatially dispersed nodes. Internet of Things (IoT) is a rapidly developing technology all over India. But, the majority of the population (around 70%) here depend on agriculture.

Around 60% of the land can be plowed and used to grow rice, potato, wheat, onion, tomato, mangoes, sugar cane, bean, cotton, cereals, etc. One unfortunate fact would be that farmers are still predominantly dependent on the traditional methods and techniques that have not evolved and were being used several years ago. This could be a major reason for the lower yield of crops. Also, several contributors might lead to lower crop yield such as seed rate, proper soil preparation, seed cultivar, lack of sufficient amount of moisture in the field, the difference in sowing time, salinity, waterlogging, not applying necessary fertilizers, protection of plants, not adopting modern technologies and machinery, misleading marketing strategies and lack of solid investment. Apart from these factors, farmers suffer huge financial losses due to the incorrect usage methods for irrigation, pest and insect control, and the prevention of plant diseases. Improper usage and miscalculation in the quantities of insecticides and pesticides and wrong weather forecast further contribute to their losses. To get a higher crop yield, monitoring would be the key task for the farmers. Because of the existing constraints on agriculture, we need to take immediate measures to develop enhanced and economically viable strategies to cultivate crops. This chapter proposes an Internet of Things (IoT) based, user-centric architecture for helping farmers make use of the advancements in technology to improve their crop yield and increase profits in selling their products.

The IoT application in agriculture is meant to empower farmers with the necessary tools to make better decisions and employ automation techniques that effortlessly combine knowledge products and several other services that will help in generation productivity, profit, and quality [2]. Using sensors parallelly with intelligent algorithms can make smart recommendations regarding field quality maintenance to yield high-quality crops.

2. Existing infrastructure in agriculture

The existing infrastructure in the agricultural domain can be categorized based on the kind of work it performs, such as processing, production, consumption, distribution and management of waste, excluding the cyber-infrastructure and the development of the workforce [3]. An instance could be that the infrastructure related to food production includes farming equipment like combine-harvesters, tractors, irrigation infrastructure and duster aeroplanes. In the processing of food, infrastructure includes processing plants and food factories. The distribution of food infrastructure comprises transportation systems such as trucks, railroads, barges, ships, etc., retail stores and supply chains. Apart from the different infrastructures mentioned, agriculture also requires energy and water infrastructure. The water infrastructure composes the pumping systems, natural and built structures for the storage of water, networks monitoring water management and governance. The infrastructure related to energy composes the electric grid, biofuel production, and carbon-based energy production such as fossil fuels.

2.1 Limitations in current agriculture infrastructure

The current infrastructure mentioned has several challenges and limitations that need addressing to achieve intelligent agriculture infrastructure. The limitations are illustrated below in Fig. 1.

2.1.1 Social limitations

Some social challenges include the aging workforce, shortage of labor, and the lack of urban community's involvement. Presently, the various kinds of crop production demand high labor. Also, it has been noted that the average

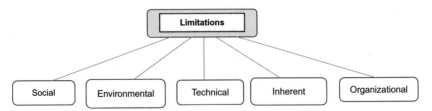

Fig. 1 Limitations of current agriculture.

age of a farmer is around 60 years. With these facts made available, it is evident that this domain will face labor shortages in the upcoming decade.

2.1.2 Environmental limitations

The pollination factors such as bees and butterflies declining rapidly is a simple example of an environmental challenge that a farmer faces in his daily life. Unpredictability in climatic conditions is also one huge factor that hinders this domain.

2.1.3 Technical limitations

Technical Limitations include the inadequacy of cyberinfrastructure since the internet bandwidth is limited in predominantly remote farming areas. The GPS services used for placing the farming equipment precisely are vulnerable to spoofing and jamming. Apart from the problems already stated, there is inadequacy in the mechanisms that share agricultural data. This data is sensitive and needs to be shared in a highly protected manner. The differential privacy, the best-of-the-breed protection methodology, is insufficient for this data because of its spatiotemporal nature. Due to this, farmers are widely reluctant to share their data. The remote sensing satellites provide coarse resolution images, not frequently monitoring the crops or the farming area, introducing a delay in the condition detection and preventing us from taking corrective measures early. Automation levels vary significantly in the farming sectors. For instance, certain farms like pecan, corn and wheat have a much higher automation level than the vegetable and fruit crops. Manual interventions are also frequently needed since the fruits or vegetables do not become ripe together at a particular time. Hence, picking them up periodically as it ripens becomes necessary.

2.1.4 Inherent limitation

Farmers manually predict when to apply fertilizers and how to manage land. These predictions are often prone to errors since they are based on approximations and assumptions.

2.1.5 Organizational limitation

The supply chain lacks visibility and transparency. This is one of the major organizational limitations.

Having mentioned the limitations that exist in the various domains within the agricultural sector, it is time that we build smart solutions for we now have the power of technology and educated youths who can contribute wisely towards the growth of this sector. A good broadband and internet connection to these areas could be among the starters.

2.1.5.1 We still get food and the necessary things to make our living. Does that mean the agricultural sector is functioning well?

In Assam a majority of state's population, almost 90% of an estimated 33.4 million in 2001, live in rural areas where the mainstay of business is production agriculture. In terms of the domestic production (SPD), the agriculture sector contributed over 40% of the state income in 2000–01. Assam is far behind in the use of modern agricultural technology to improve its agricultural productivity compared to the rest of the country. For example, the agricultural productivity index for Assam was 201 in 2010–11 compared to 179 in India.

2.1.5.2 What could be the cause?

There is still slavery in Assam and farmers wo constitute the majority of the population are faced with poverty. Their lack of knowledge about modern technology, illiteracy, lack of knowledge about market demandable agricultural commodities, irrigation system, flood, drought etc. are the primary reasons for why the productivity is still lower in that region. Let us now see what could be done to increase the productivity and make the farmer's life better.

2.2 Intelligent workforce infrastructure

To improve the agricultural industry and bring them at par with the other sectors, it is essential to: [4–6]

(1) Train farmers frequently and get them updated on the recent technologies so that they are ready to embrace any advancements that will contribute to the improvement in crop yield or crop quality

(2) Find methods to involve the next generation in this sector

(3) Find ways to involve employees from various engineering sectors such as computer science and mechanical mining into the agricultural sector and

(4) Go beyond the traditional methods to engage citizens.

To achieve this and to bring youngsters into this domain:

(4a). We need to find ways to transform agriculture into a "cool" concept in the minds of the forthcoming generations. By introducing modern technology like drones and robots and highlighting their application in various stages of farming, we can gamify the entire process to attract young minds.

(4b). Small workshops and teaching events can be held to introduce the farmers to these new technological tools and give them hands-on experience as to how to use the tools.

(4c). Augmented and Virtual reality can also be exploited when it comes to the agricultural domain. These platforms could be built in a cost-effective and scalable way to make the training process fun and realistic. Furthermore, associating the process of farming with video games can help involve young minds. This could also be a factor that triggers various ideas and solutions to existing problems in the agricultural domain.

(4d). we need to develop teleoperation facilities to engage more people in labor-surplus areas on the farm. This will help us bridge a gap in the economy by bringing unemployed people from the labor shortage areas into the areas where intense labor is needed and places where there is a workforce shortage.

3. IoT ecosystem—A complete view

This section is meant to provide an insight into the components that make up an IoT ecosystem [7]. The five major components as illustrated in Fig. 2.

- IoT devices
- Communication technology
- Data storage and processing
- Knowledge Layer
- Service Layer

3.1 IoT devices

IoT is a collection of sensors and actuators that interact through a wireless medium. They are embedded devices programmed to carry out a particular task.

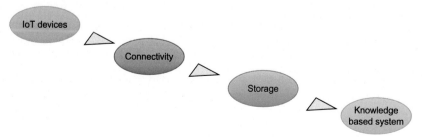

Fig. 2 IoT ecosystem—components diagram.

They consist of the FPGA or microprocessors and have input and output interfaces to take in signals and send out signals. Communication modules are responsible for handling the commutations within the network and sending the signals. The information that these sensors gather is not only sensory information. They monitor the field on various factors like soil moisture, rainfall, wetness in the leaves, speed of the wind, air temperature, and other essential factors for the healthy growth of plants.

3.2 Communication technology

For a successful deployment of the IoT systems, Communication technology plays a key role. It is responsible for the data transmission between the devices connected as a network. They can support either the IoT sensors or the backhaul network. While IoT devices transmit lesser data over very short distances, the backhaul networks transmit data over long ranges and at a very high speed with little power consumed. Some technologies even provide bi-directional communication. This includes error correction techniques, handshaking signals, data reliability, and encryption. The type of topology of the network also influences the kind of communication technology deployed within it. Some of the commonly used topologies are peer to peer, star, and bus topology.

3.3 Data storage and processing

We know that an IoT ecosystem comprises the collection of the ever-changing data and analyzing them to ensure that the environment in which the crops grow is healthy and crops give their maximum yield. The data collected is highly unstructured and can be of any form like audio, video, text—one prominent location for the storage of these data that are collected is the cloud. The data are mostly analyzed in the data using fog or edge computing. The gateways within the devices perform the necessary computations and analyze them to lessen the latency associated with the cloud and reduce the cost of computations. Users are also an important source of data that we do not want to miss out on. The information gathered through their circles, either professional or casual, are valuable resources to the domain. Their circles can also be their place for information sharing and help in leveraging the delivery of services. Many different management information systems have aided in effectively managing this information. Some examples are Far mobile, On-farm systems, Crops, Easy farm and so on. These provide a platform for managing and storing farm data.

3.4 Knowledge layer

Efficient farming mechanisms are the need of the hour. It requires methods that consume minimal resources but yield maximum output. Essential parameters need to be monitored to ensure an effective outcome. The data collected needs to be stored, and then these relevant parameters need to be analyzed. To send efficient warning messages, it is required that appropriate parameters are chosen and monitored. Data respective to each kind of crop have their set of conditions for relevant warning messages to be sent. To ensure that resources are spent appropriately, measures need to be taken considering the environmental conditions.

3.5 Service layer

This layer is associated with the end-users. Three major services are provided to the stakeholders. They comprise the fertilization schedule, irrigation schedule, and an app to analyze the crops.

4. Agricultural monitoring system based on sensors

An approach to ensure that crops receive what is essential for their healthy growth is called precision agriculture [8]. It has been devised to enhance crop productivity and improve yield. The goal is to ensure sustainability, profitability, and environmental protection. Precision agriculture is also known as satellite agriculture. Satellite is used to monitor the crop, and this farming method is site-specific. Modern farmers have evolved and made use of smart farming tools. Researchers working in precision agriculture have come up with concepts that engage modern technologies in everyday farming practices. With the deployment of these tools and ideas, we can effectively monitor the plants in terms of different parameters mentioned in the previous section. Prototype with maps of the crop fields accelerates the planning and monitoring process. They also give the farmers and the planners an idea of where to deploy the sensors and what topology to implement.

4.1 Field assessment

As mentioned as an inherent limitation, various predictions and irrigation methods are made by farmers manually based on assumptions and experience, if any. To overcome these limitations, low-resolution cameras and sensor networks are deployed. Solar power nodes that monitor the moisture level are used to collect data that are then analyzed. The captured data, such as the greenness of the crops, are then sent to the base station. The base station sets aside some time for the member node to transmit data to it. Cattle are also monitored using similar techniques.

4.2 Cattle behavior control

Monitoring and handling animals has always been a challenge. Their behaviors are not easily predictable, and their mental or positional state is difficult to calibrate. Behaviors can be a consequence of various factors such as the climate, temperature, kind of food, and similar factors. Animals are active predominantly during the daytime than at night. Their activity images such as sleeping and movement are sent for calibration. Inertial sensors collect information like the speed of movement of the animal, turning rate and so on. These help us identify their behavior.

4.3 Traditional agricultural monitoring

Tension meters and drip irrigation systems are employed to monitor the irrigation in the farmlands. Fuzzy logic controllers are employed to monitor independent crops [9]. The fuzzy logic controller has been implemented for an efficient irrigation system for fields with different crops. This increases the accuracy of the collected values and accordingly aids in making better decisions. Modern agriculture that is based on greenhouse concepts needs to be accurately controlled to monitor temperature and humidity. The atmospheric conditions of plants inside the greenhouse vary from place to place, making it difficult to maintain uniformity at all places in the farmhouse manually. So, GSM has been used to report the irrigation status to farmers' mobile handsets. The sensor-based irrigation monitoring system is divided into the bottom and upper layers, as shown in Figs. 3 and 4. Hierarchical sensor networks are placed in the bottom layer, where nodes are placed in widely separated clusters. These nodes send the data to the Base station (BS) connected by a Wireless LAN that holds the data logger software. The upper layer consists of five modules: the "acquisition module, network management module, alarm/network status display module, decision making module and business module." Real and non-real-time data are collected using the data gathering modules from the sensor network stored in a database for decision/alert notifications. Alert notifications or displaying information to the end-users are carried out by the Alarm/Network status display module. The Alarm/Network status display module acts as an access point between end-users and other modules/networks.

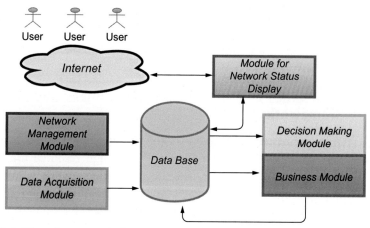

Fig. 3 IoT based smart agriculture monitoring system.

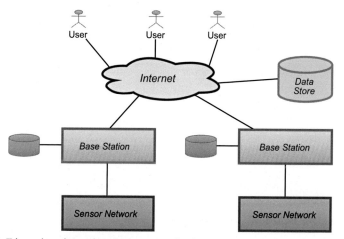

Fig. 4 IoT based traditional agriculture monitoring system.

The network management module performs the conditions of networks such as localizations, collision, and network configuration. This developed system has introduced the concept of M2M communication, where water and energy are conserved by using an intelligent sensor network and efficient routing protocol.

4.4 Intelligent IoT based irrigation system

The existing Agricultural monitoring system has employed wireless sensors for monitoring the soil condition for irrigation. Also, some of the systems have employed mobile handsets for delivery. Intelligence does not exist in any of these systems to analyze the real-time data based on experience for irrigating the field. Most of the system captures the data from the field and accordingly controls the sprinkler valve for watering the field.

5. Difficulties in sensor-based agribusiness observing frameworks

This sensor-based farming observing framework should contain a choice assert framework to expand creation, redesign resource use, and decrease natural risks [10–12]. Furthermore, this framework would be autonomous, sensible (by poor farmers), precise and remotely controllable. The time of this sensor-based farming observing framework must be as a base one yield per season (4 - a half years). Anyway, the sensor-based farming observing framework had a high likelihood in traditional agribusiness in rising nations.

5.1 Cost

Cost is the utmost significant factor on account of emerging nations. Cost can be diminished by methods for sensor-based farming framework communication networks that are not currently being used in agricultural monitoring systems. Moreover, this single framework of the sensor-based farming observing system can be utilized in multiple applications and exactness horticulture, requiring thickly conveyed camera and sensor hubs.

5.2 Reliability

Reliability on sensor and camera hubs regardless of the atmospheric conditions in most creating nations is extraordinary and rare. Subsequently, sensor hubs must be covered to shield from open-air conditions, including heat and moisture. One significant example is the accessibility and reliability of media communications and the remote infrastructure in rustic territories of creating nations being extremely troublesome.

5.3 Resources

All over the world, the vast majority of poor farmers depend on downpour to cultivate for the creation of nourishment. In addition, conventional water system philosophy isn't efficient in semiarid zones in creating nations. Thus, a sensor-based agriculture monitoring irrigation system with a decision support system is required to maximize irrigation and food production while minimizing the intake of good water.

6. Factors affecting climatic changes in savvy agribusiness

Environmental change will impact crop dissemination, creation and increment dangers related to cultivating. All-inclusive, environmental change is relied upon to diminish grain creation by 1% to 7% by 2060. In any event, 22% of the developed region under the world's most significant yields is anticipated to encounter negative effects from environmental change by 2050. Vulnerabilities in atmosphere systems could likewise impact ranchers' decisions and whether they will put resources into fundamental sources of info and assets for their territory.

Let us currently take a gander at a little nation to know why we need atmosphere brilliant farming.

The Republic of Bulgaria is arranged on the eastern Balkan Peninsula. The fundamental climatic attributes of Bulgaria are: calmly mainland and subtropical (in the south) atmosphere with four seasons and high variety in the temperature, precipitation and moistness among the nation districts. Mountains spread 60% of the nation region as the streams are short, low-water and unevenly apportioned through the nation. Because of the mainland atmosphere the mid-year in Bulgaria is hot and the winter—dry and cold. There are droughts in summers in July and August. The measure of precipitation is commonly low with varieties among the districts.

West and upper east breezes rule and, in the winter, there are solid north and upper east breezes. In light of the solid and consistent breezes, the snow spread is frequently overwhelmed from the level territories and the dirt gets frozen. In end, Bulgarian agrarian creation is downpour taken care of, significantly relies upon precipitation systems and atmosphere changes are a significant factor for rural advancement of Bulgaria.

You may think with such extraordinary conditions why intend to develop crops? Be that as it may, Farming assumes a critical job for the economy in Bulgaria. About 5% of GDP and 17.2% of complete fare of the nation in 2009 were given by agribusiness.

Presently it sure isn't anything but difficult to simply stop farming for them.

So what do the individuals in provincial territory accomplish for the living when the atmosphere isn't favorable to them?

The following segment gives them one arrangement.

6.1 Climate-brilliant horticulture

Research and strategy interfaces between environmental change and farming have propelled. "Brilliant atmosphere horticulture" has risen as a structure to catch the idea that rural frameworks can be created and actualized along with improving nourishment security and country employments, encouraging environmental change adjustment and giving moderation benefits. Atmosphere brilliant horticulture incorporates huge numbers of the field-based and ranch based economic agrarian methods that the board rehearses in writing and puts it to wide use, for example, preservation culturing, agroforestry, building up the board, and others.

6.2 Key highlights of atmosphere shrewd rural scenes

Atmosphere keen horticultural scenes work on the standards of the incorporated by the board while unequivocally coordinating adjustment and relief into their administration destinations. An evaluation of environmental change elements identified with horticulture proposes three key highlights that portray a keen atmosphere scene: shrewd atmosphere practices at the field and ranch scale; a decent variety of land use over the scene; and the board of land use communications at scene scale.

6.3 Climate-keen practices at field and homestead scale

Atmosphere brilliant scenes have contained an assortment of field and ranch rehearsals, in various land and residency types, that help both adjustment and relief destinations. These practices incorporate soil, water optimization and supplement the board alongside agroforestry, domesticated animals, farming, woodland, and field procedures. Progressively, effective water administration and assets undermined by environmental change is likewise basic for arriving at the adjustment and occupation objectives of atmosphere smart farming. Best practices for water systems, water-reaping innovation, and farm cultivating frameworks can improve water-use proficiency and preservation. Especially in semi-dry and dry areas, where water assets are a worry, interest in the water system builds creation, lessens inconstancy, and may spike extra interest in horticulture. Improved plan, development procedures, and water conveyance instruments can reduce the exceptionally high GHG emanations related to customary water system frameworks. Utilizing coordinated supplements, board standards like green composts, planting nitrogen-fixing crops, and consolidating animals' fertilizers into the dirt, diminish the measure of nitrogen lost to overflow and outflows of nitrous oxide. Applying these administration standards can serve adjustment needs by improving soil quality while diminishing ranchers' expenses and reliance on outside sources of info. Domesticated animals are especially basic for atmosphere savvy horticulture. Improved field and meadow executions including rotational nibbling, recovering vegetation and reestablishing corrupted land, will be basic for the versatility of environmental change. They additionally add to relief through carbon sequestration in profound established vegetation and soils. For better fertilizer, changing over excrement to biogas gives the additional advantages of an elective vitality source with less negative wellbeing impacts from cooking, warming, and lighting. Improved feed blends and healthful enhancements can diminish methane outflows; be that as it may, this is progressively practical at bigger activity sizes.

6.4 Diversity of land use over the scene

The second component of brilliant atmosphere scenes is an elevated level of decent variety. This incorporates land spread, land use, species, and a varietal of assorted plants and creatures. Assorted variety have fewer atmosphere relief and adjustment capacities: (1) to decrease dangers of creation and vocation misfortunes from whimsical and unforgiving climatic conditions; (2) to use zones of the scene deliberately as crisis nourishment, feed, fuel, and salary stores; and (3) to continue to insignificantly upset natural surroundings inside the scene mosaic that likewise fills in as carbon stocks.

6.4.1 Reduce hazard

A decent combination of land use and species can decrease natural dangers related to a homogeneous yield spread, regarding vermin, illnesses and powerlessness to sudden climate conditions. Improving hereditary assorted variety on ranches by expanding the number of various harvests developed or the number of assortments of similar yields, additionally gives significant atmosphere adjustment and decreases the hazards to the executive's benefits. Yield hereditary decent variety improves the odds that a few assortments will fit shifts in temperature, precipitation, and saltiness systems brought about by environmental change. Additionally, having an arrangement of assorted nourishment and salary sources, from crops, domesticated animals, trees, and non–developed terrains can pad family units and networks from climatic (and other) mishaps.

6.4.2 Provide key nourishment and feed saves

Occupation flexibility of family units and networks can likewise be upgraded through access to various wellsprings of nourishment, feed and work during scenes of unfavorable climatic conditions. Wild plant species in ranches, backwoods, savannahs and wetlands contribute to the eating regimens of huge numbers of the poor in creating nations and these nourishment sources, especially the "starvation nourishments."

6.4.3 Sustaining seasonal forest as a carbon resource

In today's supreme cultivation system, we are mostly involved with annual plant species. The topography which maintains other lands such as endless grasslands, woodlands, forests, or wetlands enhances the environmental flexibility. In turn, it also improves the living surroundings for local people and bio diversification. To cut off the carbon emissions from the topographies, preserving land areas in these types of eternal systems is mostly required.

6.4.4 Effective functions of the ecosystem

To preserve the ecological connectivity between water and nutrients and improve the living conditions of wild animals, plant groups, and microorganisms favorable to humankind, natural living means like riverside areas, woodlands and wetlands should be set and maintained. A fundamental transformation approach is required to have the relation between freshwater resources and wildlife localities when the variations in climate increase. The methods used in achieving the cultivation productions should help us maintain this relationship instead of obstructing it. For domestic purposes like household and irrigation, rainwater can be preserved and used. The constructive methods of managing the crop waste and water are necessary across the river basins to control the ailments in humans and animals.

6.4.5 Improve the advantages of environmentally smart policies on the ground

The preparatory measures of topographical elements boost farming productivity. This farming productivity depends mainly on the lands used. The surrounding lands should have a nature to hold the insects and pest predators, which are beneficial to the parts of the land nearer to the forest helping the agricultural land in sustaining the pollination. To regulate the downstream fisheries, the aliments and sediments from the farmlands can be used. While livestock, forest production, and the upstream crop can be maintained to enhance water regulation for downstream irrigation, livestock wastes can be a replacement for nonrenewable fuel sources in fulfilling the local agro amenities.

7. AI in agriculture—An introduction

AI has been practiced in almost all fields, from agriculture to medical sciences. With machine learning it is possible to train machines to the extent where they can predict the output without human intervention, and it is a subdivision of AI [13,14]. Many applications of AI can be used in the farming industry, such as agricultural robots, which work faster than the human rate of activity, crop & soil monitoring, and predictions on climatic changes.

The assistance of AI in the agriculture field results in healthier crops, monitoring soil temperature and humidity, and many such tasks. We get large amounts of data from the fields every day; analyzing such data using different machine learning algorithms helps make better decisions. Implementation of drones in farms using IoT, wireless technology integrated with sensors as per the application requirement gives the data about crop condition, spraying pesticides etc.

Decisions are made based on the data collected from sensors, and the machine learning algorithms analyze that data. Integration of IoT and AI monitors every step of crop growth.

This makes the farmers predict the yield in advance based on the information from the ML algorithm regarding climatic conditions and drastic weather changes that may result in natural disasters helping us take certain measures to save the crops and farms.

8. Machine learning techniques for smart agriculture

The ability of a machine to learn from the previous results is possible only with Machine learning. Its algorithms use statistical methods to train themselves from the dataset. The effectiveness of an algorithm depends on its performance, i.e., how well it can adjust with the increasing number of samples during training. The data which comes from the sensor contains non-linearity in data, the usage of Machine learning techniques can take such problems to care. Thus, we can expect the right decision making with less human interference. Estimation efficiency is influenced by data handling, reference model presentation, and the correlation between source and reference variables.

8.1 Wide division of machine learning algorithms

This division includes supervised and unsupervised learning. Supervised learning uses a defined set of labeled data to train a model to predict the target variable for the sample data. The purpose of the unsupervised learning approach is to find hidden patterns. It is mainly practiced in applications with no particular purpose or where the information present in the data is unclear. This is also suitable as a method to reduce the dimensionality of data with a variety of applications. These wide divisions are mostly used in smart technologies like IoT in a variety of domains. For example, WSN, IoT allows smart farming where ML techniques evolve to measure and know the big data in this field. The management deals with the estimation of yield, weed identification along with livestock and phenotype classification.

9. Artificial neural network (ANN)

An artificial neural network is an information processing system with a performance similar to a biological neural network. Chains of decision units such as perceptron's or radial basis functions are cascaded to construct this learning algorithm mainly used to recognize non-linear and complex functions. The neural network is mainly distinguished based on (1) its architecture (pattern of connections between the neurons), (2) its algorithm (method of determining the weights on the connections), and (3) its function of activation. The architecture of the ANN algorithm is designed with input units, single or multi-layer hidden units, and output units. ANN can also be used to solve the problem of classification and regression. ANN learning algorithms implementation include the radial basis function, perception algorithms, back-propagation, and feed-forward propagation.

9.1 Artificial neural networks in agriculture

The agriculture sector is often incorporated with Artificial neural networks due to its superiority over traditional systems. A neural network takes the benefit of predicting and forecasting its result based on parallel reasoning. Moreover, neural networks can be trained instead of thorough programming.

10. Automation and wireless system networks in agriculture

Embedded intelligence (EI), the result of emerging research in the automation field, made inventions and breakthroughs that made the agriculture sector adapt to it. Smart farming, smart crop management, smart irrigation and smart greenhouses are some of the outcomes of embedded intelligence in the agriculture sector. For a growing nation, it is necessary to include these growing technologies in the agriculture sector as many sectors are interdependent on agriculture.

10.1 Smart agriculture

The process of producing food and fiber to meet the fundamental need is called Agriculture. Agriculture is limited by disputes between nations and environmental limitations throughout history. Nowadays, an increase in global warming leads to changes/imbalances in the climate conditions, which in turn impacts agriculture; as a result, sustainability and the increase of agricultural production is possible only with the collection of more data and analyzing the use of them. Detection and transmission of necessary data can be possible only with the efficient use of information technology. Even though different productivity is obtained from different regions of farms due to different soil structures, it aims to distribute the inputs such as fertilizers and chemicals used to grow crops based on farms' needs. However, the concept "Sustainable Agricultural Production" aims to protect the environment and the natural resources for the past 2 decades, which gives special importance to using these inputs in as little amount as possible and with as much care as possible. In this situation, variation in agricultural production needs to be measured, and the inputs need to be applied only after considering the results. To be successful in this process called precision, agriculture reliability and continuous data is needed.

10.2 Smart farming (SF)

Incorporating communication and information technologies into equipment, sensors and machinery for use in agricultural production systems, grant to generate a large volume of data and information with insertion of automation into the process, result in Smart farming. New and leading technologies like the Internet of Things (IOT) and cloud computing are expected to accelerate this development by introducing more artificial intelligence and robots into farming. Smart farming relies on making decisions from analysis of various farm data transmitted and concentrated in a remote storage system. SF originated with the combination of computer science and software engineering and arrived with computing technologies and the transmission of data from agriculture, with the overall computing environment. These elements of computing are embedded in objects and linked with each other and the internet. The SF field includes other terms with similar meanings, such as smart agriculture. Farm management information systems in the agriculture field presented ideas for technologies like precision agriculture and management information systems, which helped in overlapping technologies and interfaces accordingly for SF. The use of sensors in agriculture made the tools of SF achievable. A sensor is an electromagnetic device that collects measurable quantities from the environment and converts them to signals that an instrument can read. Sensors read measurements like temperature, humidity, light, pressure, noise levels, presence or absence of certain objects, mechanical stress levels, speed, direction, and object size.

10.3 Smart IoT agricultural revolution

Intelligent Agricultural Revolution refers to using and integrating the latest technology such as IOT more widely in agriculture to increase the quantity and quality of harvests of domestic crops. Through this system, farmers can monitor and control agricultural activities from anywhere using a smartphone through a specially created application. Through IOT and smart agriculture, we invite young farmers involved in agriculture. It does not involve brute strength but is entirely driven by the technology at our fingertips. For example, the drone can be used to spray insecticides, analyze and monitor the results of soil cultivation of plants quickly and without using a workforce. In addition, sensors or sensor-based Internet of Things (IOT) can also transmit data or information related to the plant immediately (real-time) for further action by the farmers.

10.4 Smart system design methodologies

This section gives the details about the design and implementation of the Smart Agriculture System. People interested in agricultural development for more production and faster results about the humidity of the soil, air pressure, temperature, and location of sensors can use this system. Smart Agriculture System uses the communication between Arduino and Raspberry Pi, which implements the Machine-to-Machine technology. Communication using devices like wireless sensors is preferred. With the help of the internet, data is transferred by sensors to Android-based mobile devices. As a result, the Smart Agriculture System is a combination of software and hardware components.

11. Hardware components in the smart agriculture system

11.1 Humidity sensor

Humidity sensors sense the relative humidity of the environment in which it is placed. They measure both moisture and temperature in the atmosphere and express relative humidity as a percentage of the ratio of moisture and temperature in the air to the maximum amount held in the air at the current temperature. As the hotness of air increases, the relative humidity changes because it holds more moisture.

Most humidity sensors use capacitive measurements to determine the amount of moisture in the air. This type of sensor reads measurements by placing two electrical conductors and a non-conductive polymer with a film between them to create an electric field between them. Moisture from air assembles on the film and creates a voltage difference between those two plates. This change is then converted into a digital measurement.

11.2 CO$_2$ sensor

A carbon dioxide sensor or CO$_2$ sensor is used to measure the quantity of carbon dioxide gas in the air in which it is placed. The most common principle of CO$_2$ sensors uses infrared gas sensors (NDIR) and chemical gas sensors. The CO$_2$ sensors measure CO$_2$ levels by observing the number of infrared radiations being observed by CO$_2$ molecules. Similar to CO$_2$ sensors, a carbon monoxide sensor or a CO sensor detects the presence of carbon monoxide (CO) gas to prevent the environment from getting poisoned.

11.3 Moisture sensor

The Moisture sensors are used to detect the amount of moisture in soil or judge if there is water around the sensor. As soil moisture measurement plays an important role in agriculture, it helps farmers manage their irrigation system more efficiently. By knowing the exact condition of soil moisture in their fields, farmers reduce water usage and increase the yield and quality of the crop.

11.4 Rain drop sensor

The Raindrop sensors are used to detect rain. A raindrop sensor is a board on which lines of nickel are coated. The simplicity of raindrop sensors made it easy to detect rain. It can also measure the rainfall intensity and is also used as a switch when raindrops fall on the raining board. It uses the principle of resistance. Raindrop sensors also measure the moisture via analog output pin and provide a digital output when the moisture exceeds the threshold. Rain sensors come under the classification of weather sensors. It takes advantage of detecting water beyond a humidity sensor can detect.

11.5 Ultrasonic sensor

Ultrasonic sensors use sound waves to measure the amount of water left inside a water tank. This helps farmers in making further arrangements for spare water tanks avoiding risk.

11.5.1 Implementation using raspberry PI 2 model B

A sensor network has tiny objects installed in different wireless sensor network monitoring areas to measure various physical data to finish a specific task. Improvement of growth in various crops depends on environmental parameters like light intensity, soil moisture, relative humidity, soil temperature, usage of fertilizers and pH, etc. Any minor changes to these parameters can cause problems like improper growth of crops and diseases in plants, etc., resulting in less crop yield.

Fig. 5 shows the proposed smart agriculture, consisting of a transmitter section (i.e., mobile controlled robot) and a monitoring section. The transmitter section consists of Raspberry Pi 2 Model B, various sensors such as humidity sensors, pH sensors, Thermohygro sensors, CO_2 sensors, Soil moisture sensors, Obstacle sensors, Power supply section (i.e., solar plate), ZigBee transmitter, a Wi-Fi modem, Water sprinkler, Dc motor for spraying insecticides, an LM380 audio power amplifier, speaker, Liquid crystal display (LED), and camera. At the same time, the monitoring section consists of an Android smartphone, ZigBee receiver and a Laptop with the application language. Raspberry Pi 2 Model B recommends Raspbian as its

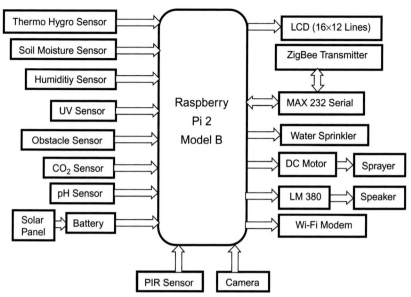

Fig. 5 Implementation of the proposed smart agricultural system.

operating system, free OS based on Debian, optimized for Raspberry Pi hardware. It consumes low power with a high-performance controller for interfacing various sensors and performing the required task based on the written program.

12. Use cases

12.1 Solar fertigation: Internet of Things architecture for smart agriculture

Using the industry 4.0 framework, Smart agriculture is a good Internet of Things (IoT) application domain. Furthermore, change in climate affects the natural resource exploitation policies becoming a serious matter in agriculture and food production. Therefore, monitoring environmental parameters and agricultural processes continuously helps in resource optimization and maximization of food production. This work proposes the conceptual model and the design of Solar fertigation, an IoT system specifically designed for smart agriculture. In particular, the envisioned solution can detect some of the most meaningful terrain parameters to feed a decision-making process that drives automated fertilization and irrigation subsystems. In addition, Solar fertigation is powered by a photovoltaic plant to attain energy self-sustainability.

12.2 Wireless sensor based crop monitoring system for agriculture using Wi-Fi network dissertation

The process is coupling the sensor devices with wireless technologies to monitor the important parameters of Indian Agriculture such as temperature, humidity and moisture. Their idea is to have wireless sensors connected through Wi-Fi to a Central Monitoring Station through General Packet Radio Station. In addition to that, it also connects to the Global Positioning System (GPS) to send messages to the central monitoring station. Further, they also has an external sensor such as soil moisture, pH and leaf wetness. Depending on the values from the sensors, such as soil moisture, it will turn the water sprinkler on or off. If the pH sensor senses any information, it will be sent to the base station to inform the farmer using a GSM modem to take further action.

12.3 Secure smart agriculture monitoring technique through isolation

Using mobile and fixed sensors and mobile devices like smartphones and tablets, the farmers can collect data in various formats regarding crops, soil, and weather, allowing them to monitor their crops and access their data effortlessly. The data collected are sent to the core cloud platform, where they are processed and analyzed using specific algorithms. The result generated is sent back to farmers to improve their agriculture process and allow remote actuating of the irrigation system. The same devices, sensors, and actuators are also shared by other stakeholders like disaster early warning systems for efficient real-time management, and therefore they must be controlled with security.

12.4 Realizing social-media-based analytics for smart agriculture

When a user uploads a query in social media where other users can post their solutions, along with possible diseases names, images, and features of the solution, like the time taken by disease to get cured and the cost of resolution, the text used by the users can be unstructured; thus the features can be mentioned anywhere in the text. In this case, NLP techniques are used to extract those features to identify the disease based on the names suggested by community members. It also considers the image posted by the user and evaluates the merits of the solutions posted by the members to select a list of possible solutions. Later, we can extract the solution, and a ranking model sorts the solution by considering trusted DB and relevant extracted features of the solutions.

13. Conclusion

The growth in the global population is compelling a shift towards smart agriculture practices. This coupled with the diminishing natural resources, limited availability of arable land and increase in unpredictable weather conditions, makes food security a major concern for most countries. As a result, the use of Internet of Things (IoT) and data analytics (DA) can enhance operational efficiency and productivity in the agriculture sector. As a result, there is a paradigm shift from using wireless sensor networks (WSNs) as a major driver of smart agriculture to use IoT and DA. The IoT integrates several existing technologies like radio frequency identification, WSN, and end-user applications. We presented the IoT ecosystem and how the combination of IoT and DA is enabling smart agriculture. Furthermore, we provide future trends and opportunities categorized into technological innovations, application scenarios, business, and marketability.

References

[1] J. Guth, U. Breitenbücher, M. Falkenthal, P. Fremantle, O. Kopp, F. Leymann, L. Reinfurt (Eds.), A detailed analysis of IoT platform architectures: concepts, similarities, and differences, in: Internet of Everything, Springer, Singapore, 2018, pp. 81–101.

[2] J.-C. Zhao, J.-F. Zhang, F. Yu, J.-X. Guo, The study and application of the IOT technology in agriculture, in: 2010 3rd International Conference on Computer Science and Information Technology, vol. 2, IEEE, 2010, pp. 462–465.

[3] X. Yu, W. Pute, W. Han, Z. Zhang, A survey on wireless sensor network infrastructure for agriculture, Comput. Stand. Inter. 35 (1) (2013) 59–64.

[4] E. Peres, M.A. Fernandes, R. Morais, C.R. Cunha, J.A. López, S.R. Matos, P.J.S.G. Ferreira, M.J.C.S. Reis, An autonomous intelligent gateway infrastructure for in-field processing in precision viticulture, Comput. Electron. Agric. 78 (2) (2011) 176–187.

[5] T. Kalaivani, A. Allirani, P. Priya, A survey on Zigbee based wireless sensor networks in agriculture, in: 3rd International Conference on Trendz in Information Sciences & Computing (TISC2011), IEEE, 2011, pp. 85–89.

[6] K.R. Dabre, H.R. Lopes, S.S. D'monte, Intelligent decision support system for smart agriculture, in: 2018 International Conference on Smart City and Emerging Technology (ICSCET), IEEE, 2018, pp. 1–6.

[7] S. Bansal, D. Kumar, IoT ecosystem: a survey on devices, gateways, operating systems, middleware and communication, Int. J. Wirel. Inf. Netw. (2020) 1–25.

[8] K. Lakshmisudha, S. Hegde, N. Kale, S. Iyer, Smart precision based agriculture using sensors, Int. J. Comput. Appl. 146 (11) (2016) 36–38.

[9] S. Pooja, D.V. Uday, U.B. Nagesh, S.G. Talekar, Application of MQTT protocol for real time weather monitoring and precision farming, in: 2017 International Conference on Electrical, Electronics, Communication, Computer, and Optimization Techniques (ICEECCOT), IEEE, 2017, pp. 1–6.

[10] P. Nayak, K. Kavitha, C.M. Rao, IoT-enabled agricultural system applications, challenges and security issues, in: IoT and Analytics for Agriculture, Springer, Singapore, 2020, pp. 139–163.

[11] A. Saidu, A.M. Clarkson, S.H. Adamu, M. Mohammed, I. Jibo, Application of ICT in agriculture: opportunities and challenges in developing countries, Int. J. Comput. Sci. Math. Theory 3 (1) (2017) 8–18.

[12] S. Ghani, F. Bakochristou, E.M.A.A. ElBialy, S.M.A. Gamaledin, M.M. Rashwan, A.M. Abdelhalim, S.M. Ismail, Design challenges of agricultural greenhouses in hot and arid environments—a review, Eng. Agric. Environ. Food 12 (1) (2019) 48–70.

[13] R. Divya, R. Chinnaiyan, Reliable AI-based smart sensors for managing irrigation resources in agriculture—a review, in: International Conference on Computer Networks and Communication Technologies, Springer, Singapore, 2019, pp. 263–274.

[14] G. Bannerjee, U. Sarkar, S. Das, I. Ghosh, Artificial intelligence in agriculture: a literature survey, Int. J. Sci. Res. Comput. Sci. Appl. Manag. Stud. 7 (3) (2018) 1–6.

About the authors

C-15884

Veeramuthu Venkatesh received B.Tech degree in Electronics and Communication Engineering from NIT Trichy, M.E degree in Embedded System, Anna University, India in 2007 and completed Ph.D. from SASTRA University Thanjavur, India in 2018 respectively. He joined SASTRA University, Thanjavur, Tamil Nadu, India as a Lecturer in the Department of Computer Science Engineering since 2008 and is now Assistant professor-III, His research interests include Wireless sensor networks, Context aware computing Information fusion and IoT. So he has published 50+ Research articles in National & International journals and 2 IEEE conference papers.

Pethuru Raj have been working as the chief architect and vice-president in the Site Reliability Engineering (SRE) division of Reliance Jio Infocomm Ltd. (RJIL), Bangalore. My previous stints are in IBM global Cloud center of Excellence (CoE), Wipro consulting services (WCS), and Robert Bosch Corporate Research (CR). In total, I have gained more than 17 years of IT industry experience and 8 years of research experience.

Finished the CSIR-sponsored Ph.D. degree at Anna University, Chennai and continued with the UGC-sponsored postdoctoral research in the Department of Computer Science and Automation, Indian Institute of Science (IISc), Bangalore. Thereafter, I was granted a couple of international research fellowships (JSPS and JST) to work as a research scientist for 3.5 years in two leading Japanese universities. Published more than 30 research papers in peer-reviewed journals such as IEEE, ACM, Springer-Verlag, Inderscience, etc. Have authored and edited 16 books thus far and contributed 35 book chapters thus far for various technology books edited by highly acclaimed and accomplished professors and professionals.

Focusing on some of the emerging technologies such as IoT, data science, Blockchain, Digital Twin, Containerized Clouds, machine and deep learning algorithms, Microservices Architecture, fog/edge computing, etc.

R. Anushia Devi received B.E. degree in Computer Science from Anna University, M.E degree in Computer Communication, Anna University, India and completed Ph.D. from SASTRA University Thanjavur, India in 2020 respectively. She joined SASTRA University, Thanjavur, Tamil Nadu, India as a Lecturer in the Department of Computer Science Engineering since 2007 and is now Assistant professor-III, her research interests include Information security, Data hiding, Information fusion and IoT. So she has published 20+ Research articles in National & International journals and 2 IEEE conference papers.

CHAPTER ELEVEN

5G Communication for edge computing

D. Sumathi[a], S. Karthikeyan[a], P. Sivaprakash[a,b], and Prabha Selvaraj[a]
[a]VIT-AP University, Amaravati, Andhra Pradesh, India
[b]PPGIT, Coimbatore, India

Contents

Abstract

Edge computing is a computational exemplar that facilitates the functionalities of the cloud to the edge servers in the mini clouds so that the high computational tasks could be performed along with the storing of a huge amount of data in the near vicinity to the user equipment. Nowadays, the progress in the development of IoT devices increases exponentially, and since these devices must be interconnected to transfer the information to the cloud and also it accesses the data from the cloud. Therefore, the objective of edge computing is to satisfy the necessity of information transfer. Due to the deployment of edge computing in domains, security has been enhanced since there is no need for the data to travel all over the network. It has been observed that the data does not reside in one data center and it has been distributed. 5G and edge computing are found to be two intricate technologies that are determined to augment the performance of various applications considerably and the processing of a vast amount of data is done in real-time. Speed is enhanced through the innovation of 5G whereas the reduction in latency is done in edge computing through the computational capabilities in the network. Hence, several industries in

Advances in Computers, Volume 127
ISSN 0065-2458
https://doi.org/10.1016/bs.adcom.2022.02.008

307

various domains expect to make use of 5G technology to provide new business models, use cases, applications that make a tremendous shift in technology which results in the development of business. The development of 5G wireless technologies needs to involve edge computing architecture. In this regard, this chapter provides the limelight of the concepts of edge computing and the working methodology. Next, the chapter discusses in detail the importance of edge computing in 5G, the taxonomy of edge computing in 5G, and the functional components of edge computing. A section will provide the evolution of 5G. A brief explanation of the architecture of edge computing and 5G will also be done. Finally, an exploration of recent advancements in edge computing for 5G will be done.

1. Introduction

An exponential increase in the growth of IoT devices to transfer information from and to the cloud leads to the development of edge computing. The objective of edge computing is to transport various cloud resources like storage, networking, and compute that are required by devices, users, and applications. Edge computing is defined as an exemplar technology that facilitates the extension of cloud potential at the edge of the network by the edge server that is located in mini clouds. Through this facility, computational-intensive tasks are done, and thereby a huge amount of data is stored in close vicinity to User Equipment (UEs) [1–3]. Edge computing is chosen so that the communication requirements of the future generation applications could be met out. The common characteristics of edge computing, when compared with data center computing, are described as given below:

- Sharing of resources by multiple users and applications
- Through virtualization and abstraction of the resource pool, more benefits are obtained
- API is utilized in providing interoperability

Various advantages that are achieved through the deployment of edge computing are adaptability, data throughput, scalability, robustness, and reliability. Recently, it has been observed that the crucial requirements of the IT industry in various areas like security and privacy, real-time business, optimization of data, agile bonding, and to reach out to the needs of the high bandwidth and low latency in the network, edge computing has been deployed to move the resources, storage abilities, and resources to the edge of the network.

2. Architectures of edge computing

Cloud services could be extended to the edge of the network with the help of the introduction of edge devices among cloud computing and terminal devices [4,5]. This is deployed with the edge computing architecture. The structure of edge computing could be segregated into three layers namely the terminal layer, cloud computing layer, and edge layer.

Terminal layer: It comprises several devices such as cameras, smart cars, and sensors, etc. These devices belong to both data consumers and as well as data providers. The objective of the devices is to gather the raw data that comes from various places and it is transferred to the upper layer for further computation process.

Edge layer: The primary layer in the architecture that plays an essential role in accessing the data obtained from the terminal layer. Devices such as routers, switches, gateways, and access points are installed in this layer. Transmission of data is made feasible since it is located near to the user. Real-time analysis and intelligent processing are done efficiently when compared with cloud computing.

Cloud layer: Servers and storage devices that perform well plays a crucial role in analyzing the huge amount of data. Besides this, certain operations like business decision support and consistent maintenance are also performed. Data that come from the edge layer are stored permanently for further process. Based on the control policy, the cloud module is also used to do modify the installation method dynamically.

2.1 Edge computing reference frame

In 2018, an edge computing white paper 3.0 has been released by ECC that has been instigated by Huawei in association with various institutions like Shenyang Institute of Automation, the Chinese Academy of Sciences, the China Academy of Information and Communications, and other well-known enterprises [6]. The principle behind the reference frame is the model-driven engineering methodology. The four objectives that have to be achieved over with this model are

- Demonstration of the reusable knowledge model system and need to complete the interdisciplinary industry.
- The decoupling of the software interface has been done in addition to the reduction in the system heterogeneity.

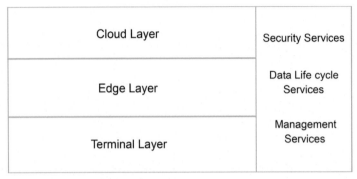

Fig. 1 Edge computing reference frame 3.0

- Aids in the life cycle of progress service, processing of data, and security.
- Cooperation could be achieved in the digital and physical world.

Fig. 1 shows the architecture of the ECC edge computing reference. From various inferences, the functionality of each layer could be demonstrated.

Services that are provided in this framework are security, lifecycle, and management services. The entire framework has been linked with these services. The objective of the data lifecycle services is to impart the management that could handle various operations like preprocessing, investigation, dissemination, and accomplishment of data along with the storage and visualization. Control of the architecture and bestowing the information lies with the management services. Through the deployment of security services, the whole life cycle of data has been brought under surveillance with the help of the business orchestration layer. Also, the deployment and optimization of the data service are done. Assurance and reliability of the whole architecture are made feasible with the adaptation of the security services on the particular architecture of edge computing. The edge layer consists of two entities namely edge node and edge manager. The edge node transforms the business of edge computing. All edge nodes are supervised by the edge manager. Edge nodes could be categorized as edge sensors, edge gateways, edge controllers, and edge gateways. Transferring of information in the layers from top to bottom and vice-versa is made possible with the development of local edge resources. Several service frameworks such as real-time computing, lightweight computing intelligent distributed and gateway systems have been provided.

2.2 The architecture of edge computing

The architecture of edge computing from the perspective of communication service providers has been discussed in this section. The functional components of edge computing could be identified as given below by referring Fig. 2.

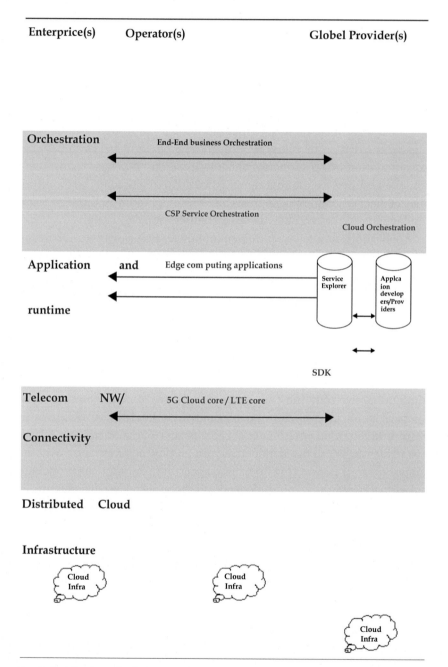

Fig. 2 Architecture of edge computing

Distributed cloud infrastructure: Cloud data centers that are located globally and in access locations are combined in the network and it is found to have functioned with the help of an essential significant planning and administration system. The selection of the corresponding infrastructure depends on the corresponding applications and use cases.

Connectivity: As soon as the installation is completed, the configuration of connectivity must be done. Applications that are to be executed need necessities based on mobility, latency, throughput, and bandwidth so that execution must be done on several hardware setups. New innovative solutions are required for the routing of data that are required for applications. Management of communication among the applications in the server and employment of these mobile network solutions must be done.

Application runtime execution environment: Service provided by edge computing is the runtime execution environment that could be able to function virtual network functions and non-telco workloads. Applications could be hosted on the environment and they must synchronize with the requests made by the development communities. To execute these applications, several features and operational needs are deployed in addition to the diversified platform mechanisms. Customization of the environment could be done by the application developer based on the requirements.

Dynamic orchestration and management: A centralized planning and administration facility must be enabled to be aware of the topology that has been configured in the network and the availability of resources in the distributed cloud environment. The objective of the orchestration layer is to render the coordinated individual planning and administration over the numerous organization functions. Managing the platforms for workloads and virtual network functions depends on the service level agreements.

Service exposure: Server can disclose the core competencies that are existing internally in the operator or to another partner. Several capabilities such as optimization, security, connectivity, data, and analytics could be added more significance to both internal and external users. Moreover, a comparison of the edge computing reference architectures is given in Table 1.

It has been observed from various inferences that there is tremendous growth due to the progress in innovative technologies in terms of smart devices, sensors, and various other gadgets. Hence, there is a huge expansion of data which might be heterogeneous. Due to this, enterprises acquire solutions that depend on several architectures. This in turn results in augmenting the access speed and as well as visualization of data along with the other

Table 1 Comparison of edge computing reference architectures.

Architectures	Distributed cloud infrastructure	Connectivity	Security	Data encryption	Blockchain	Standards
FAR-Edge RA	Yes	Yes	Yes	No	Yes	No
Edge computing RA 2.0	Yes	Yes	Yes	No	No	Yes
Industrial Internet Consortium RA	Yes	Yes	Yes	No	No	Yes
Edge computing reference architecture 3.0.	Yes	Yes	Yes	Yes	No	No
EdgeX Foundry	Yes	Yes	Yes	Yes	No	Yes

information from diversified systems irrespective of their interconnection. As the development of edge computing continues, consequently the solutions are used to accumulate and accomplish the data. Recently, the growth in innovative technologies targets the business associated with IoT solutions like Assisted Reality (AR) which results in substantial consequences in providing applications in various domains such as logistics, transport, and manufacturing. A combination of 5G, mobile edge computing, and IoT will lead to the generation of more IoT devices recently.

3. 5G and edge computing

The performance of applications could be amplified with the help of two indistinguishable technologies namely 5G and edge computing. Processing enormous data in real-time could be also enabled additionally. The increase in speed is done by 10 times in 5G when compared with the 4G, but mobile edge computing decreases the latency through the computing competencies into the network. The main features of 5G are

➢ A huge volume of data generation due to the extraordinary growth of mobile devices that results in ultradense networks.

➢ QoS requirements are enforced to execute interactive applications that need high throughput and ultralow latency.

➢ Interoperability of a varied range of user equipment, QoS necessities, type of networks, and so on.

5G comprises three predominant technologies for the functioning of user equipment in terms of high network capacity [7]. They are described as given below:

- mmWave communication makes use of high-frequency bands that ranges from 30 to 300 GHz [8] and also high bandwidth [9].
- Transmission range and interference could be decreased by the utilization of mmWave. Cells that are small in size permit the user equipment to perform the communication process.
- A huge number of antennas are used by the base stations so that the transmission could be done directly which results in less interference. This permits the neighboring nodes to connect concurrently.

3.1 Importance of edge computing

Applications that are enriched with high interaction are aided with certain features such as high throughput and low latency [10]. During the real-time packet delivery among driverless cars, the delay in data transmission is less than 10 ms [11]. During the accessibility from the cloud, the end-to-end delay might be more than 80 ms [12,13] and it is considered to be unbearable. The introduction of edge computing satisfies the submillisecond needs of 5G applications and thereby the energy consumption is reduced by 30–40% [14] that contributes to less energy consumption during the data access from the cloud [15]. The next section provides the taxonomy of edge computing.

3.2 Taxonomy of edge computing

It could be categorized into subsequent aspects namely objectives, access technologies, computing devices, computing models, empowering technologies, computational order, and applications. It is denoted in Fig. 3.

Moreover, the role of edge computing in 5G could be explained in terms of the taxonomy. It could be determined as the goals, computational platforms, characteristics, utilization of 5G functions, performance metrics, and several roles of edge computing in 5G.

3.2.1 Goals

The purpose of edge computing in 5G is to satisfy various constraints such as

- ➢ Improvement in data management: Due to the extensive growth of user equipment, the handling of a huge amount of data becomes a complex

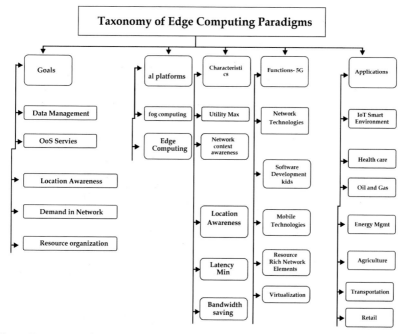

Fig. 3 Taxonomy of edge computing

process since accessing data leads to high latency [16]. Hence, the data could be managed locally by edge servers.

➤ Enhancement in QoS services: This service is intended to provide delivery of content rich in multimedia that requires less latency and high bandwidth [17]. Customization of QoS could be provided to develop new and custom-made applications [18].

➤ Examining network demand: Prediction of the essential resources to accommodate the user demand in a local vicinity. The efficient allocation of resources could be done fairly based on the prediction of the network demand.

➤ Handling of location-awareness: Location-based service providers are capable of providing services and data to the edge clouds.

➤ Augmentation in resource organization: Optimization of network resource usage has been done to enhance the networks since the resources are inadequate. This is one of the challenging issues since it needs to serve many applications that are diversified in nature and also demands and user requirements that vary based on time.

3.2.2 Computational platforms

Deployment of the computational platforms could be done either in a single mode or in combination mode. This kind of service depends on several network scenarios and the requirements of an application/service. The three main computational platforms are described as given below:

✓ Cloud computing: A massive amount of data is gathered from several user equipment and transferred to the cloud. Subsequently, decisions or data are also transferred from the cloud to the user equipment. The provision of real-time services is found to be a little complex due to the distance between the cloud and the user equipment.

✓ Edge computing: It has the facility of collecting, storing a huge amount of local data and information from the various user equipment in that area. It has been observed that edge computing connects easily with the user equipment due to its proximity. There are three types of edge computing. They are named local devices, localized data centers, and regional data centers.

➢ Local devices: These are deployed for specific home applications. Cloud storage is joined with applications in such a way that there is no need to move the applications into the cloud.

➢ Localized data centers: These are capable of providing considerable processing and storage capabilities. Also, the deployment of data centers is fast in present environments. These are made accessible as they are preconfigured and assembled on site. 1–10 Racks data center is mainly used for broad applications that need less latency, high bandwidth, security, and availability.

➢ Regional data centers: Data centers with more than 10 racks could be positioned near the user and data source. Hence, processing and storage abilities could be enhanced due to the scalability.

✓ Hybrid computing combines both edge and cloud computing. Real-time data could be processed at the edge computing and nonreal-time data are processed in the cloud computing. Throughput is increased with edge computing.

3.2.3 Characteristics

Network context awareness: The whole network context information could be obtained by enabling the edge servers. Certain information like the location of the user, traffic load, and allotted bandwidth facilitates edge servers to adjust and act according to the impulsive network constraints and user equipment. Hence, the edge servers handle a huge amount of traffic so that the performance of the network could be enhanced. Microlevel information also supports in providing precise services to traffic flows so that the user needs are satisfied.

Location awareness: Through this feature, the edge server can collect the data from the sources in its close vicinity.

Low latency and close vicinity: Through this, the response delay is decreased. Three major factors that contribute to the response time are identified as the propagation delay, communication delay, and computational delay.

✓ Propagation delay: It is based on the dissemination distance.
✓ Communication delay: It depends on the data rate.
✓ Computational delay: It is determined by the computational time.

Usually, in cloud computing, the end-to-end delay is more than 80 ms [12].

3.2.4 5G Functions

Technologies that enable the edge computing could be described as given below:

Network function virtualization (NFV): Network functions are performed in virtual machines. Managing a huge amount of data so that it results in malleable and scalability in networks [19,20]. Demands from the network could be managed in any of the two entities namely edge or at the cloud that could protect the data and information in transferring to the cloud.

Access to new radio (NR): 5G is a novel standard that connects several devices so that scalability and low latency could be achieved so that the networks could be extended shortly. Accessing radio access technologies includes access to several other technologies like NR [21].

Software-defined network (SDN): The network could be partitioned into data and control planes for providing the networks actively and flexibly that aids in the deployment of new services and to simplify the network management [22]. The objective of the data plane is to forward the traffic based on the decisions done by the control plane and the control plane manages the policy on the cloud. Edge servers operate and manage the real-time responses that are generated during the network functioning.

Multiple input/multiple output technology (MIMO): 5G New radio encompasses the deployment of MIMO technology to cover the network and capacity. The transmission gain and efficiency could be enhanced by the deployment of the massive MIMO through the implantation of several antenna components. This works based on the Shannon principle [23]. In this, the energy efficiency and the capacity of the network are boosted.

Edge server takes the task of computation with massive MIMO that results in less latency and energy consumption [24].

Communication among the devices: Throughput of the system, efficiency of the energy could be improved through the direct communication that would be enabled among the devices without the necessity of the base stations [25,26]. Edge servers carry out the computational tasks that the UEs are authorized with the computational abilities. Through this direct communication facility, edge computing ensures the process to be a success [27,28].

3.2.5 The functioning of edge computing in 5G

Storage: Through this technology, a huge amount of data is being offloaded to the edge clouds. The special capability of this is to provide the distributed local storage for the substantial data. Storage is compatibly less when compared with cloud computing. Data like metadata and computing strategies are stored in the edge server with the deployment of several types of storage policies [29]. Temporary data storage is also used for holding the transient information about the group of devices that are connected [30,31].

Computation locally: Edge clouds do the computation process irrespective of its complex nature. It is observed that the simple computation is done in the conventional cache and access technologies whereas the local computation and data processing functionalities are done in the edge servers intelligently and autonomously [32]. Through the performance of smart tasks and responding locally results in the cost reduction and delay that occurs due to the transmission of the needed data to the cloud.

Data analysis: To make a decision, an extensive analysis of data is required. To decrease the latency for the transmission and receiving of data and responses from and to the cloud, data analysis could be done locally. To do this, raw data is collected from several applications and it is analyzed to produce the necessary information needed for the decision-making process [33,34].

Decision-making process: Decisions depend on the processed data [35]. The advantages of the decision process in the edge cloud have resulted in the diminishment of the requirement of components, data, and the exchange of information or data frequently. Therefore, the availability of the system and the bandwidth has been improved.

Local operation: Remote monitoring or controlling the devices is enabled through this technology [36].

Progress in the local security: One among the functionalities of the edge server is that it can distribute the software patches, detection of malware, issuing of security credentials, and its management apart from communicating in a secured manner. Moreover, it can perform countermeasure attacks. Due to the juxtaposition of edge computing, detection of malevolent objects is done rapidly and real-time responses could be activated so that the impact of the attacks could be lessened. Therefore, the service disruption is minimized. The scalability and the edge computing abilities could expedite the implementation of blockchain among the user equipment but, with restriction in capabilities [37].

It could be observed from various inferences that the performance of the applications could be improved and thus it empowers the massive data transmission in a real-time environment. When the speed of 5G is combined with the processing capabilities of edge computing, then the applications that need low latency acquired the focus. Hence, applications that make use of AR/VR, AI, and robotics need intensive decisions from computing resources. However, several business applications have the potential of availing the benefits from both technologies.

3.2.6 Integration of 5G and edge computing
The working of IoT devices could be still made efficient by providing a high level of connection on the network edge. Revolution gets initiated when the processing and storing of data is done in the devices that are located at the edge that has been connected to the devices in the network. Data transmission could be done uninterruptedly and effectively with the help of an adequate network connection. For example, data that are collected from the sensors that have been implanted in an autonomous car might be transferred to the other vehicles for further process. To perform this transmission rapidly, there needs a necessity for the 5G technology. Due to the availability of the 5G infrastructure, data centers and other IoT devices that are located in the edge can form a distinguished processing area where the data could be constructed and collected. Furthermore, data analysis also could be done locally with a marginal latency. Therefore, the network edge acts as a loop of organized 5G networks that are interrelated. This device has the capability of handling the data and also it arranges the information that has to be transmitted to the centralized servers based on the preferences.

Due to the increase in the speed and bandwidth of the network, the backhaul across the network might jeopardize the innovation and thereby the performance of the applications degrades. Therefore, hosting applications at the edge might perform well with high bandwidth and less latency due to the provision of 5G. Several other advantages of edge computing are

Progress in agility: CoSPs have been facilitated with the trial periods on the new services and upgrade based on their performance. Upgradation could be done without visiting the site. This is made possible with the replacement of the customer premises equipment with the edge computing platforms. A rapid recovery facility is also ensured during the process of the network failure since the CoSP has the dynamic provision ability to choose the essential service needed.

Context awareness: Hosting applications in the edge possess this facility. Suppose, if an AR application is hosted in the edge, it deals with data and many images associated with that corresponding location. Decisions could be made in a real-time context since they might be delayed if it has to contact the cloud for deciding.

Manage without connection: Edge computing enables the application to run despite the feeble connection that exists between the cloud and edge.

Through the special feature of rapid data processing, performance could be enhanced. Also, the cloud-based services could be made easier for the users. Extensive growth of IoT devices and 5G implementation devices are evolving to perform the computation process with the help of edge computing technologies. The most important attribute is that the complexity arises during the bandwidths balancing among the several networks since there is a need to interchange the data to data centers and other devices that are located in the edge. Edge computing proves to show its performance through the deployment of the distributed computing technology for the location and the capacity of the network. There are issues during the data transfer. In this case, data gets transferred among the network through the gateways. Another issue is that the backup of the data. Organizations must adhere to the policies of data management during the backup and accessing of data.

3.2.7 Security and privacy

Traditional security principles and practices will not be substantial for the functioning of the 5G network. There is a necessity for cutting-edge security measures that results in the innovation of the new network model. Apart from that, several other protective and proactive security measures are in need for the protection of the loads of data that is been transferred in milliseconds among the IoT devices. Due to the advancement in the endpoints number, the risks that might be encountered are also assumed to be high in number. Also, data security is considered a challenging task and it will increase vulnerabilities. Protection to the network could be devised in several ways. Methods such as strong authentication mechanisms at various levels and security for the gateways could be done. Additionally, the SDN and NFV endpoints must be secured by following several security measures properly. The integrity of networks could be done so that the security challenges that arise due to the evolvement of 5G could be reduced.

4. 5G and edge computing use cases

The main objective of 5G is to provide an extensive bandwidth. To perform this process, there is a need for an enhancement in the infrastructures so that it could handle the issues such as crucial traffic prioritization. This ensures that the throughput could be high and the latency could be less. Cloud architectures that are used in massive data centers play a significant role in providing more flexibility and efficient use of hardware resources through the signaling functions. To avail the similar advantages for radio applications, without creating an impact on the computational power and the speed of the light, a transition is required to move to the edges of the network. This empowers in providing the high bandwidth that is needed for the radio equipment to communicate among the applications. A comparison of several architectures has been described in Fig. 4 to illustrate the evolution of edge computing.

A two-tier architecture is described in the pre-edge computing in which the clients get connected to the service providers and the network. To decrease the investments done by the providers, the network access could be revolutionized and thereby the providers turn into a transport service provider belongs to another category of EaaS (Edge-as-a-Service).

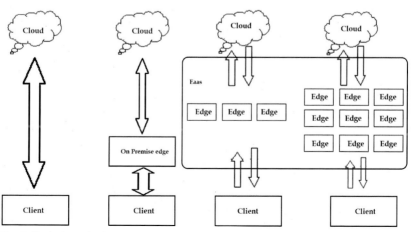

Fig. 4 Evolution of edge computing

This section describes many use cases and characteristics that are aided by 5G.

Smart city: Intelligent decisions could be made with the help of sensors and actuators that could be deployed across the city so that the city becomes a smart city. Transmission of data is done at a high rate and low latency.

Smart grid network: Through this grid, changes could be detected, understood intelligently, and planning of the behavior in response to the changes. To do this process, a clear synchronization has to be done for keeping the grid at a stable condition during the time of impulsive loads or demands.

Virtual reality, gaming, and augmenting reality: Information could be provided through live streaming so that it gives the facility of shrouding the significant information. One could get completely engrossed in the virtual world.

Remote surgery and examination: Through these contemporary technologies, the operator could be able to perceive and sense the physical world even from a distant location. One such application is remote surgery in which the latency communication is very low and the reliability is high which might result in life in a risky situation.

4.1 Industry 4.0

It is one among the use cases of 5G. To support Industry 4.0 connection needs, 5G could be deployed. A huge increase in production and several innovation strategies could be enabled through the investment in the dense networks which focus on the manufacturing department. Moreover, the manufacturing is strengthened and hence an extensive analysis of data results in handling the intricacy and instability of the global markets leads to highly competent capabilities. There is a possibility of data flow from the customers to the producers through the products that act intelligently which enables product development rapidly. Cyber-Physical Systems is a combination of real and virtual systems that leads to the rearrangement of supply chains and value creation. From various inferences, it could be denoted that Industry 4.0 is identical with the amalgamation of CPS in domains such as production and delivery. Moreover, the usage of IoT and its services have been found extensive usage in the industry.

4.2 5G Communication technology in Industry 4.0

To perform the wireless networking and to make the factory intelligent, Industry 4.0 does the process. Evolution progresses in the fourth revolution in the world and it is denoted by Industry 4.0. Industrialized elements such as

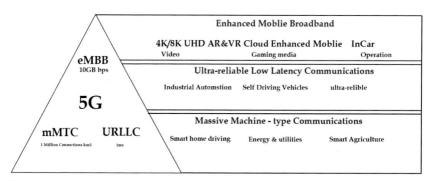

Fig. 5 5G use cases

people, machines, and equipment are found to be in close association with the help of 5G so that the manufacturing process could be done with efficacy and coordination. Fig. 5 exhibits the deployment of 5G use cases. The implantation of AI and Machine Learning results in making accurate decisions in various fields like agriculture, healthcare, and supply chain logistics. 5G makes its footprints in the industry with the help of network slicing in three different ways and they are described as given below.

1. eMBB (Enhanced Mobile Broadband)
2. URLLC (Ultra-Reliable Low Latency Communications)
3. Massive Machine type communication

This provides extensive coverage when it is compared with 4G and also offers high data rates. Through the sustenance of the comprehensive range of stakeholders that includes Communication Network Operators (CNO), providers of communication and industrial technology, Tier 1 manufacturers, and system integrators, 5G lays its foundation with Industry 4.0. The objective of using URLLC is to satisfy the end-to-end (E2E) latency and reliability. This needs various components that include effective data resource distribution, fast data turnaround, innovative channel coding mechanisms, and uplink transmissions that are grant-free. The connection among numerous IoT devices per square kilometer could encompass a wide area and could penetrate more in indoor networks.

The implementation of 5G in the smart factory has been considered in this work. The process of deployment of 5G in the smart factory is shown in Fig. 6.

An example of the smart factory under the 5G communication is shown in Fig. 6 which the data collected is massive and wireless transmission is used to transmit the data to various other departments. Sensors that are deployed in various equipment transmit the data to the respective departments for further process.

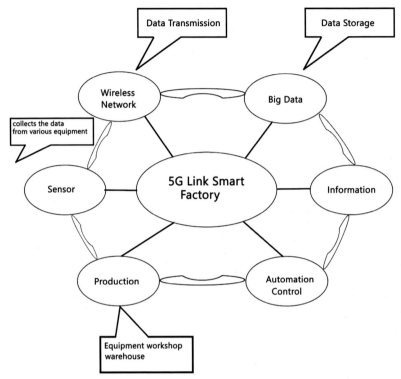

Fig. 6 5G links smart factory

The manufacturing process is found to be an intelligent task due to the continuous connection of 5G network which has been implanted its footprint in various departments such as procurement, logistics, design, warehousing, and production. Smart factory involves the process of decision making in loading and unloading of materials, logistics, and storage mechanisms. Due to the availability of sensors at various places, the information is being transmitted within a short period so that the decisions could be taken rapidly in terms of any fault tolerance actions. Data thus collected from various parts results in a massive database. Further, the best solution could be achieved by the deployment of industrial robots so that the supercomputing abilities of cloud computing for self-governing learning are used. D2D technology under 5G is used for communication among objects which in turn decreases the end-to-end delay in services and thereby the network load gets by-passed. Moreover, the response is swift in these scenarios. The secret behind the smart factory's success is the utilization of Enhanced Mobile Broadband (eMBB) technology for the collection of a massive amount of data for analysis and to take the corresponding action based on the data. Human-computer interaction is carried out in the system and AI plays a vital role in the optimization of the processes and products. Various activities that are carried out in the smart factories are

Cloud services: Huge data storage, management of data, and computing are the activities that depend on the cloud computing architecture.

Ubiquitous computing: Data from various entities such as software, hardware, and other personnel are collected for further process.

Accretion of knowledge: Data analysis abilities could be enhanced due to the available mechanisms so that the information could be consolidated and reused further.

Application improvement: Industrial APP development environment is provided in which the functions and resources function.

A smart factory is known for its functionality of the application platform for several intelligent technologies and is likely to the cooperation with the other state-of-the-art technologies to exploit the resource utilization to the maximum, efficacy in production and economic benefits. Additionally, it also aids in the process of determining the variations and anomalies that occurs during the manufacturing process. Due to this action, the supply of energy, personnel, and production could be fine-tuned to assure the production normally which results in the efficacy of the energy. The deployment of ERP leads to the automatic computation for procuring the raw materials and as well as supplier information. Production efficiency shall show tremendous progress due to the state-of-the-art technology that has been used to interconnect all the departments like sales, production, and design. Additionally, the optimization of resources is also obtained and thereby results in the improvement of the quality of the product.

5. Challenges during the deployment of edge computing in 5G

Among the many issues, the most predominant is the security and privacy of data. Through this deployment, the data cannot travel across the network. Yet, two factors result in the vulnerability at the edge of the network.
- The dynamic nature of the network results in the divergence of the data and the network components.
- Scalability is a significant factor due to the augmentation in the devices that communicates with each other.

Enhancement in adopting the security and privacy policies has to be strengthened due to the significance of the data. The possible solutions are

- Applications that execute on the edge cloud must not be aware of the data that is being processed. Hence the information that is used over here, must be encrypted.
- Additionally, the raw information could be detached before reaching the edge cloud so that it assures privacy [38,39].

Quality of Experience (QoE): This is one of the most significant parameters used in the evaluation of customer satisfaction with the service provider. The main issue is to obtain a balanced deal between the high availability and the QoE of the application. Availability must be high even when the user equipment is not in the vision of the edge server and the QoE of the application could be high when the UE is close to the edge server so that the jitter and delay are reduced. The state of the network is maintained with the help of edge computing. Due to this, the arrangement among the QoE and the availability could be obtained. Aggregation of signaling messages could be done to decrease the signaling overhead. Hence, the network congestion is reduced which results in the enhancement of the scalability and the high throughput.

Protocol standardization: To deploy edge computing in 5G, there arises a necessity of protocol standardization. For functioning properly, a group of the systematic body is needed which takes the responsibility of forming the rules and policies for the deployment. Yet, two main challenges must be addressed.

- Agreeing on a common standard is a complex task due to the flexibility and varied customization by several vendors.
- Due to the tremendous increase in the assorted UEs, it needs diversified interfaces to perform the communication with the edge cloud.

Hence, Standards Institute (ETSI) [40] is identified to take care of the communication among the edge servers with the conglomerated UEs, several layers, and computation patterns in a multivendor environment.

Heterogeneity in communication: Due to the dynamic nature of various factors such as data rate, transmission range, it becomes a complex task to determine a solution that might be compatible in a diversified environment. Solutions based on the software might be developed to carry out the workloads in the edge nodes and at several hardware levels. Parallelism in task and data level partitions the workload into self-governing and minor jobs that could run parallel in several layers in edge clouds [41].

6. Conclusion

The evolution of the next-generation cellular network results in a significant enhancement in the quality of service. An emerging technology that empowers the progress of 5G through the deployment of edge cloud capabilities so that many of the issues that arise in the conventional cloud could be avoided. The taxonomy denotes the several features of edge computing. Through this deployment, various services like processing of data locally, progress in data rate, and availability could be achieved. A separate section has been contributed to discuss the use cases that are supported by 5G. Finally, the challenges that could arise during the setting out of edge computing in 5G.

References

[1] W.Z. Khan, E. Ahmed, S. Hakak, I. Yaqoob, A. Ahmed, Edge computing: a survey, Future Gener. Comput. Syst. 97 (2019) 219–235.

[2] N. Hassan, S. Gillani, E. Ahmed, I. Yaqoob, M. Imran, The role of edge computing in internet of things, IEEE Commun. Mag. 56 (11) (2018) 110–115.

[3] E. Ahmed, A. Ahmed, I. Yaqoob, J. Shuja, A. Gani, M. Imran, M. Shoaib, Bringing computation closer toward the user network: is edge computing the solution? IEEE Commun. Mag. 55 (11) (2017) 138–144.

[4] J. Ren, H. Guo, C. Xu, Y. Zhang, Serving at the edge: a scalable IoT architecture based on transparent computing, IEEE Netw. 31 (5) (2017) 96–105.

[5] H. Bangui, S. Rakrak, S. Raghay, B. Buhnova, Moving to the edge-cloud-of-things: recent advances and future research directions, Electronics 7 (11) (2018) 309.

[6] S. Carlini, The drivers and benefits of edge computing, in: Schneider Electric Data Center Science Center, 2016.

[7] B. Yang, Z. Yu, J. Lan, R. Zhang, J. Zhou, W. Hong, Digital beamforming-based massive MIMO transceiver for 5G millimeter-wave communications, IEEE Trans. Microw. Theory Tech. 66 (7) (2018) 3403–3418.

[8] N.C. Luong, P. Wang, D. Niyato, Y.-C. Liang, Z. Han, F. Hou, Applications of economic and pricing models for resource management in 5G wireless networks: a survey, IEEE Commun. Surv. Tutor. 21 (2018) 3298–3339.

[9] Z. Pi, J. Choi, R. Heath, Millimeter-wave gigabit broadband evolution toward 5G: fixed access and backhaul, IEEE Commun. Mag. 54 (4) (2016) 138–144.

[10] A. Ateya, A. Muthanna, I. Gudkova, A. Abuarqoub, A. Vybornova, A. Koucheryavy, Development of intelligent core network for tactile internet and future smart systems, J. Sens. Actuator Netw. 7 (1) (2018).

[11] M. Simsek, A. Aijaz, M. Dohler, J. Sachs, G. Fettweis, 5G-enabled tactile internet, IEEE J. Sel. Areas Commun. 34 (3) (2016) 460–473.

[12] S. Choy, B. Wong, G. Simon, C. Rosenberg, The brewing storm in cloud gaming: a measurement study on cloud to end-user latency, in: 2012 11th Annual Workshop on Network and Systems Support for Games (NetGames), IEEE, 2012.

[13] Z. Zhao, K. Hwang, J. Villeta, Game cloud design with virtualized CPU/GPU servers and initial performance results, in: ScienceCloud '12: Proceedings of the 3rd workshop on Scientific Cloud Computing, ACM Press, 2012.

[14] W. Shi, S. Dustdar, The promise of edge computing, Computer 49 (5) (2016) 78–81.

[15] W. Hu, Y. Gao, K. Ha, J. Wang, B. Amos, Z. Chen, P. Pillai, M. Satyanarayanan, Quantifying the impact of edge computing on mobile applications, in: Proceedings of the 7th ACM SIGOPS Asia-Pacific Workshop on Systems—APSys '16, ACM Press, 2016.

[16] X. Sun, N. Ansari, Green cloudlet network: a distributed green mobile cloud network, IEEE Netw. 31 (1) (2017) 64–70.

[17] Z. Zhao, L. Guardalben, M. Karimzadeh, J. Silva, T. Braun, S. Sargento, Mobility prediction-assisted over-the-top edge prefetching for hierarchical VANETs, IEEE J. Sel. Areas Commun. 36 (8) (2018) 1786–1801.

[18] E. Chirivella-Perez, J. Gutiérrez-Aguado, J.M. Alcaraz-Calero, Q. Wang, NFVMon: enabling multioperator flow monitoring in 5G mobile edge computing, Wirel. Commun. Mob. Comput. 2018 (2018) 2860452.

[19] J. Gonzalez, G. Nencioni, A. Kamisinski, B.E. Helvik, P.E. Heegaard, Dependability of the NFV orchestrator: state of the art and research challenges, IEEE Commun. Surv. Tutor. 20 (4) (2018) 3307–3329.

[20] I. Sarrigiannis, E. Kartsakli, K. Ramantas, A. Antonopoulos, C. Verikoukis, Application and network VNF migration in a MEC-enabled 5G architecture, in: 23rd IEEE International Workshop on Computer Aided Modeling and Design of Communication Links and Networks (CAMAD), 2018, pp. 1–6.

[21] S. Parkvall, E. Dahlman, A. Furuskar, M. Frenne, NR: the new 5G radio access technology, IEEE Commun. Stand. Mag. 1 (4) (2017) 24–30.

[22] Y. Li, M. Chen, Software-defined network function virtualization: a survey, IEEE Access 3 (2015) 2542–2553.

[23] C.E. Shannon, A mathematical theory of communication, Bell Syst. Tech. J. 27 (3) (1948) 379–423.

[24] S. Wang, X. Zhang, Y. Zhang, L. Wang, J. Yang, W. Wang, A survey on mobile edge networks: convergence of computing, caching and communications, IEEE Access 5 (2017) 6757–6779.

[25] Y. He, F.R. Yu, N. Zhao, H. Yin, Secure social networks in 5G systems with mobile edge computing, caching, and device-to-device communications, IEEE Wirel. Commun. 25 (3) (2018) 103–109.

[26] H. Wang, J. Wang, G. Ding, L. Wang, T.A. Tsiftsis, P.K. Sharma, Resource allocation for energy harvesting-powered D2D communication underlying UAV-assisted networks, IEEE Trans. Green Commun. Netw. 2 (1) (2018) 14–24.

[27] X. Chen, L. Pu, L. Gao, W. Wu, D. Wu, Exploiting massive d2d collaboration for energy-efficient mobile edge computing, IEEE Wirel. Commun. 24 (4) (2017) 64–71.

[28] Y. He, J. Ren, G. Yu, Y. Cai, D2D communications meet mobile edge computing for enhanced computation capacity in cellular networks, IEEE Trans. Wirel. Commun. 18 (3) (2019) 1750–1763.

[29] J. Zhang, W. Xia, F. Yan, L. Shen, Joint computation offloading and resource allocation optimization in heterogeneous networks with mobile edge computing, IEEE Access 6 (2018) 19324–19337.

[30] N. Bessis, C. Dobre, Big Data and Internet of Things: A Roadmap for Smart Environments, vol. 546, Springer, 2014.

[31] J.A. Silva, R. Monteiro, H. Paulino, J.M. Lourenco, Ephemeral data storage for networks of hand-held devices, in: 2016 IEEE Trustcom/BigDataSE/ISPA, IEEE, 2016.

[32] N.H. Ndikumana, T.M. Tran, Z. Ho, W. Han, D.N. Saad, C.S. Hong, Joint communication, computation, caching, and control in big data multi-access edge computing, IEEE Trans. Mob. Comput. (2019) 1359–1374.

[33] Y. Li, J. Xue, W.Z. Wang, T. Li, Edge-oriented computing paradigms, ACM Comput. Surv. 51 (2) (2018) 1–34.

[34] E. Ahmed, I. Yaqoob, I.A.T. Hashem, I. Khan, A.I.A. Ahmed, M. Imran, A.V. Vasilakos, The role of big data analytics in internet of things, Comput. Netw. 129 (2017) 459–471.

[35] M.M. Hussain, M.S. Alam, M.M.S. Beg, Feasibility of fog computing in smart grid architectures, in: Proceedings of 2nd International Conference on Communication, Computing and Networking, Springer Singapore, 2018, pp. 999–1010.

[36] C. Xu, K. Wang, P. Li, S. Guo, J. Luo, B. Ye, M. Guo, Making big data open in edges: a resource-efficient blockchain-based approach, IEEE Trans. Parallel Distrib. Syst. 30 (4) (2019) 870–882.

[37] R. Yang, F.R. Yu, P. Si, Z. Yang, Y. Zhang, Integrated blockchain and edge computing systems: a survey, some research issues and challenges, IEEE Commun. Surv. Tutor. 21 (2019) 1508–1532.

[38] E. Ahmed, M.H. Rehmani, Mobile edge computing: opportunities, solutions, and challenges, Future Gener. Comput. Syst. 70 (2017) 59–63.

[39] K. Kaur, S. Garg, G. Kaddoum, M. Guizani, D. Jayakody, A lightweight and privacy-preserving authentication protocol for mobile edge computing, arXiv (2019). preprint arXiv:1907.08896.

[40] K. Jain, S. Mohapatra, Taxonomy of edge computing: challenges, opportunities, and data reduction methods, in: Edge Computing, Springer International Publishing, 2018, pp. 51–69.

[41] T. Taleb, S. Dutta, A. Ksentini, M. Iqbal, H. Flinck, Mobile edge computing potential in making cities smarter, IEEE Commun. Mag. 55 (3) (2017) 38–43.

About the authors

Dr. D. Sumathi is presently working as an Associate Professor Grade 2-SCOPE at VIT-AP University, Andhra Pradesh. She received the B.E Computer Science and Engineering degree from Bharathiar University in 1994 and M.E Computer Science and Engineering degree from Sathyabama University in 2006, Chennai. She completed her doctorate degree in Anna University, Chennai. She has an overall experience of 22 years out of which 6 years in the industry, 15 years in the teaching field. She has guided many projects during her tenure. She is an active member of the AIR-Centre of Excellence at VIT-AP. She is a life member of ISTE. Her research interests include Algorithm Analysis, Cloud computing, Network Security, Data Mining, Natural Language Processing, Machine Learning, Deep Learning, and Theoretical Foundations of computer science. She has published papers in international journals and conferences. She has published 6 patents. She has developed various projects during her tenure. She is guiding various students to do research projects in

NLP, ML, and DL. She has given various talks in educational institutions. She has organized many international conferences and also acted as Technical Chair and tutorial presenter. Further, she has published 5 patents. She has published book chapters in CRC Press, IGI Global, Springer, IET, De Gruyter.

Dr. S. Karthikeyan is working as an Associate Professor in the School of Computer Science Engineering at VIT-AP University, Amaravati, Andhra Pradesh-India. He has 11 years of industry and academia experience. He is alumnus of PSGTech, Coimbatore and VIT University. Currently he is heading High Performance Computing (HPC) Research Lab at VIT, AP Campus. In addition to this, he is a **NASSCOM Associate Analyst certified Trainer** and handling various Data Analytics courses in recent days. He has been teaching for 11+ about years and have guided and motivated over 100+ students on independent research projects over the last five years and personally supervised Undergraduate Research interns in the last five years. Also, currently training students on the programming, technology side IoT and AI and ML and helping them get industry ready. My research interest includes Cloud and Big data Analytics and AI & ML, Hadoop, Apache Spark. Also Dealing with IoT projects, Drones, NoSQL (MongoDB, Cassandra, Neo4j), R Programming and Full Stack developer. Also, he is interested in teaching, research article writing, developing IoT solutions and making as patent. He published several articles, conference papers, journals, and authored book chapters. He also organized and attended 70+ FDPs, 25+ workshops, 10+ refresher courses, 20+ conferences, and 21+ technical talks in his domain and gained expertise. He received 10+ awards namely **Best Researcher Award, Best researcher with Patent Award, Academic Excellence awards, Most Selfless service award, Above & Beyond Award, Best Paper award, Best performer Award, best Team Leader Award from various prestigious organizations in India**. Also he **holds 10+ Indian design and Product patents in IoT and Machine learning areas and also 4 more** are under examination in IPR. Life member in international professional bodies such as ISTE, IAENG, ISRD, IFERP. Also, he is

Former Founder of Google Developer Club VIT AP and Founder of Rudra Open Community Organization (NGO) and Microsoft Academy and Learn Programme at VIT.

Mr. P Sivaprakash is presently a Research Scholar cum Assistant Professor, Department of Computer Science & Engineering, PPG Institute of Technology, Coimbatore, affiliated to Anna University-Chennai, Tamilnadu, India. He received the B.E., Computer Science & Engineering and M. Tech Information Technology degree from Anna University, Chennai. His research interests include Cloud Security, Network security, Big data Analytics, High performance computing and cloud computing. He has an abundant knowledge in networks and big data analytics. He has completed a certified course in Cisco certified Network Associate.

Dr. Prabha Selvaraj is working as an Associate Professor SG1, SCOPE, VIT-AP University, Andhra Pradesh. She has completed her B.E CSE, M.E CSE and PhD in Anna University, Chennai and has 20+ years of experience in teaching and industry. Her research interests include Data Mining, Security, Wireless Sensor Network, IoT, Information Retrieval System, Image processing, machine learning and blockchain. She has published 35+ papers in various national and International journals and conferences and seven book chapters in IGI, Wiley and IET in the area AR, Big data analytics, IoT, WBAN, Block chain and Image Processing. She has published seven patents. She has received Top best faculty award in CSE by EET CRS and Best Educators Awards 2018 by IARDO, Distinguished Leader in Engineering Discipline 2019 by Higher Education Leadership Awards. She is a reviewer in few national and international journals and conference.

CHAPTER TWELVE

The future of edge computing

Swaroop S. Sonone[a], Kavita Saini[b], Swapnali Jadhav[c], Mahipal Singh Sankhla[d], and Varad Nagar[d]

[a]Department of Forensic Science, Dr. Babasaheb Ambedkar Marathwada University, Aurangabad, Maharashtra, India
[b]School of Computing Science and Engineering (SCSE), Galgotias University, Delhi, Uttar Pradesh, India
[c]Government Institute of Forensic Science, Aurangabad, Maharashtra, India
[d]Department of Forensic Science, Vivekananda Global University, Jaipur, Rajasthan, India

Contents

Advances in Computers, Volume 127
ISSN 0065-2458
https://doi.org/10.1016/bs.adcom.2022.02.009

333

Abstract

Edge computing is dispersed computational model that accompanies computing and data storages closer to the sites wherever it is required, to increase response rates & preserve bandwidths. Edge computing is modifying form data is being managed, treated, & produced by billions of machines throughout the globe. The rapid emergence of Internet linked gadget with the Internet of Things (IoT) among with newer technologies which need real-time computational powers, continue to drive edge computational system. The future of edge computing holds the key assets as it can be modified to advance any of the technology currently present. We will be able to transform the computing technology by constructing newer network architectures. Drawbacks of traditional cloud computing have given birth to edge computing. Edge computing can transform the present technologies efficiently and productively. Edge includes elasticity to hold existing and upcoming AI (Artificial Intelligence) requests & the facts that computations at edge dodges network latency & enables quicker response. Soon, we can see that edge computing is becoming the only one and most relied upon computing technique for advanced machines. The capability of edge computing is tremendous, it is just needed to be understood and applied in day-to-day life.

1. Introduction

In the recent past, there has been an incredible expansion in the communication sector especially wireless communications [1,2]. The data rates have also been grown speedily. Accompanying data rates, there arose numerous novel wireless facilities, which need exhaustive computation in a brief time [3–5]. Edge is the concept to store and use it and bring the computational sources nearer to the gadgets which produce and consumes the data. The predominant counterpart to such an approach is cloud computation, where the alleged sources are present at the data centers [6]. Edge fetches computation and storing sources closer to the handlers & data resources, consequently avoiding costly and slower links to detached cloud computational substructures.

Edge is a computing model that allows edge–server in smaller (mini) clouds or called edge clouds to extend the cloud abilities at the edge of the networks to execute computationally rigorous errands and stock a huge quantity of numbers at closer proximity to handlers. Edge is the start or end of the objects which the network covers. It could be any computational or networking gadget. A smartphone could be the edge among the cloud & the body. Edge establishes a novel idea into the computational landscapes. It offers the services & facilities of the cloud near to the users and is recognized by the quick processing & fast application response timings. The present advanced Internet abled gadgets like surveillances, virtual realities, & real-time traffic tracking requires quick processing & fast responsive gadgets [5]. Handlers/users generally run such applications on their source controlled mobile-phone while the main services and processings are executed on the cloud server. Leveraging the service of clouds by mobiles results in higher latency & mobility problems [7]. Edge satisfies the above-mentioned application demands by carrying the processing to the edge of the networks. Edge has an above hand over the cloud in varied dimensions making it a possible and strong contender to apply in all fields. Edge computing can be used for almost every possible connected technology. It can sufficiently and powerfully replace the existing technologies and may even contribute to the upcoming automation. The emergence of edge over cloud computing has opened the barricades to be utilized in all areas of human welfares.

2. Emergence of edge computing

Technical advancements in devices and gadgets computing are allowing an innovative current of real-time & universal implementations, like intellectual aid, traffic monitoring, augmented reality, vehicle tracing, and cooperating filmed streamings [8]. Edge has lately earned incredible consideration in academics as well as industries. The novel model goals to locate sources for storing and computing a bit nearer to the end operators, i.e., to the edge of the networks. Edge is referred to as the type of tech which can execute computing at the networked edge. The principal incentive to do so is the shortcoming of present-day cloud computational substructures when processes larger volume information for latency dangerous application. Edge is predicted as a next-gen communicating network to meet the demands of next-gen systems. The instant emergence of edge computing has given birth to various fields where it can be successfully implemented and applied. It can transform an existing technology by integration with the upcoming mechanization. Edge has converted itself into a potential option to go whenever a new challenge puts out its head (Fig. 1).

3. Drawbacks of out-of-date cloud computing

Cloud computation is a computing model that provides on-demand facilities to the users by the connection of computational sources which consists of storing facilities, computational sources, and many more. Cloud computation further concentrates on the effective optimizations of

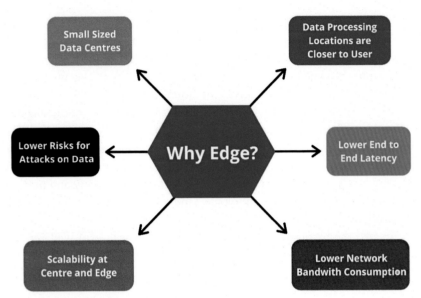

Fig. 1 Necessity of edge computing.

distributed sources among various handlers [9]. Cloud caused a technical change and standard turn into the ICT (Information & Communication Technology) division in the past years. Cloud consummated a huge acceptance in nearly each area of social lifetime [10–13]. Conventional cloud computations, that has a central computational model which offers incessant admittance to extremely proficient information hubs, has been espoused to permit UEs to divest computing and storing into the data hubs [14]. It is significant to keep in mind that edge does not aims to be a substitute for the cloud, but to accompany it [15]. Cloud cannot offer any guarantees concerning the communication part because data centers are positioned long from the customer & characteristically, neither the cloud supplier nor the operator has regulations over the transiting networks. Edge computation resolves the issues in the order that it enhances scalability in the system dimensions, i.e., a greater number of users could be aided with lower latency link when attaching supplementary edge websites for example at wireless gateways of the users. Though, due to the inadequate sources at distinct edge locations, edge computation can't provide the exact all flexibility like the cloud [16]. Although cloud is retrieved by the wired backhaul networks, one distinctive feature of the edge is that customers are characteristically linked to the edge sources by the wireless passes. In addition to the wireless accessible technology, it is also possible to differentiate the edge networks by its communicating outlines. Commonly, the present computational world involves human and varied objects which are interconnected [17,18]. There is no doubt that cloud is an efficiently organized computational model but as the manufacturing development of technologies, it flops to familiarize to the progressive interacting models. Servers of the clouds are incapable to handle such a huge quantity of information with its conventional networked structures. Additionally, fog computing experiences difficulties in the matter [19] such as customizations of the apps of fog server as per the local require-ments, accessibility of suitable fog sources to fulfill the need locally, the amount and locations for deployment of fog servers to meet the require-ments perfectly. Though fog is superior than cloud computing still it has few safety restrictions. Fog node are vulnerable to invaders & hacks. Therefore, because of the beforementioned threats of fog & cloud, the edge ideology has come in reality [20]. Because of the problems of cloud, a dis-persed computational model called fog computational model was intro-duced where the computing of information is used to happen closer to the network edges. But because of few restrictions of fog computational model similar kind but a developed was introduced known as edge comput-ing that was able to solve all of the prime problems and difficulties of cloud and fog to a greater amount [21]. This cloud computational issue could be solved by the three edge computing models which are further discussed in Section 5 [22–25] (Fig. 2).

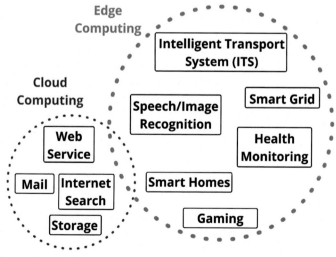

Fig. 2 Comparison of cloud computing & edge computing.

4. Significance of edge computing

Edge is an adjunct of cloud where the computational facilities are transported nearer to last handler at the network edge. Edge idea has been advanced to report the issues of higher latencies in delaying delicate facilities & implementations which are not correctly used among the cloud computational model. Such apps have the subsequent demands: (1) lower and expectable latencies, (2) position consciousness, and (3) flexibility provision. Though edge gives numerous benefits over the cloud, researches onto the upcoming fields are still in its early stages. Edge is an independent computational paradigm which comprises of various dispersed varied gadgets which interacts with the networks and execute computational task like the storing and processes [26]. Edge is a progressed stage of cloud. Edge reduces the weight of a cloud by offering the sources & facilities in edge networks. Though, edge complement cloud by increasing the operator services for the delay sensitive implementations [27]. Edge approves a distinct paradigm which carries cloud abilities close to UEs to decrease the latency. Fig. 3 showcases the cloud computing and edge computing models.

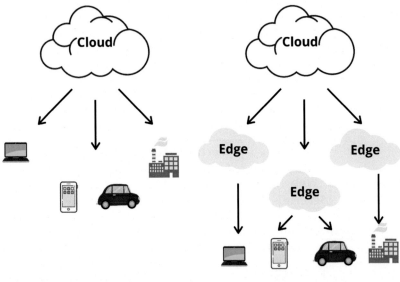

Cloud Computing Model   **Edge Computing Model**

Fig. 3 Cloud and edge computing models.

Edge could either be used as single computational medium, or a cooperative medium with varied components, consisting the cloud [28]. Today edge is the essential as the conventional cloud computational platform is not appropriate for decidedly communicating implementations which are exhaustive in computation and has greater Quality of Service demands, having lower latency and higher outputs [5]. This is due to cloud which might be far away from the end users, that expands the consumption of energy also.

To put it in another way, cloud server is stereotypically situated at a central system, and edge server having mini clouds are situated at a network edge [6,29].

Additionally, wireless gateway can offer an extra added related data to the applications, to some degree which is not obtainable in cloud [4]. Edge establishes a novel idea into the computational space. The presently advanced Internet abled implementations like surveillances, simulated reality, and real-time tracking or monitoring needs quick processes & quicker response timings. Operators generally runs such application on their source inhibited mobile phones while the central facilities and processes are executed on the cloud server. Supporting service of the cloud by mobile gadgets results into higher latency and mobility linked problems [7,30]. Edge computational tech completes the forementioned applications by providing the processes into the network edges [29]. It will be an insightful scene to observe the integration of upcoming paradigms of Internet of Things (IoTs), Blockchain technology, Artificial Intelligence with the edge computing. The field of research is open to work on these technologies with the combination of EC.

5. Edge computing technologies

Edge-cloud associated techs are gaining consideration from all over the world. Though, the idea & growth presently are in a significantly budding stages and numerous difficulties are present to be diagnosed from every field especially from the industrial and academic perspectives. Many of these present edge structures consists of devoted physical edge servers which employs with the edge devices for computing and storing, or consist simple stevedores which gives a restricted virtualization support at the edge node [5]. Edge provides a chance to perform as a discretion allowing intermediary among the operator's information and cloud dependent facilities, particularly when handlers have admittance to edge structures which are in its trust

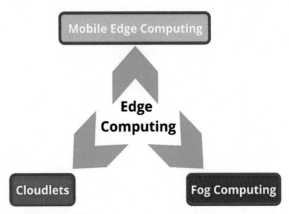

Fig. 4 Edge computing models.

area or those which are handled by the trusted suppliers [6]. Following are the three basic and fundamental sides of edge computing (Fig. 4).

5.1 Fog

Fog computational idea was given by CISCO, that permits the applications to execute straight at a network edge by millions of connected smart gadgets. Edge & fog application specifically focuses on the subcontracting computation from static or mobile operator gadgets to compute nodes in closeness. Though, these applications don't completely exploit the assistances of edge due to sources of adjoining operating systems which aren't taken into account as a probable computing node. Prominent progress in the arena of edge computation emphasis on handler's application capable to external-ization computing on computational node at network edge known as fog nodes [19,31]. Fog computing offers information, computational storing and application service to operators. This enables novel diversity of use and facilities. Fog computing is not at all likely to substitute cloud technol-ogies, though its capability to decrease dormancy and rise safety and extra conformation encounters.

The chief duty of fogging is to deliver data close to the users at edge network. A significant advantage of edge computation is that it enhances time required to act & decreases responsive time to millisecond, and preserv-ing system sources. Fog is an important development in the computing tech-nology. Emergence of fog has emphasized the dominance of a dispersed system for computation which is more elastic and nimbler than the conven-tional central model. Such dexterity and elasticity are essential with Big Data implementations enchanting the forms of the Internet of Things and its lower or no latency demands [29].

Fog has lesser latency, lesser likelihoods of outbreaks on data paths. It helps real-time communication and flexibility that is improved than cloud computational technology. Fog is a newer model and experiences numerous difficulty which differs from the problems it receives through cloud. These problems consist administration of varied gadgets, structural problems, privacy, flexibility and security issues. Fog consists of various gadgets, with numerous kinds of information gathered. Interoperability among varied objects is a tiresome work. If the quantity of linked gadgets surpasses, this may increase scalability problems for the fog computing. Additionally, for correct administration of sources and load balancing, an effective source handler is needed. Building such source handler for varied gadgets and information is not an easy job. It is also perilous to execute proper regulation and administration of devices, particularly those running real-time implementations.

Additionally, the regulation of road-traffic and billing mechanism is a must needed. One of the prime difficulties in fog is to plan a fair-minded billing structure for the facilities provided. Fog facilities are given through different pricing plans and schemes, and the handlers expects higher QoS with lowest prices [32]. The billing model must be fair minded & balanced to fascinate more readers and produce high revenue. Because of inaccessibility of any ordinary billing model for fog, it is still an open study subject. Fog comprises arrangement of luxurious gadgets and system. It is significant to execute predeployment testing of fog medium utilizing few simulation tools. Though, there is no such standard simulation model/tool present at the time for fog, that makes it an open study subject as well. Lastly, the safety against malicious threats and security attacks are also a prime study challenge for fog computing [33].

5.2 Cloudlets

Alike fog, cloudlets too represent the central tier of the three-tier structure: Mobile device–Cloudlets–clouds. Cloudlets are seen as a data center in the box with a goal to take cloud facilities near to the mobile handlers. Not just understanding to important technical development, mobile gadgets, like smartphones are still source lacking when related to other immobile systems like computer and server. This is mainly due to the smaller size, smaller memory, and short battery life. On other hand, there is an important upsurge in advancement of varied mobile apps. Maximum of the developing applications, like virtual reality, responsive media, voice recognition, natural language processing, need higher quantities of sources for processing with least latencies [7].

To match these requirements, cloudlets are manufactured with virtual characteristics to particularly give computing sources to mobile operators. Mobile devices acting as small client, can offload computing work by a wireless network to the cloudlets, which are positioned one flight away. Therefore, a cloudlet's incidence in mobile device's surrounding is essential, as the end-to-end response time with implementing applications should be smaller and expectable [8].

5.3 Mobile edge computing

Rise in the popularity of Smart Mobile gadgets like mobile phones and tablet endorses the growth of MCC or commonly known as mobile cloud computing, and it improves the source-poor mobile hardware & decreases the growth price of mobile applications by mixing the cloud computing into the mobile environment [17]. The ETSI or European Telecommunications Standards Institute has presented the idea of mobile edge computing where mobile operators can use the computational facilities from the base-stations [11,34]. MEC offers unified combination of various implementational service providers and merchants toward mobile users, initiatives and other upright parts. It is a vital part in the upcoming 5G structure which assists diversity of advanced applications and facilities where ultra-low latency is essential [35].

MEC can resolve the significant calculations of content caching, task scheduling, collaborative processing & other difficulties in huge scale systems; thus, mobile edge computing has concerned the attention of related researchers in numerous nations meanwhile it was planned [12]. Mobile edge computing tech is near to handler and can procedure information from varied Internet of Things gadgets at the similar time to accelerate the advancement of new businesses. Thus, mobile edge computing tech has superior act in endorsing smart cities and big data analysis, creating future cities smarter [16].

6. Possible advancements in digitization using edge computing

Edge is the future. It can be applied to any sector without any doubt. There is possibility of addition of edge computing in any field of technology. Few of the applications of edge computing in the upcoming digital world are as follows (Fig. 5).

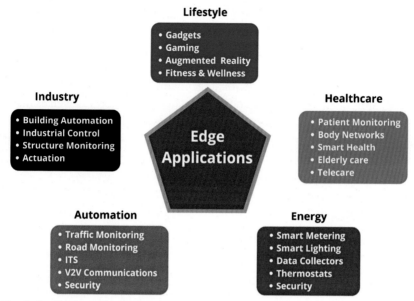

Fig. 5 Applications of edge computing.

6.1 Edge computing in network architecture

Edge solution is generally multifaceted dispersed structures surrounding & balance between the work among the edge layer, the edge cloud or edge network, & the enterprise layer. Additionally, along with the edge devices there are edge local servers [21]. Edge is Internet of Things at large amount due to the high increasements in gadgets & information being produced by such gadgets, there will be blockages & latency problems with present structures. Edge computation discourses these problems, by shifting the processing to the network edge. This shift to the edge of network, way from the data centers & near to the users, cut down on time taken to transfer information & take opinions, when related with conventional dispersed cloud computing. Edge allows contracting out computing to compute node in the near vicinity of the data sources, that centrals to lower latencies & enhanced bandwidth usages [25]. Edge computing plays significant role in IoT applications where instant processing of data is required [16].

6.2 Remote monitoring

New digital services have emerged that are much more proactive with faster response to critical incidents and execution of maintenance activities. Imagine a vendor service representative appearing at one of your sites to fix a problem you didn't even know you had yet! New generation administration tool takes advantages of newly introduced technologies like cloud computing, data analytics, AI (Artificial Intelligence) and mobile communication. Compared to the traditional, license-based, install a third-party server on every site approach, next generation management tools offer greatly simplified deployment and maintenance of the tools as well as management of the physical infrastructure. This is especially valuable when you are managing multiple edge computing installations with widely dispersed assets. These new digital remote monitoring services are built on top of this new management platform and share its benefits [28].

6.3 Healthcare

There have been countless software-enabled improvements in healthcare. Today, apps, medical devices, and data allow machines and doctors to monitor and treat patients near and far. As medical devices improve, they will be able to sense more things about your body and respond appropriately in real-time [6]. In order for these healthcare devices to work, they need edge computing and powerful AI, which isn't there quite yet. This technology can enable emergency calls and response before heart attacks, shows vital signs monitoring and responses.

One of the important uses of edge in healthcare is noninvasive cancer cell monitoring and responses. It can give a smart and personalization health nudges along with the electrolyte imbalance monitoring and notifications [29].

Edge have been positively applied in current times & is now generated utilized in various medicinal devices. Edge allows operators to regulate & act to health associated information produced by different servers. Various structure which utilizes cloud, fog, and edge computing has been used to gain the assistances of concerted computing models. The utilization of smart sensors in Intensive Care Unit & closed loop networks is able of saving numerous lives. Health consultants having admittance to the cloud servers can directly analyze patient illness and help them instantly [5,7].

6.4 E-Commerce

E-Commerce can be profited extremely through edge. For lower bandwidth of wireless systems, there may be greater postponements for mobile phones. Today, online shopping is becoming prevalent, so to advance the operator usage it has become an essentiality. At an edge node information can be coordinated in the backend with the clouds. Edge thus can advance user experience & latency for time sensitive implementations. Users can get the exact product from market through advancements in network structures. Real-time usages for wearables and accessories achieved through the edge computing [30].

6.5 Markets/business

Like the E-commerce sector, the markets have also great potential of advancing using the edge computing. There is great upsurging requirement into the companies to exchange the information. For the consideration, the social healthcare area is where administrative organizations, pharma industries, logistics and insurance companies are needed to cooperate among themselves for ideal efficiency and offering the on-time facilities. It is just happening due to the assistance of edge. A shared platform is offered by edge where all these companies exchange their information & get latest data in real-time when the situation demands [35].

6.6 Security

Edge has introduced facial identification by the Internet of Things (IoT) system that has an effective implementation at airport for inspection at security stations and in various platforms too. Different kinds of sensitive substitutes make the security system and everyday living very secure. Truthfully, it is greatly appropriate in cybersecurity mediums as it mainly concerns with the information sources not by any platforms. Edge security is an implementation of safety practices at network nodes which are outside the system core [16].

Edge needs the similar fundamental safety characteristics like the network cores; the complete network should be observable to managers. Automatic regulating apparatuses should be utilized by these handlers. Complete data is required to be encoded. There should be regulations on admittance to operate information & network sources. Edge gadgets can take the form of localized data centers, miniature data centers, or practically any tiny gadget with computational power closer to the handler/operator. IoT is going to progressively depend on & disposition by the edge network.

6.7 5G Communications

5G is next-gen mobile network which seeks to attain considerable enhancement on superiority of facilities like high output & lower latency. Edge is developing tech which allows the growth to 5G by carrying cloud abilities closer to the handlers (or operator equipment) to tame the internal issues of conventional clouds, like higher latencies & the absence of safety. MEC has developed an essential portion in the rise of 5G tech in the cellular networks as edge executes computing near to the mobile users [36].

Edge advances system performance to assist & position dissimilar situations, like remote surgeries. Though the positioning of edge in 5G offers many advantages, the integration of edge & 5G carries new problems which should be solved in the upcoming times. Edge can be located at varied premises like enterprises, in industrial structures, in houses & transporters, consisting rails, flights and automobiles. The edge substructure can be regulated or presented by interactive facility providers or varied kind of service providers. Several other uses cases which need different implementations to locations at various places. In these situations, a dispersed cloud is beneficial that can be seen as an implementation setting for requests over numerous locations, counting connectivity regulation as a solution [19]. Nowadays, there are only new market for the kinds of applications that enables 5G like virtual reality, mass scale IoTs, AUVs/drones, robotics, etc. Edge can offer developers a setting to make the 5G appliances which don't exist today even without "full 5G" being available yet [33].

6.8 Smartphone advancements

Smartphone advancements in the sectors of gaming, application development & real-time data management and transfer are only possible due to the edge computing. Business experts says that this is a developing drift which will be a game changer in the mobile phone markets. As the relocation of the data centers near to the users, edge dramatically decreases latency that results into significant data transfers at a faster and larger quantity.

In gaming sector, a small delay of a second is capable to completely disturb the experience of user. Gaming is a real-time experience, & number of people are playing tournaments from their gadgets every day. But when thousands of these people are playing a graphic integrated game at a same time, the greater the probability of latency is tremendously increased. Edge networks can offer a faster and seamless experience [11,30].

7. Opportunities for edge in future

Edge computing depend on a diversity of technologies that have nurtured its growth. Edge is founded on the concept of locating smaller servers known as edge servers or sourced networking gadgets in the close proximity of handlers. By this way, few of computing & data storing load is carried from cloud to edge servers. Operator device generally consists of wireless sensor systems, smartphones, handheld devices, & numerous IoT gadgets which needs real-time responses [3].

Positioning computing & storing sources at network edge can allow a great quantity of appliances which require real-time responses. Examples of these applications consist road-traffic regulation and steering, which counts (1) traffic reporting & computing of roads for a particular area close to the edge, (2) data sorting & segregation, which executes prefiltering of information & content at edge prior to sending it to clouds to decrease the data size, and (3) virtual reality, real-time communicating videoplays, & health regulatory networks which could create quick responses by utilizing edge nodes, thus advancing the user experience for time bounded appliances [4].

In spite of difficulties which rises when realizing edge computing, there are various chances for institutional researches. Opportunities can anything, just it is necessary to see and utilize it. Benchmarks, standards, and marketplace edge can be utilized in day-to-day practice & can be available openly if tasks, relations and dangers of all revelries involved are expressed. Frameworks & languages are many options to execute applications in the cloud paradigm. In light weighted libraries & algorithms, not like server's edge nodes that won't assist heavy weight software as the hardware restraints. Edge can benefit industries & academic collaborations.

Edge offers a special chance for institutions to focus again on its research activities widely in applied distributed computing, particularly inside the cloud and mobile computations [31]. Leading academic organizations that have dependable industry & government relationships have yet produced more expressive and impactful study. Researches in an edge space can be drifted by an open association of industry associates like as mobile handlers and user, device developers & cloud providers, along with interested academic allies to share benefits of both [6]. Fewer edge computing applications are discussed as follows.

7.1 Multimedia and edge computing

Multimedia, particularly, video is one of the prime clients of total Internet bandwidths. It has been stated that in the year 2015, video data was of 70% of the total Internet traffic. These stats are fortuned to grow to 82% in the year 2020 [19]. In the upcoming Internet of Things situations, numerous multimedia producing devices like closed-circuit TVs & visual sensor systems will produce huge volumes of multimedia data [31]. As the multimedia needs higher bandwidths, processing, and storing, so regulating these large volumes in means of interactions, processing, & storing is a real problem. Edge is intended to help in these conditions to decrease the complete end-to-end bandwidth utilization, supply, effective processing, and storing for multimedia [29].

Today, multimedia world is utilizing more newer techs by the implementations of edge applications such as video analytics, smart speakers, smart TVs, etc. For example, the smart speaker has the inherent arrangement to track the voice teaching and play the melody. Edge computing has provided a solution for low quality of video-conference & it connects the servers at the edges of the gadgets which has resolved all the annoying difficulties of video conferencing. Edge intelligence powered by edge computing has drifted all the entertaining mediums from a smaller screen of mobile phones to huge screen of televisions by presenting smart TVs [33].

7.2 Energy efficiency and edge

At this stage, the world will require an advanced computational solution to save our finite resources and prevent climate change. Farming in drought-stricken areas can be accomplished with drip-monitoring and measurement systems. Before, it was prohibitively expensive to outfit a 1000-acre farm with sensors and connect each to a cloud system. With edge computing, network connectivity isn't as big of an issue. These systems can make independent decisions that balance ground moisture with available water resources [1]. Energy efficacy is one of the compulsory & chief issue today due to environmental effects, energy requirements, and pricing.

The Information Communication Technology area is one of the prime energy clients. ICT division is also regarded as a chief CFC (Chloro Fluoro Carbon) contributor, a prime greenhouse gas, emitting about the 2% of global CFC emissions [29]. The greenhouse gases emissions by cloud data centers are calculated to be 1034 metric tons in 2020, which indistinctively raises the climatic concerns and calls for accurate answers. In near past, various proposals have been offered to employ edge computing for refining energy competence of cloud facilities and operating devices. Mobile edge computing allows the offloading of computation intensive & energy consuming applications from mobiles to servers, thus decreasing the energy usages of operative gadgets. Lastly, edge has been studied as an incentive for refining energy competence of cloud applications [3].

7.3 Smart living

Edge computing has taken autonomous cars pretty far. However, this technology will be far more life-changing when it has the ability to connect to other cars, buildings, and structures. These smart ecosystems will compile the benefits of autonomous driving and turn cities into AI-powered machines [29]. Communication & conversation among smart appliances, like controllers, sensors, and actuators, is an essential and general spectacle in all areas of smart living and universal surroundings [5]. Smart objects & cloud interactive model utilized in varied smart solution, like cognitive gateways, showcases varied restrictions & flaws in the cloud interactions, particularly impulsive postponement and jitter [5]. Edge technology provides a solution for such issues which delay the ideas and actions of smart living solution. Development in the smart gadgets and detectors are focusing to fulfill the smart lifestyle goals. EHOPES a common abbreviation of Smart Energy, Health, Offices, Protection, Entertainment, and Surroundings presents the basic parts of smart life. Fig. 6 shows the usage of edge in smart houses [5]. Smart homes consist of smart TVs, smart lights, robotically controlled vacuum cleaners and various home appliances that run on the basis of IoTs.

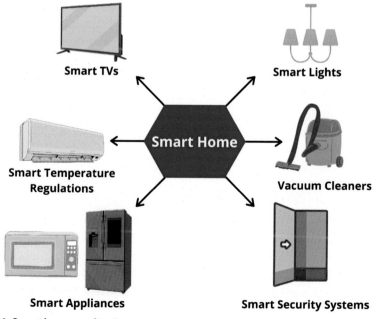

Fig. 6 Smart home applications.

Objects are regarded to be ineffective when linked with the cloud by just adding a wireless unit to such gadgets. The efficiencies of such smart homes are increased drastically when segregated wireless detectors & regulators are located in the walls, floor, and pipe of such homes along with connected appliances. The huge quantity of data produced by detectors; regulators & things must be treated nearby at edge of the smart homes in place of transferring it to clouds for effective utilization of such smart home [4]. This is applied by edge with an edge way gadget located in the house along with the things & running a specific OS called edge operating system limited just for edge. Smart houses furnished with a great amount of IoT gadgets belongs to an upcoming application area of edge. IoT implementations intended for the regulating & metering of smart houses will enable users to get automatic and accurate readings of various meters and allow invoices accordingly without any delay. Such IoT application are made for distant regulation & metering of different efficacies like water, power, and fuel. The information gathered from IoT gadgets can be transferred to the edge servers for the further processing rather than transferring it to the clouds, that can central to the real-time data analytics [19]. Edge & IoTs can be utilized in smart power administrations. These appliances automatedly notices the usages and supply patterns. Distributing nodes are utilized by the edge for real-time detection & processing. Cloud-computing is used as a cooperative tool to make these appliances robust and dynamic for huge quantities of data in the positioning of wider area energy systems. Edge in IoTs can help in making smart towns. Edge can be efficiently utilized in light regulation networks of roads & streets, water and air superiority regulation, finding alternative ways in emergency situations of accidents or disasters, & automated watering gardens in towns. Edge in IoT setting assists traditional logistics & provides newer captivating potentials which makes the flow administration of systems automated and relaxed [31].

7.4 Communication efficiency

5G and edge computing technologies has the capability to help organizations deliver a widespread range of exciting products and services. As IoT devices and their algorithms become more powerful, they need to be equipped with larger processors and storage, and consequently higher power requirements. This is not always possible within the limitations posed by the form factor of the device, and hence designers need to leverage the cloud for better compute and storage. With their distributed assets and unique network resources, Communication Service Providers are in a unique position to create value from the emerging distributed cloud and edge computing paradigm. The convergence and availability of 5G and edge computing should also usher in several new and interesting services that had not been possible earlier due to inadequate bandwidth for devices on the move, as well as portable devices that required high computing resources but that could not leverage cloud computing due to latency issues. Furthermore, network providers and digital ecosystem enablers can also leverage new business opportunities as they can fulfill the infrastructure requirements and provide better connectivity solutions [5,37].

7.5 Resource management

Edge computing solutions give the benefits of cost savings & simplifications of edge endpoint administrations. By applying an autonomous life-cycle administration method, one administrator can manage deployments to thousands of endpoints, with management tasks carried out based on intent and with no intervention needed [3].

7.6 Environment monitoring

Environmental regulation, consisting data gathering and reporting of metrics such as: outdoor temperature and humidity, water quality, pH value and temperature, dissolved oxygen, ammonia, nitrogen, and nitrite [30,35]. The more intelligent the edge becomes, the more important it is to understand the performance of the application at the edge and how it impacts the overall business performance. The growth of edge computing is being fuelled by the Internet of Things. Vast amounts of data are being generated by sensing devices that capture information about the physical environment—everything from humidity and light to chemicals and vibration. The collaborative use of edge and cloud computing in IoT can enhance the quality of existing monitoring systems. An automated system will collaborate with sensors and actuators. Applications have been developed for monitoring critical entities that exert a major effect on the environment. These entities include monitoring of gas concentration in air, water levels in lakes and underground, lighting conditions, soil humidity, and changes in land position. Environment monitoring is crucial in many fields, such as agriculture, forestry, and food safety [2,38] (Fig. 7).

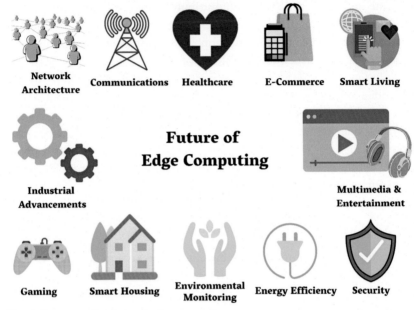

Fig. 7 Future applications of edge computing.

8. Conclusion

Edge computing is gradually developing and also paving the way for more effectual dispersed computing. The main aim of edge computing is to offload computations near the edge devices. It will also be necessary for researchers to address the above challenges in order to mitigate them in the future of edge computing. Many organizations are adopting the use of edge computing since it is useful and also productive. More innovative applications of edge computing are emerging to make human's life safer and more comfortable by supporting multiple devices such as secure smart homes, automated vehicle insurance, safer remote surgeries, etc. Therefore, edge is growing as an upcoming tech to satisfy the current requirements of increased data production. Currently, edge computing has pleased all structures of transmissions, substation, energy productions, energy usages, energy supply and transmittance involved in all network related fields. Lastly, we can conclude that the hope of upcoming future is on the shoulders of edge computing. Integration of varied and novel technologies will help into progressing humanity into safe hands and efficient usage of resources.

References

[1] F. Shi, J. Xia, Z. Na, X. Liu, Y. Ding, Z. Wang, Secure probabilistic caching in random multi-user multi-UAV relay networks, Phys. Commun. 32 (2019) 31–40.

[2] L. Fan, N. Zhao, X. Lei, Q. Chen, N. Yang, G.K. Karagiannidis, Outage probability and optimal cache placement for multiple amplify-and-forward relay networks, IEEE Trans. Veh. Technol. 67 (12) (2018) 12373–12378.

[3] J. Zhao, Y. Liu, Y. Gong, C. Wang, L. Fan, A dual-link soft handover scheme for C/U plane split network in high-speed railway, IEEE Access 6 (2018) 12473–12482.

[4] X. Liu, F. Li, Z. Na, Optimal resource allocation in simultaneous cooperative spectrum sensing and energy harvesting for multichannel cognitive radio, IEEE Access 5 (2017) 3801–3812.

[5] S. R. Jena, R. Shanmugam, K. Saini, S. Kumar, " Cloud computing tools: inside views and analysis", International Conference on Smart Sustainable Intelligent Computing and Applications under ICITETM2020 Elsevier, Pages 382–391.

[6] K. Saini, P. Raj, Handbook of Research on Smarter and Secure Industrial Applications Using AI, IoT, and Blockchain Technology, IGI Global, 2021. ISBN13: 9781799883678, ISBN10: 1799883671, EISBN13: 9781799883685.

[7] J. Ren, Y. He, G. Huang, G. Yu, Y. Cai, Z. Zhang, An edge-computing based architecture for mobile augmented reality, IEEE Netw. 33 (4) (2019) 162–169.

[8] M. Villari, M. Fazio, S. Dustdar, O. Rana, R. Ranjan, Osmotic computing: a new paradigm for edge/cloud integration, IEEE Cloud Comput. 3 (6) (2016) 76–83.

[9] I. Stojmenovic, S. Wen, The fog computing paradigm: scenarios and security issues, in: 2014 Federated Conference on Computer Science and Information Systems, IEEE, 2014, pp. 1–8.

[10] W. Shi, J. Cao, Q. Zhang, Y. Li, L. Xu, Edge computing: vision and challenges, IEEE Internet Things J. 3 (5) (2016) 637–646.

[11] U. Shaukat, E. Ahmed, Z. Anwar, F. Xia, Cloudlet deployment in local wireless networks: motivation, architectures, applications, and open challenges, J. Netw. Comput. Appl. 62 (2016) 18–40.

[12] I. Yaqoob, E. Ahmed, A. Gani, S. Mokhtar, M. Imran, S. Guizani, Mobile ad hoc cloud: a survey, Wirel. Commun. Mob. Comput. 16 (16) (2016) 2572–2589.

[13] W. Bao, D. Yuan, Z. Yang, S. Wang, W. Li, B.B. Zhou, A.Y. Zomaya, Follow me fog: toward seamless handover timing schemes in a fog computing environment, IEEE Commun. Mag. 55 (11) (2017) 72–78.

[14] L.M. Vaquero, L. Rodero-Merino, Finding your way in the fog: towards a comprehensive definition of fog computing, ACM SIGCOMM Comput. Commun. Rev. 44 (5) (2014) 27–32.

[15] W. Li, Z. Chen, X. Gao, W. Liu, J. Wang, Multimodel framework for indoor localization under mobile edge computing environment, IEEE Internet Things J. 6 (3) (2018) 4844–4853.

[16] Z. Ning, X. Kong, F. Xia, W. Hou, X. Wang, Green and sustainable cloud of things: enabling collaborative edge computing, IEEE Commun. Mag. 57 (1) (2018) 72–78.

[17] J. Gedeon, F. Brandherm, R. Egert, T. Grube, M. Mühlhäuser, What the fog? Edge computing revisited: promises, applications and future challenges, IEEE Access 7 (2019) 152847–152878.

[18] D.N. Le, R. Kumar, B.K. Mishra, J.M. Chatterjee, M. Khari, (Eds.), Cyber Security in Parallel and Distributed Computing: Concepts, Techniques, Applications and Case Studies, John Wiley & Sons, 2019.

[19] N. Hassan, S. Gillani, E. Ahmed, I. Yaqoob, M. Imran, The role of edge computing in internet of things, IEEE Commun. Mag. 56 (11) (2018) 110–115.

[20] P. Pace, G. Aloi, R. Gravina, G. Caliciuri, G. Fortino, A. Liotta, An edge-based archi-tecture to support efficient applications for healthcare industry 4.0, IEEE Trans. Industr. Inform. 15 (1) (2018) 481–489.

[21] J. Pan, J. McElhannon, Future edge cloud and edge computing for internet of things applications, IEEE Internet Things J. 5 (1) (2017) 439–449.

[22] H.T. Dinh, C. Lee, D. Niyato, P. Wang, A survey of mobile cloud computing: archi-tecture, applications, and approaches, Wirel. Commun. Mob. Comput. 13 (18) (2013) 1587–1611.

[23] N. Abbas, Y. Zhang, A. Taherkordi, T. Skeie, Mobile edge computing: a survey, IEEE Internet Things J. 5 (1) (2017) 450–465.

[24] E.M.E. Computing, I. Initiative, Mobile-Edge Computing: Introductory Technical White Paper, ETSI, Sophia Antipolis, France, 2014, pp. 1–36.

[25] H. Li, G. Shou, Y. Hu, Z. Guo, Mobile edge computing: progress and challenges, in: 2016 4th IEEE International Conference on Mobile Cloud Computing, Services, and Engineering (MobileCloud), IEEE, 2016, pp. 83–84.

[26] Y. He, F.R. Yu, N. Zhao, V.C. Leung, H. Yin, Software-defined networks with mobile edge computing and caching for smart cities: a big data deep reinforcement learning approach, IEEE Commun. Mag. 55 (12) (2017) 31–37.

[27] Y. Hao, P. Helo, A. Gunasekaran, Cloud platforms for remote monitoring system: a comparative case study, Prod. Plann. Control 31 (2–3) (2020) 186–202.

[28] Y. Cao, S. Chen, P. Hou, D. Brown, FAST: a fog computing assisted distributed analytics system to monitor fall for stroke mitigation, in: 2015 IEEE International Conference on Networking, Architecture and Storage (NAS), IEEE, 2015, pp. 2–11.

[29] S. R. Jena, R. Shanmugam, R. Dhanaraj, K. Saini, "Recent advances and future research directions in edge cloud framework", Int. J. Eng. Adv. Technol. (IJEAT) ISSN: 2249-8958, Volume 9 Issue 2, December, 2019 DOI: 10.35940/ijeat.B3090.129219.

[30] K. Saini, V. Agarwal, A. Varshney, A. Gupta, E2EE for data security for hybrid cloud services: a novel approach, in: IEEE International Conference on Advances in Computing, Communication Control and Networking (IEEE ICACCCN 2018) Organized by Galgotias College of Engineering & Technology Greater Noida, October 12–13, 2018, 2018, https://doi.org/10.1109/ICACCCN.2018.8748782.

[31] M. Liu, F.R. Yu, Y. Teng, V.C. Leung, M. Song, Distributed resource allocation in blockchain-based video streaming systems with mobile edge computing, IEEE Trans. Wirel. Commun. 18 (1) (2018) 695–708.

[32] P. Wang, C. Yao, Z. Zheng, G. Sun, L. Song, Joint task assignment, transmission, and computing resource allocation in multilayer mobile edge computing systems, IEEE Internet Things J. 6 (2) (2018) 2872–2884.

[33] Z. Chen, L. Jiang, W. Hu, K. Ha, B. Amos, P. Pillai, M. Satyanarayanan, Early imple-mentation experience with wearable cognitive assistance applications, in: Proceedings of the 2015 Workshop on Wearable Systems and Applications, 2015, pp. 33–38.

[34] J. Gedeon, C. Meurisch, D. Bhat, M. Stein, L. Wang, M. Mühlhäuser, Router-based brokering for surrogate discovery in edge computing, in: 2017 IEEE 37th International Conference on Distributed Computing Systems Workshops (ICDCSW), IEEE, 2017, pp. 145–150.

[35] M. Syamkumar, P. Barford, R. Durairajan, Deployment characteristics of "the edge" in mobile edge computing, in: Proceedings of the 2018 Workshop on Mobile Edge Communications, 2018, pp. 43–49.

[36] J. Zhang, B. Chen, Y. Zhao, X. Cheng, F. Hu, Data security and privacy-preserving in edge computing paradigm: survey and open issues, IEEE Access 6 (2018) 18209–18237.

[37] S. Mohril, M.S. Sankhla, S.S. Sonone, R. Kumar, Blockchain IoT concepts for smart grids, smart cities and smart homes, in: Blockchain and IoT Integration: Approaches and Applications, 2021, p. 103.

[38] Singh, A., Sonone, S. S., Sankhla, M. S., Parihar, K., & Saxena, M. Blockchain for IoT edge devices and data security. In Handbook of Green Computing and Blockchain Technologies (pp. 141–169). CRC Press.

About the authors

Swaroop S. Sonone is currently working as an Assistant Professor in Department of Forensic Science, Dr. Babasaheb Ambedkar Marathwada University, Aurangabad, Maharashtra, India and he has completed his Bachelor's degree in Forensic Science and Master's degree in Forensic Science, specialized in Digital and Cyber Forensics from Government Institute of Forensic Science, Aurangabad, Maharashtra. His broad area of interest includes digital and cyber forensics, mobile forensics, computer forensics, and multimedia forensics. He is currently working on digital transaction security, cybercrime vulnerabilities, mobile forensics, cyber security, digital evidence, and their legal aspects in courtroom. He has been participating and presenting his work for more than 20 national and international conferences and workshops. He has published more than 15 papers in *Scopus Indexed Journals* and several papers/chapters are under progress. He has delivered talks at various places including Commissioner of Police Office, Aurangabad, Sessions and District Court, Aurangabad about Forensic Science and its significance in society. He has given briefings about the instrumentations and their workings to visitors at Government Institute of Forensic Science, Aurangabad. He has also worked as Secretary at Council for Research Applied to Forensic Technology and Sciences, the research hub of Government Institute of Forensic Science, Aurangabad.

Dr. Kavita Saini is presently working as Professor, School of Computing Science and Engineering, Galgotias University, Delhi NCR, India. She received her PhD degree from Banasthali Vidyapeeth, Banasthali. She has 18 years of teaching and research experience supervising Masters and PhD scholars in emerging technologies. She has published more than 40 research papers in national and international journals and conferences. She has published 17 authored books for UG and PG courses for a number of universities including MD University, Rothak, and Punjab Technical University, Jallandhar with National Publishers. Kavita Saini has edited many books with International Publishers including IGI Global, CRC Press, IET Publisher, Elsevier and published 15 book chapters with international publishers. Under her guidance many MTech and PhD scholars are carrying out research work.

She has also published various patents. Kavita Saini has also delivered technical talks on Blockchain: An emerging technology, web to deep web, and other emerging areas and handled many special sessions in international conferences and special issues in international journals. Her research interests include Web-Based Instructional Systems (WBIS), Blockchain Technology, Industry 4.O, and Cloud Computing.

Swapnali Jadhav born on April 10, 1998, natively from Sangli District has completed her Bachelor's in Forensic Science and Master's degree in Forensic Science, specialized in Forensic Chemistry and Toxicology from Government Institute of Forensic Science, Aurangabad, Maharashtra. She has been participating and presenting her work for more than 15 national and international conferences and workshops. She has published more than 10 papers in *Scopus Indexed Journals* and several papers/chapters are under progress. Her broad areas of interest include forensic chemistry, nanotechnology, toxicology, forensic medicine, etc. She is engaged in the research of effects of various chemicals on human body. She has hand on experience on variety of sophisticated instruments like UV-Visible Spectrophotometer, IR Spectroscopy, etc. She has completed her internships under Council of Scientific and Industrial Research-North East Institute of Science and Technology (CSIR-NEIST) and Council of Scientific and Industrial Research-Indian Institute of Integrative Medicine (CSIR-IIIM) in year 2021 and 2022, respectively. She has delivered talks at various places including Commissioner of Police Office, Aurangabad, Sessions and District Court, Aurangabad about Forensic Science and its significance in society.

Mahipal Singh Sankhla is born on May 19, 1994, at Udaipur, Rajasthan. Currently, he is working as an Assistant Professor in the Department of Forensic Science, Vivekananda Global University, Jaipur, Rajasthan. Prior to this he has served as Assistant Professor in the Department of Forensic Science, Institute of Sciences, SAGE University, Indore, M.P. He has completed BSc (Hons.) Forensic Science and MSc Forensic Science. Currently he is pursuing PhD in Forensic Science from Galgotias University, Greater Noida, U.P. He has done training in Forensic Science Laboratory (FSL) Lucknow, CBI (CFSL) New Delhi, Codon Institute of Biotechnology Noida, Rajasthan State Mines & Minerals Limited (R&D Division) Udaipur. He was awarded "Junior Research Fellowship-JRF," DST-Funded Project at "Malaviya National Institute of Technology—MNIT," Jaipur and "Young Scientists Award" for Best Research Paper Presentation in Second National Conference on Forensic Science and Criminalistics and "Excellence in Reviewing Award" in International Journal for Innovative Research in Science & Technology (IJIRST). He is edited of 4 books and published 10 book chapters in various national and international publishers. He has published more than 120 research and review papers in peer review international and national journals. He has participating and presenting his research work for more than 25 national and international conferences and workshops and organized more than 25 national and international conferences, workshops, and FDP.

Varad Nagar is born on March 17, 2002, at Varanasi, Uttar Pradesh. Currently pursuing BSc (H) Forensic Science from Vivekananda Global University, Jaipur, Rajasthan, India and Pursuing Foundation Degree from IIT Madras in Data Science and Programming. He has been participating and presenting his work for more than 10 national and international conferences and workshops. He has published more than 12 papers in *Scopus Indexed Journals* and published 10 book chapters in various national and international publishers and several papers/chapters are under progress. He has hand on experience on variety of sophisticated instruments like UV-Visible Spectrophotometer, IR Spectroscopy, SEM, etc.

CHAPTER THIRTEEN

Edge computing security: Layered classification of attacks and possible countermeasures

G. Nagarajan[a], Serin V. Simpson[b], and R.I. Minu[c]

[a]Department of Computer Science and Engineering, Sathyabama Institute of Science and Technology, Chennai, Tamil Nadu, India
[b]Department of Computer Science and Engineering, Thejus Engineering College, Thrissur, Kerala, India
[c]Department of Computer Science and Engineering, School of Computing, SRM Institute of Science and Technology, Kattankulathur, Tamil Nadu, India

Contents

Abstract

Edge computing is a widely accepted approach in cloud based Internet of Things (IoT) environment to overcome the issues related to traditional cloud computing. Edge offers a fast computational response to IoT applications. The emerging IoT applications cannot

Advances in Computers, Volume 127
ISSN 0065-2458
https://doi.org/10.1016/bs.adcom.2022.02.010

be served by a centralized server placed at cloud. The latency of traditional execution pattern is not acceptable in some applications which perform the computations in sequential pattern with high time constraints. The deployment of edge node at the edge replaces the computation process from the distant cloud server to the edge node. Placing of resources at the edge of a network can be done in different ways. Based on the execution process at the edge, the computing scenarios are categorized into fog computing, mobile edge computing and mobile cloud computing. The end node can be placed as a one hop neighbor of the edge node and also as a member in MANET-IOT network which is connected to edge node. The edge computing also opens a wide range of chances to attackers to intrude the network. This chapter discusses the possible attacks in the edge computing, classifies the attacks based on modified OSI model for cloud computing and analyzes the countermeasures present in edge computing environment.

1. Introduction

Internet of Things (IoT) has a vital role in all emerging applications. IoT made it possible to include the nonliving things in communication. That removed the problems of inaccurate prediction of the state-of-object. An object became capable to claim the accurate state-of-object without any interference with the help of IoT. But, such claims demanded a huge work to be done in background to infer the state-of-object by the self-assessment process. It required enough resources and computation facilities for the assessment. The number of entities increased the demand of resources. Also, all the communications in the world have a sequential pattern. Such sequences also have dependencies on previous communications. Thus, each communication must happen during the intended time. The computational limitations must not disturb the flow of communication. Having large resource strength at each object was not practically possible [1].

At this juncture, cloud computing offered a solution for the resource constraints. All the computations can be executed by the resources placed at the cloud. The network is required to provide a communication channel between the cloud and end node. Cloud computing could stand as an optimal solution for this problem only till the emerging of applications that requires quick responses. The applications like high speed autonomous driving vehicles and augmented reality cannot be served effectively by cloud due to the high latency and jitter. Also newly emerging applications demand context aware routing. A large server placed at a distance for serving "n" number of end nodes cannot offer a context aware routing for each end node. Edge computing has been introduced at this juncture. This is a concept of placing the resources at the edge of Internet rather than placing the same at a distance [2].

Edge computing is capable to offer the end nodes, the required service with low latency and jitter. The edge nodes are usually placed one hop away from the end nodes. Thus it can provide a context aware routing and mobility support for all the end nodes. Edge also reduced the bandwidth utilization by doing the required computations at the edge of the network. Also, the implementation of edge nodes reduced the energy utilization at the central cloud server. In short, the edge computing has removed all the shortcomings of present network. The deployment of large number of edge nodes at the edge of the network became an open door to the attackers in the IoT network. Several attacks are possible in the edge based IoT networks. This chapter classifies the attacks in the edge computing network and discusses the existing countermeasures for defending against such attacks [3].

This chapter is organized in the following manner; the next section illustrates the four layer architecture of edge computing. Section 3 presents a layer-wise analysis of security threats present in the edge computing environment. Section 4 discusses about the edge based existing solutions for the security issues present in real world IoT applications. Section 5 gives an overview of open research challenges in edge computing. Section 6 summarizes the chapter and also discusses the scope of future works in edge computing.

2. Four layer architecture of edge computing

The cloud computing has several limitations due to its centralized architecture. IoT requires fast processing for real time applications. Also, it demands a context aware routing because the requirement is highly dependent on the environment. Traditional cloud computing cannot provide such services efficiently. But, the basic contribution of cloud network toward the low-resource computation cannot be neglected easily. Thus, some paradigms have risen to make the cloud computation stronger. The most effective way to accomplish the requirement is edge based computing. The computing facility has brought to the edge of the network. Thus the data need not be taken up to the cloud server for the processing. The edge becomes capable to offer all the services that were previously offered directly from the cloud. The cloud server will be notified about a finished computation whenever it is necessary.

The four layer architecture of edge computing is shown in Fig. 1. The initial layer is connected to physical world. The duty of IoT layer is to sense

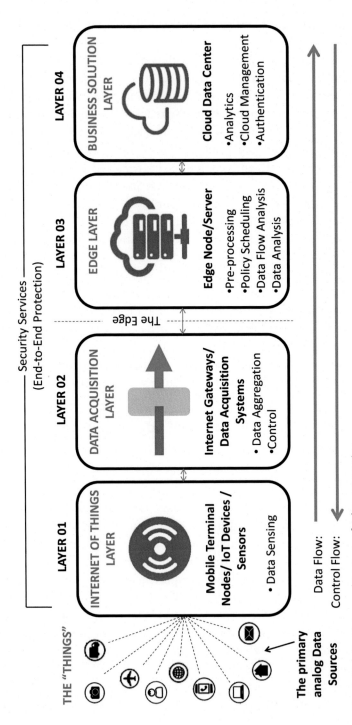

Fig. 1 The four layer architecture of edge computing.

the data and transfer the same to data aggregation system in layer 02. Layer 02 is an Internet gateway between the edge layer and IoT devices. Edge layer is placed in between the cloud service and IoT applications at the edge of the network. The edge layer is responsible to do the preprocessing, policy scheduling, data analysis and data flow analysis.

3. Security attacks in edge computing: Layered classification and analysis

International Organization of Standardization (ISO) has structured the network activities into a seven layer abstraction, called Open System Interconnection Model (OSI Model). Each layer in OSI Model is an abstraction of a set of activities and the participating components. Each layer contributes to their neighboring layers. Based on this layered classification, ISO could define the peer-to-peer interaction concept of network communication. Cloud technologies have been grown to the outer boundaries of OSI model. Thus the conventional OSI model failed to address the overall cloud functions within the designed boundary. A modified OSI model has been raised based on the proposal by CISCO for cloud platform. Rather than layering the network into seven, the modified OSI model keeps a five layer classification based on the cloud operations. The modified OSI model for cloud computing is shown in Fig. 2.

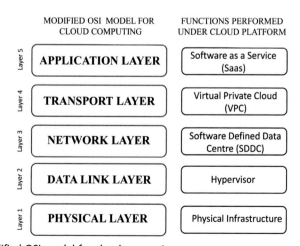

Fig. 2 Modified OSI model for cloud computing.

3.1 Physical layer

This layer is analogous to the layer 01 of traditional OSI model. But, it is more complex. The components in this layer are not visible to users. The infrastructure is completely handled by cloud service providers. The physical devices present in the datacenters are the foundation of cloud platform. Cisco UCS, HP converged system, and VCE vBlock are the known examples for the infrastructures used to build cloud environment. Actual data transmission is happening through this bottom layer. Binary data will be transmitted through this layer as analog/digital signals. Attackers can perform signal related attacks in this layer. They can try to retrieve the data from the signals. To defend such cases, other layers jointly contributes security paradigms by various mechanisms like encryption, authentication, etc. In such cases, Signal Jamming can be done to perform simple DoS attack. Through this jamming attack, attackers will not be able to get the data, but they can make the resources unavailable for a while [4].

3.2 Data link layer

Hypervisor is installed on top of physical layer. It provides virtualization. The hypervisor present in data link layer allows the users to access the physical devices by creating Virtual Machines (VMs). It helps the user to use the newly installed infrastructures without any delay. Such newly installed physical devices can be made available by VMs. This layer is responsible for quick response and resource rich execution. Hyper-V, KVM, Xen, NSX, ACI, etc. are the examples for the hypervisors used in industries. Data link layer provides Logical Link Control and Media Access Control. Media access control can be done either in centralized or in distributed manner in edge computing. This layer is mainly affected by spoofing attacks to get access to the communication. Logic Link Control includes error control, flow control and acknowledgment services. Attackers can perform DoS attacks by simply making the bandwidth unavailable for communication. Inference Attack is a passive traffic analysis attack present in data link layer. Attacker observes the data transmission through the edge nodes and tries to get knowledge about the network and network entities. This attack does not have the intention to get the messages from ongoing transmission. An attacker can do more with the knowledge about the network than the communicating messages. The most commonly used path and actively communicating entities can be identified by inference attack [5].

3.3 Network layer

The cloud accounts are created in this layer under software defined data center (SDDC). Network layer handles the computing and networking process by utilizing the resources under this layer. In SDDC, all services will be provided to the user by infrastructure virtualization. The data center will be automated by software. In SDDC, the resources and other infrastructures are managed by intelligent software systems. vSphere, OpenStack, AWS, Azure and GCP are used to manage the cloud infrastructure in this layer. In this layer, the data will be treated as tiny packets. The routing of data packets are handled in this layer. Thus, this is the most attacked layer among the layers introduced in modified OSI model for cloud computing. A plenty of attacks are present in edge environment which mainly focuses on packet routing. The attack can happen from both inside and outside the network. Reputation based attacks/Smearing attacks are the most common attacks in network layer. A node always tries to increase the reputation among its neighbors in order to participate in possible communications. The attackers can falsify the reputation about a node either for making the targeted node to participate in a communication or for keeping that node away from the communication. DoS attacks (e.g., Flooding Attack) and Identity based attacks (e.g., Sybil Attack) are also present in this layer [6,7].

3.4 Transport layer

The actual workload creation, data storage and process execution are happening in this layer. This is accomplished with the help of Virtual Private Cloud (VPC). This layer can also be called as Native Service Layer or Machine Instance Layer. The users in this layer will get the service even if they do not have access to network layer. This layer functions as IaaS (Infrastructure as a Service). The VPC provides access to developer or administrator through RDP, SSH or any other protocols. Transport layer is responsible for flow control, error control and segmentation process. A communication can be established either as a connection oriented service or as a connection less service. This layer has a set of protocols for controlling the flow of data through the network. An attacker can disturb/disrupt the flow by exploiting the vulnerabilities present in transport layer protocols. Most commonly used transport layer protocols are transmission control protocol and user datagram protocol. Replay attack is an active DoS attack present in transport layer. Attacker captures all the packets coming from the edge nodes and forward to the intended recipient edge node as the packets originally come from the source node. By this attack, the attacker is able to read or modify the message, if the message is not encrypted at source. A properly encrypted message cannot be tampered by this attack. In such cases, this kind of attack can contribute only delay to the network. The replay attacker always tries to hide the presence of attack in the network. This attack can be performed as a combination attack of one or more attacks. For example, replay attack can be done along with any cryptanalytic attack to decrypt the messages. So, the replay attack can be extended to message modification attack with the help of any cryptanalytic attack [8,9].

3.5 Application layer

This is the end user layer. It provides measurable business values to a set of constituents. The user access right is defined in application layer. This layer is analogous to the layer 07 of traditional OSI model. The front end of web applications, interactions through mobile applications and connectivity among IoT devices come under this layer. The software code running on this layer will do the necessary to make the cloud resources available to the end users. Blogs, Wikis, CRMs and HPCs are the examples of front end applications used in cloud environment. It is the only one layer which has direct interaction with the end user. Thus the data handled in this layer will always be in user understandable format. Protocols like, HTTP, SMTP and FTP are used in this layer. The attackers in this layer aims to retrieve the actual data. It can be done either by using some intrusive software (e.g., Malware Injection Attack) or by directly collecting the packets which are being transmitted during the communication (e.g., Collusion Attack). This layer also determines the way the data needs to be transferred to the network. The data can be sent either with compression or without compression. All such confidentiality aspects lie in this layer. Mostly the data will be encrypted for making it secure from the attacks. The attack in this layer also tries to discover the encryption algorithm to retrieve the actual data from the cipher text [10].

The attacks present in the network can also be classified based on this layering concept. Since each layer has peer-to-peer connectivity, each layer holds set of protocols. Thus, the attacks are mostly affected within the boundary of a layer. The layered classification of attacks present in edge computing is illustrated in Fig. 3.

Both active and passive attacks are possible in every layer. Attackers aim to intrude the network by analyzing the vulnerabilities present in each layer.

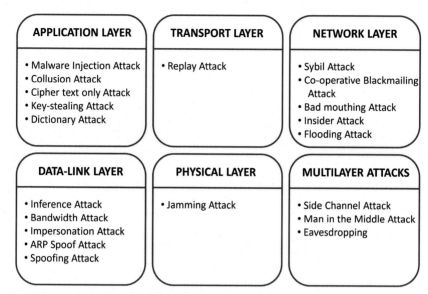

Fig. 3 Layered classification of attacks present in edge computing.

The layers are classified based on the duties which have to be jointly accomplished. All such duties are the work done required to fulfill a communication. Thus the attacks happening in each layer is dependent on layering pattern. All the above described layers are interconnected and all layers receive input from adjacent layers. Due to this relation, some attacks are possible in more than one layer. Such attacks are classified into multilayer attacks. These attacks can be performed irrespective of layers. This includes both active and passive attacks. Side Channel Attack is an active DoS attack. This is a special kind of attack that focuses on the subsidiary information of network. This attack does not consider the vulnerabilities of the system. This will only consider the additional information about the channel communication, such as signal leakage, variation in response time, power consumption and any other extra source of information. The attacker will find the way to make use of this information to perform denial of service. For example, response time will be small for highly sensitive information. An attacker can find the time of occurrence of such communication, if that occurs regularly. Performing a jamming attack during the estimated time will affect the next communication and that may cause enough damage to the IoT application. The side channel attack can be performed in any layer of network. The attacking strategies will be different based on the subsidiary information obtaining from that layer. Man in the Middle attack is an active multilayer attack. This can be performed in between any peer to peer communication. The attacker will be present in between the edge servers and end devices. Both of them cannot identify the presence of attacker. The necessary communication will be carried out between the entities without knowing the presence of attacker in middle. The attacker will get the data from the network. Eavesdropping attack is a passive multilayer attack. It comes under information leakage attack. This attack will not harm any ongoing communication directly. But it could overhear the data transferred through the edge nodes. A good encryption scheme can successfully prevent this attack [11].

4. Edge based existing solutions for the security issues present in real world IoT applications

The increased use of IoT devices also increased the security issues in the cloud platform. The emergence of edge computing also opened a wide door to the attackers to intrude the network. Since the IoT devices are mostly implemented in real world applications, the vulnerabilities of the network are equally dependent on the physical infrastructure as well as real world reasons. This section discusses the issues present in smart city, industrial environment, smart campus, vehicular network and healthcare system. This section also discusses the existing edge based solutions in cloud environment.

4.1 Smart city

A large number of edge computing units have been deployed for handling the communications/service requests in smart city. Almost all digital equipment are now connected to create a smart environment. The main goal of network engineer is to increase the performance of edge computing units (ECU). But, a linear increase could not be achieved due to privacy and security issues. Thus, the overall security maintenance has become challenging. In order to increase the overall performance of edge computing units, Xu et al. [12] proposed a Trust oriented IoT service placement (TSP) method. They have proposed this work for addressing the security issues in IoT based smart city environment. The authors try to increase the performance of strength pareto evolutionary algorithm (SPEA2) to get an improved placement strategy.

4.2 Industrial environment

The combination of IoT network and traditional industry is called Industrial Internet of Things (IIoT). IIoT receives wide acceptance from all industries due to the smart services that could be exclusively offered by IIoT. IoT network has a great role in reducing the man power requirement at industry. That could make the work easier than the conventional execution pattern. But the increased use of IoT devices also results in the reduction of Quality of Service (QoS). This is happening because of the improper network management and the security breaches. Thus industrial IoT demands a large scale improvement at defining security boundaries. Wang et al. [13] proposed an edge based unified trustworthy environment establishment for IIoT. The main aim of the proposed work is to identify the malicious service providers and service consumers. It also eliminates malicious messages and chooses only trusted service providers. Edge computing is incorporated with the cloud platform to perform trust evaluations based on the collected service records. The security is mandatory in industrial IoT. The security breaches in any level will badly affect the expected profit.

4.3 Smart campus

The most popular and widely used network paradigm in campus is mobile social networks. Thus mobile social network is used for the fast content delivery in smart campus. Conventional notice boards have been replaced with this new paradigm. The interesting and significant feature of mobile social network is the reliability. The message can be delivered both in "one to one" or "one to many" fashion. It is possible to ensure the delivery in both cases. Also, one important feature is the response scheduling. The response which we would expect to get can be scheduled. Sending reminder and nested communications are other salient features of mobile social networks. The wide acceptance of social network is mainly due to its incomparable services offered for accomplishing the needs of campus. An institution can use either an in-house platform or can utilize any globally accepted social network. Both cases are vulnerable to security attacks. Also the quality of experience needs to be enhanced for utilizing the same for institution purposes. Xu et al. [14] proposed an edge computing enabled mobile social network for smart campus. The main goal of proposed work is proper resource allocation. In order to find the optimal edge node, a reverse auction game technique is used. The game analysis results in a Bayesian equilibrium which is used to perform optimal edge node selection for each mobile user. Smart campus is a place where the crowd density will be larger than normal IoT environment. But, the crowd density will have a predictable parallel line pattern with time duration. Thus a minimum edge node and maximum utilization strategy can be applied to smart campus.

4.4 Vehicular network

The vehicular network is the most advanced application of Internet of Things. It requires dynamic processing. The traditional cloud platform is not capable to handle vehicular network due to the increased latency. The vehicular network can be implemented only with the fifth generation (5G) network, since it requires fast processing and immediate response from the service providers. The vehicular network can be used to avoid congestion and accidents. It can also increase the traffic efficiency. The main reason for choosing 5G network is the capacity of 5G network to perform device to device (D2D) communication. In D2D communication, the network devices can exchange information without utilizing network infrastructure. The direct D2D communication avoids the risk involved with traditional cloud network. Since the D2D communication does not demand a reliable connection with nearby base station, the applications which use D2D communication can also be used in 4G and 3G networks. Security attacks on vehicular network can risk the life of passengers. Thus such attacks have to be identified and avoided initially before it hits badly. Such attacks can be avoided only by improving the authentication system. The issues with authentication system may lead to privacy leakage. Zhang et al. [15] proposed an edge based secure authentication system for vehicular network. They have designed a protocol that could be expected to run over 5G network. The system model includes edge computing vehicles (ECVs), normal vehicles, road side units (RCUs) and cloud platform. Since the proposed scheme adopts D2D communication, the protocol can also be run over 4G and 3G networks. The edge based execution removes the issues associated with conventional execution.

4.5 Healthcare system

The emergence of wearable devices changed the concepts of conventional healthcare system. We have several sensor based wearable devices which can monitor our heartbeat, blood pressure, temperature, etc. The most significant achievement is that, an immediate medical care can be offered with the help of these sensor based wearable devices. Self-assessment might not be accurate, since public is still not having enough knowledge about human body. So, the healthcare system is highly dependent on modern technologies. Wearable devices can monitor our health and sensor nodes can generate the data. A cloud based processing unit can find the health condition based on the sensed data. All these processes are automated with the help of IoT. The hospitals can offer emergency assistance to the patients even if they are at home. Patients may be unaware about their serious internal health issues. The most modern assessments can predict the future health condition of a patient. Such findings are helpful to give medical assistance to the patient before he/she fell into the worst condition. The data saved at cloud datacenter will be huge. An accurate prediction and self-learning can be achieved only by considering such a big data. But, the data processing becomes complex due to big data. Also privacy and security issues are present in this field. The medical assessment of a person is completely private and such records need to be handled properly. No one wishes to reveal their health conditions to the public. Also, anyone can find the details of a human body from the sensed data. Illegal organ trade is the major concern which may happen if the medical records reach at improper hands. Thus the privacy and security of cloud based healthcare system is really important. Liu et al. [16] proposed an edge based privacy preservation technique for wearable devices. The system model includes patients, wearable devices and edge computing units. The proposed technique includes identity authentication and data access control. The method considers time aware and space aware contexts. In time aware context, access control is maintained by encryption algorithm and bloom filter is applied to preserve the privacy. In space aware context, hash code based authentication is used to preserve the privacy. Both contexts aim to preserve the privacy of sensitive data.

5. Discussion

This section discusses the open research challenges present in edge computing. All the edge paradigms have been introduced to overcome the shortcomings of cloud computing. The modern IoT applications could run smoothly with the help of these edge paradigms. This chapter discussed the vulnerabilities of edge paradigms in detail. The main reasons for all the discussed security attacks are resource constraints and the storage limitations of edge nodes.

5.1 Analysis of existing defense mechanisms

Several mechanisms have been proposed in edge environment to avoid and eliminate the security attacks. The application layer is mainly affected by malware based attacks. Application layer is defended against malware attacks by architectural enhancements. Application layer also has the responsibility to keep the confidentiality of network data. Thus the data will be encrypted. All attackers aim to find the encryption algorithm based on the available information to retrieve the original data. Research is still going on in this field to develop more secure encryption schemes. All security paradigms present in application layer and transport layer try to increase the strength of authentication schemes. Network layer in edge computing is affected by identity based routing attacks. All existing works in this layer hold a robust trust based mechanism to identify the attack. Data link layer and physical layer are prone to signal level attacks. Such attacks are identified and eliminated by employing signal level mechanisms. Multilayer attacks are also discussed in this chapter. It is not possible to define an exact solution to multilayer attacks, since it shows different behavior in different layers. Thus, such attacks have to be analyzed and treated according to the characteristics of each layer.

5.2 Open research challenges

Edge computing is an emerging area required to run the advanced IoT applications. Edge computing needs to be secured for eliminating the existing vulnerabilities. The existing solutions discussed in this chapter could overcome the vulnerabilities up to an extent. Still some more enhancements are needed due to the following reasons.

5.2.1 Rapid increase in the number of network components

The introduction of 4G LTE network increased the speed (5–12 Mbps) and capacity of network. But, 4G LTE network is not sufficient to run the IoT applications. Thus an enhancement has been done on 4G LTE network known as LTE-Advanced, which has a speed of 300 Mbps. Low level IoT applications have been introduced during this phase. Such applications could run in LTE-Advanced network with the help of cloud server. The main difference between the traditional network and IoT network is the increased connectivity among living and nonliving objects. For achieving the connectivity, all the objects must be installed with an IoT device. In this scenario, having highly efficient device at all communication end points is not practically possible. Thus IoT uses low-weight end nodes with an access to the cloud server. All the computations will be carried out at cloud server. Upon increasing the number of components, the response time of the cloud server increased rapidly. Thus the speed of the network expected to increase for the smooth running of IoT applications. The emergence of 5G network with a speed of 20 Gbps has solved the existing problems. But, the use of remotely placed cloud server again continued with a threat of having response delay due to the unexpected network issues. The number of components increased day by day and the speed of 5G network was sufficient to handle all the end devices by cloud monitoring. The emergence of quick response demanded services like vehicular network and augmented reality, again pointed toward the known threat associated with remote cloud servers. The edge computing has been introduced in this stage to overcome the threat. The edge servers are capable to provide quick response to the IoT applications, since such servers are placed near the edge of the network. But the deployment of such servers needed in every point to accommodate the increased number of network components. Thus a highly equipped device cannot be used due to the economic constraints. So the capacity of edge server has been limited due to the economical reasons. That in turn increased the security issues associated with edge computing. The IoT devices are increasing day by day. Thus there is a need for some robust mechanisms to prevent the security attacks on edge nodes.

5.2.2 Heterogeneous nature

The most challenging task in IoT environment is the incorporation of heterogeneous components. The range of components can be varied from a simple sensor node to highly efficient edge device for advanced IoT applications. The edge server needs to accommodate all types of components irrespective of their capacity. Device specific security breaches must also be considered by edge devices. The security attacks associated with one type of device may become stronger in another device due to the device specific security breaches. Thus, handling security in heterogeneous environment is a crucial task. Also, the edge servers have limitations due to the low–resource capacity. The edge servers are equipped with the capacity to perform all network related computations. But, it will not have enough capacity to incorporate security related advanced algorithms. Thus, it is a highly difficult task to ensure the security in edge environment under these resource constraints.

5.2.3 Possibilities of identity based attacks

Identity based attack is the most challenging attack in edge environment. It is happening in more than one layer of network. In application layer, it is present in the form of authentication attacks. The attacker tries to find the matching credentials to intrude the network by using the identity of other nodes. The spoofing and impersonation attacks present in data link layer are also identity based attacks. Such attacks are also present in network layer. Sybil attack present in network layer can generate multiple identities by using a single compromised node. All these identity based attacks are active attacks. The main reason for such impersonation attack in edge environment is the lack of centralized monitoring system among the edge servers. The cloud server will not be updated regularly. All the computations can be performed at the edge of network. Only log information will be passed to cloud server whenever the application demands a cloud server update. Thus, the identity based attack became a major threat among edge nodes. New mechanisms have to be incorporated with edge computing to make it stranger against security breaches.

6. Conclusion and future works

The cloud computing has become efficient to serve all high level IoT applications, only after the introduction of edge computing platforms. This chapter discussed the security attacks and possible defense mechanisms present in edge environment. The security attacks on edge computing are happening mainly due to the limitations of resource limited edge devices. Two major classifications of attacks have been contributed in this chapter. A general classification of attack has been done based on the attacking strategy and a detailed layer based classification has been done later, based on the modified OSI model for cloud computing. The attacks happening in each layer have relation with the functionalities of that layer. Thus, all those attacks are occupying inside the virtual boundary of each layer. Multilayer attacks have also been discussed in this chapter. Some common factors or strategies made such attacks possible in more than one layer. The defense mechanisms are categorized in this chapter, based on the classification of attacks. The existing mechanisms contribute toward one or two types of attacks expected to be present in the edge environment. A robust mechanism is needed in edge computing for eliminating all the vulnerabilities associated with edge nodes. The future works in edge computing can be extended toward handling the increasing heterogeneous network components and also toward the successful elimination of identity based attacks.

References

[1] M. De Donno, K. Tange, N. Dragoni, Foundations and evolution of modern computing paradigms: cloud, IoT, edge, and fog, IEEE Access 7 (2019) 150936–150948.

[2] S. Gong, A. El Azzaoui, J. Cha, J.H. Park, Secure secondary authentication framework for efficient mutual authentication on a 5G data network, Appl. Sci. 10 (2) (2020).

[3] R.I. Minu, G. Nagarajan, Bridging the IoT gap through edge computing, in: Edge Computing and Computational Intelligence Paradigms for the IoT, IGI Global, 2019, pp. 1–9.

[4] B. Li, T. Chen, G.B. Giannakis, Secure mobile edge computing in IoT via collaborative online learning, IEEE Trans. Signal Process. 67 (23) (2019) 5922–5935.

[5] T. Shanshan, M. Waqas, S.U. Rehman, M. Aamir, O.U. Rehman, Z. Jianbiao, C.-C. Chang, Security in fog computing: a novel technique to tackle an impersonation attack, IEEE Access 6 (2018) 74993–75001.

[6] J. Yuan, X. Li, A reliable and lightweight trust computing mechanism for IoT edge devices based on multi-source feedback information fusion, IEEE Access 6 (2018) 23626–23638.

[7] R. Smith, D. Palin, P.P. Ioulianou, V.G. Vassilakis, S.F. Shahandashti, Battery draining attacks against edge computing nodes in IoT networks, Cyber-Phys. Syst. 6 (2) (2020) 96–116.

[8] K. Fan, M. Liu, G. Dong, W. Shi, Enhancing cloud storage security against a new replay attack with an efficient public auditing scheme, J. Supercomput. 76 (2020) 4857–4883.

[9] K. Mahmood, X. Li, S.A. Chaudhry, H. Naqvi, S. Kumari, A.K. Sangaiah, J.J.P.C. Rodrigues, Pairing based anonymous and secure key agreement protocol for smart grid edge computing infrastructure, Future Gener. Comput. Syst. 88 (2018) 491–500.

[10] X. Li, S. Liu, F. Wu, S. Kumari, J.J.P.C. Rodrigues, Privacy preserving data aggregation scheme for mobile edge computing assisted IoT applications, IEEE Internet Things J. 6 (3) (2019) 4755–4763.

[11] Z. Wang, A privacy-preserving and accountable authentication protocol for IoT end-devices with weaker identity, Future Gener. Comput. Syst. 82 (2018) 342–348.

[12] X. Xu, X. Liu, Z. Xu, F. Dai, X. Zhang, L. Qi, Trust-oriented IoT service placement for smart cities in edge computing, IEEE Internet Things J. 7 (5) (2020) 4084–4091.

[13] T. Wang, P. Wang, S. Cai, Y. Ma, A. Liu, M. Xie, A unified trustworthy environment establishment based on edge computing in industrial IoT, IEEE Trans. Industr. Inform. 16 (9) (2020) 6083–6091.

[14] Q. Xu, Z. Su, Y. Wang, M. Dai, A trustworthy content caching and bandwidth allocation scheme with edge computing for smart campus, IEEE Access 6 (2018) 63868–63879.

[15] J. Zhang, H. Zhong, J. Cui, M. Tian, Y. Xu, L. Liu, Edge computing-based privacy-preserving authentication framework and protocol for 5G-enabled vehicular networks, IEEE Trans. Veh. Technol. 69 (7) (2020) 7940–7954.

[16] H. Liu, X. Yao, T. Yang, H. Ning, Cooperative privacy preservation for wearable devices in hybrid computing-based smart health, IEEE Internet Things J. 6 (2) (2019) 1352–1362.

About the authors

Dr G. Nagarajan received the B.E. degree in Electrical and Electronics Engineering from MS University and the M.E. degree in Applied Electronics from Anna University, in 2000 and 2005, respectively, and the M.E. degree in Computer Science and Engineering from Sathyabama University and and the Ph.D. degree in Computer Science and Engineering from Sathyabama University, in 2007 and 2015. He is a Faculty Member of the Department of Computer Science and Engineering, School of Computing, Sathyabama Institute of Science and Technology, Chennai, India. His current research interests include Computer Vision, IoT, 5G, Edge/Fog Computing, Artificial Intelligence, Machine Learning, and Wireless Sensor Network. He has published more than 70 research papers in peer-reviewed journals such as IEEE Conference, ACM, Springer-Verlag, Inderscience, and Elsevier. He also has contributed 15 book chapters thus far for various technology books. Finally, he has authored and edited 3 books thus far and is focusing on some of the emerging technologies such as the IoT, Edge/Fog Computing, Artificial Intelligence (AI), Data Science, Blockchain, Digital Twin, 5G, etc.

Mr. Serin V. Simpson has received his B.Tech degree in Information Technology from University of Calicut in 2012. He has received his M.Tech degree in Computer Science and Engineering from University of Calicut in 2015. Currently, he is doing Ph.D. programme in Computer Science and Engineering in Sathyabama Institute of Science and Technology. His research area includes Edge Computing and Network Security. He has 1 year of industrial experience and 6 years of teaching experience.

Now, he is working as Assistant Professor in the department of Computer Science and Engineering at Thejus Engineering College, affiliated to APJ Abdul Kalam Technological University. He has published papers in the research areas of Edge/Fog Computing, Cloud Computing, Wireless Sensor Networks, IoT, Mobile ad-hoc Networks, Network Security and Cluster based communication.

Dr R.I. Minu received the B.E. degree in Electronics and Communication Engineering from Bharathidasan University in 2004 and the M.E. degree in Computer Science and Engineering, and Ph.D. degree in Computer Science and Engineering from Anna University in 2007 and 2015, respectively. She is a Faculty Member of the Department of Computer Science and Engineering, School of Computing, SRM Institute of Science and Technology, Kattankulathur, India. Her current research interests include Computer Vision, Machine Vision, IoT, 5G, Edge/Fog Computing, Artificial Intelligence, Machine Learning, and Semantic Web. She has published more than 40 research papers in peer-reviewed journals such as Elsevier, Springer-Verlag, Taylor & Fransics, and IEEE Conferences. She has authored and edited 3 books thus far and is focusing on some of the emerging technologies such as the IoT, Edge/Fog Computing, Artificial Intelligence (AI), Data Science, Blockchain, Digital Twin, 5G, etc.

CHAPTER FOURTEEN

Blockchain technology for IoT edge devices and data security

M.P. Anuradha[a] and K. Lino Fathima Chinna Rani[b]
[a]Department of Computer Science, Bishop Heber College, Affiliated to Bharathidasan University, Tiruchirappalli, Tamil Nadu, India
[b]Department of Computer Applications, Bishop Heber College, Affiliated to Bharathidasan University, Tiruchirappalli, Tamil Nadu, India

Contents

Advances in Computers, Volume 127
ISSN 0065-2458
https://doi.org/10.1016/bs.adcom.2022.02.011

379

Abstract

The Internet of Things (IoT) is expanding rapidly across various trade verticals in real time applications. In IoT-architecture, there is a need to focus offline devices with its extensive-data due to its data processing restrictions and limited storage volume; but, the IoT-edge computing overcomes the above issues and offers a reliable architecture for managing the IoT devices with data. Due to the integration of computing and transaction processing systems, Blockchain in IoT (BIoT) provides high end security, synchronization, intellectual property management, uniqueness, affordability, and data privacy in real-time development. That basic initiative about BIoT-Edge computing technology including IoT layered architecture, IoT security threats and attacks and the operational functions of Edge computing along with architecture models are the foundations that are to be discussed in this proposed chapter. Future research directions of data integrity frameworks with edge computing platform in Blockchain of things are also focused.

1. Introduction
1.1 What is IoT?

The Internet of Things [IoT] is defined as a system of unified computational strategies. It is mechanically collaborated with digital technologies, things, creatures or individuals that are provided with unique identifiers and the capability for the transmission of information through network; but, not demanding the intercommunication of man-to-man or man-to-system. A "Thing" in IoT can be any device with any type of built-in-sensors with the capacity to accumulate and transfer data over a network without physical involvement. The rooted technology in the Internet of objects which is used in the progression of decision making supports them to interrelate with internal states and the outside settings.

In a nutshell, the Internet of Thing is a perception that attaches all the devices to the internet and lets them interconnect with each other over the internet. IoT is a massive network of connecting devices, all of which connects and shares data about how they are implemented and the situations in which they are executed. By doing so, each of the devices will be observed from the involvement of other devices, as humans do. IoT is trying to enlarge the interdependence of humans that is inter-reliant, subsidize and cooperate with things [1].

IoT enabled network is to deliver rich understanding in the field of health organizations, execution routines or monitoring of IoT-devices, and the execution settings of certain devices. The depth of data can mechanize the variation in sensitive data and stipulate the execution performance proficiently. When one or more IoT devices collected the data, the mechanism behind the IoT assists in finding fault prediction, avoiding the operational cost, enlightening consumer security, diminishing the loss because of manufacturing defects, and efficiently protecting the humans. Cloud computing technology is required for all the IoT technologies to reveal the competences of the IoT architecture. With quick scalability and huge data maintainability, ML (Machine Learning) is used to distinguish risky procedures in the information representation [2].

1.2 Basics of edge devices

An IoT network environment comprises web-enabled smart devices which are used for embedded systems such as: sensors, processors and data channels

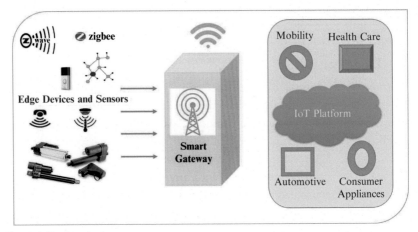

Fig. 1 IoT-edge devices.

used for accumulating data. They help to transfer and execute the procedures in the collected information.

Fig. 1 explains that the devices of IoT can send their sensor data, then gather those of which by linking to an IoT gateway or other edge devices where data is either directed to the cloud for analysis or executed locally. Occasionally, these devices interconnect with the other devices and perform in the information which they get from one another [3].

The connectivity, networking and communication protocols employed in these web-enabled devices basically depend on specific IoT applications. The IoT-client and systems can get cloud proficiencies through the edge devices. Therefore, The IoT- edge devices act like a bridge to resolve basic issues related to the centralized cloud environment. Though the cloud environment is authoritative, they make some sort of suspension for the IoT data transmission. While incorporating cloud environment proficiencies with IoT devices, edge computing executes the data quickly, avoids interruptions and ensures safety measures besides some apprehensions [4]. IoT edge devices solve the requirement of the larger IoT platform via removing the round-trip period, which is used for cloud execution.

2. IoT layered architecture

Numerous IoT services fetch various layers from IoT architecture representations. The basic four-layer architecture model consists of the perception layer, network layer, application layer and service support layer.

2.1 The perception layer

The bottom layer at the contractual IoT architecture environment is called the perception layer. In this layer, the sensors and associated devices originate to perform as they gather various volumes of data based on the requirement of the project. These can be edge devices, sensors, and actuators which cooperate with their network environment. This layer either acts as an in-charge to assemble information through IoT things or information taken from the network settings like Wireless Sensor Networks [WSN], interconnected IoT devices etc. It also computes the data collected from the above resources. This is considered as the physical layer resides in the standard IoT framework. The IoT devices such as the gateways, sensors, RFID tags, etc., are located in this lowest layer. Therefore, it is termed as the device layer or the recognition layer.

2.2 The network layer

The data which are collected by WSN, IoT, Interconnected IoT devices is required to be carried and managed. That's the work of the network layer. It links these device policies to other smart objects, servers, and network devices. It also controls all the data transmission process in the network layer and also is accountable for executing the Perception Layer's acknowledged information. Moreover, it takes responsibility for communicating information to the application layer via various network settings like wireless as well as wired networks with Local Area Networks (LAN). The foremost channels for information broadcast include FT TX, 3G (or) 4G, Wi-Fi, Bluetooth, Zigbee, UMB, infra-red, etc. Massive data will be processed and there must be a need to adopt a wide-ranging middleware to accumulate and execute this enormous data.

2.3 The application layer

This layer acts as the upper-most layer of the IoT Framework. This layer encompassing the application user interface takes accountability to organize data and access to the services. It takes authority for transferring several services to diverse customers. This layer offers a request of specific provision to network users. The requirements that are carefully chosen is based on the data which is collected from the IoT's object sensors.

2.4 The service support layer

The service support layer is located in the subsequent layer from the upper-most layer. The standard architecture models contain an additional layer by

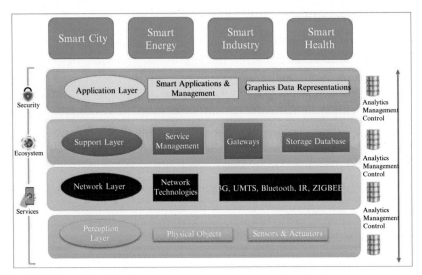

Fig. 2 IoT layered architecture.

the name of support layer, which is placed in the midst of the application layer and network layer. The ITU-T (International Telecommunications Union—Telecommunication Standardization Sector) recommends a four-IoT-Layered architecture which is described in Fig. 2. The fourth layer is intended to enhance the level of security in the standard IoT Framework. If the gathered sensitive information is directly sent to the network layer, there may be a chance of threat increases. An innovative layer is projected in order to avoid errors in three-layer architecture. Here, the support layer performs two tasks. First, it authorizes the quality of data received from the various authenticated users and implements the protection mechanism against the attacks. Here, several methods are to be followed to validate the authenticated clients and their data. The common security mechanism is performed by authentication, which is executed by using secret keys and passwords. The second role of the support layer is directing the processed data towards the network layer. Here, the communication channel for transferring the data from support layer to network layer may be wireless-network and wired network settings. The security competences are classified into two types (i) generic (ii) specific. These are dispersed among the four layers and shown in Fig. 2.

3. IoT security threats and attacks

The attack against the protection of IoT is also abbreviated as Interconnection of Threats (IoT). Certainly, IoT devices like sensors, actuators, etc., are predominantly susceptible to physical threats, software threats and hardware threats [5]. Existing IoT software and hardware are constructed with expertise's solutions through an extensive number of employees. Few of these IoT-software are a diverse combination of mechanisms renewing through current results for implementing in well-constructed IoT platforms. They assure that these mechanisms are executed in a protected technique.

Comprehensive risk and threat analysis mechanisms with management tools for IoT platforms are required to prevent the attack. To develop the qualification strategies for IoT attacks involves the study of various threat types and the disciplinary measures executed if the attacks are taking place. This chapter commences with the view of IoT attack classification and the analysis of security attacks supporting to spot an actual assessment of the IoT enabled networks. In addition, it permits the authentic user to regulate the qualification strategies.

3.1 Classification of attacks based on IoT-architecture

As has been elaborated in Section 1, a concrete IoT architecture representation is analyzed here as well. Generally, the IoT framework is composed of four layers that are illustrated in Fig. 2. There are some other vigorous attacks that may occur in all the fundamental layers such as: perception, network, and service layers. Furthermost, the main protection of security correlations in the IoT four-layered framework are briefly explained in Table 1.

3.2 Attacks—sensing (or) perception layer

To entirely protect the IoT devices, there needs a plan as well as constructed methods which must be hooked with IoT devices. It defines that IoT

Table 1 Threat categorization according to IoT architecture.

Security correlations	Application and interface layer	Service-support layer	Network layer	Device layer
Uncertain web edge	*	*	*	
Inadequate authentication	*	*	*	*
Insufficient network security services		*	*	
Transport encryption deficiency		*	*	
Confidentiality apprehensions		*	*	*
Insufficient cloud security interface	*			
Inadequate mobile security interface	*		*	*
Apprehensive protection formation	*	*	*	
Uncertain software and hardware	*		*	
Low security on physical devices			*	

devices themselves are to verify the uniqueness of data, sustain legitimacy, encrypt the information for maintaining honesty and restrict the quantity of deposit information for assuring confidentiality. The device security model needs to be firm enough for avoiding illegal access. Nevertheless, it must be adaptable for the protection of effective exchanges to the society as well as diplomacy.

3.2.1 Node attack
The authentic information can be extracted by an attacker on the devices as a substitute for demolishing the authorized data.

3.2.2 Sinkhole attack
If the sensors of the IoT-devices in the network are unnoticed for a long-time, then these devices are turned to be vulnerable to sinkhole attack. In this occurrence, the threatened system pulls out the authenticated data through all the neighboring systems in the network.

3.2.3 Selective-forwarding attack
Vulnerable nodes may select data–packets and illegally access data. After the obtaining of these data-packets, they can be released. Thus, SFA judiciously can screen specific data-fragments and permit the remaining packets. The dropped data-packets may contain essential data for additional execution.

3.2.4 Witch attack
It can be arising after an unauthorized IoT system and gets hold over the failure of an IoT-node. If the authentic node fails to work, then the accurate connection to the next node will deviate through the vulnerable node for further transmission which ultimately leads to data loss.

3.2.5 HELLO flood attacks
The vulnerable IoT-nodes originate a HELLO flood attack via transmitting the HELLO texts to each and every adjacent node. They can be accessed on a regular basis. Hence, there may be a chance to consider this node as a nearby IoT-node for each node in the network. Sequentially, this unauthorized system can transmit a HELLO text to each of its neighboring IoT-nodes and obstruct the valid nodes. This attack acts as the basis of non-accessibility of properties for authentic operators by dispensing a massive junk message to obtain provision.

3.2.6 Physical damage

Invaders can violate the procedural facts by their malicious activities. They are implemented through demolishing the IoT-devices. Though IoT-device inclusions are frequently non-tamperproof, the IoT-devices may be wide open and the physical parts can be examined through investigations and pin-headers. Hardware protection involves outlining inviolability of IoT-devices. Due to this, it is hard to tamper confidential information like user data, passwords and encryption mechanism. The exterior access can be dominated in most of the IoT-devices. Consequently, an intruder could replicate the whole system and spoil the authentic code and information.

3.3 Attacks—Network and service support layers

These layers, combinedly, represent the IoT service management structure. Moreover, it is accountable for IoT-devices which are used by the authentic users, spreads on policies combined with rules, and merges digitization with the IoT-devices. In this layer, Role based access control is used to find out the authenticated users and devices by using security mechanisms. In order to achieve authentication, there needs to maintain a sustainability for an audit track variation made by all authenticated users as well as devices, so that it will be intolerable for disproving the activities engaged in the IoT-network. This data control can also be assisted to recognize the malicious actions if it is detected. There are some harmful attacks may occur in the network and service support layer. These are as follows:

3.3.1 Man-in-the-middle (MITM) attack

It is an illustration of snooping probably in the IoT environment. Though device confirmation contains the verification of exchangeable devices of physical features, self-identity stealing can occur in this attack.

3.3.2 Replay attack

In this attack, the information is gathered from either the personal uniqueness of data interchange or from the other IoT authorizations. These sensitive data can either be deceived and transformed or reiterated. This attack is considered principally a dynamic method of man-in-the-middle attack.

3.3.3 Denial of service attack

The IoT-devices in IoT-network are considered to be resource controlled and these are exposed to various resource attacks extensively. Intruders could lead messages and information to an explicit system for the utilization of resources specifically.

3.3.4 Sybil attack

A malicious node pretends to be an active IoT-node in the network by using various uniqueness. Therefore, a network unknowingly permits a malicious node to implement and execute more than one time, in order to avoid redundancy. In wireless sensor networks (WSN), the attacker may use this malicious system to diffuse data via a negotiated system. They bring down the network system under their control.

3.3.5 Sinkhole attack

Intruders may outbreak incorporated system by attacking the information stream through neighboring systems. The whole system may deceive as well as simulate the information which was reached already in its terminal. In the WSN, the attacker can use this vulnerable system to fascinate the network traffic, and attack the sensor's data.

3.3.6 Sniffing attack

Intruders can utilize the sniffer systems and services to reach the network information and extract the sensitive data for further occurrences.

3.4 Attacks—Middle-ware layer (or) service support layer

This layer offers an interface as well as services to the application layer. Intruders may break the facility to disturb the application layer. The threats in database and system server can destroy the data as well as the operational node productivity. The cloud threats primarily intended on data virtualization that can produce an enormous danger for the user's confidentiality. The goal of middleware layer threat is to abolish the eminent applications as well as user's concealment.

3.4.1 Flooding attack in cloud

Cloud flooding attack is one of the methods of denial-of-service attacks. In this attack, intruders continuously transmit their requirements to a cloud service which can weaken the cloud applications in such a way it attacks the QoS (Quality-of-Service). It is a malicious attack on a network resource which avoids authentic users to utilize the resources and naturally employs by starting an overwhelming number of fake requests for services.

3.4.2 Malware injection

The intruder can modify the information and access the control over the resources as well as execute the vulnerable program via inserting illegal service code or simulating mechanism that is hooked on the cloud. It defines that, intruders replicate the data and upload a victim's application code, therefore, malicious code replies to the victim's request if some application is needed. At last, an attacker may accomplish the sensitive data services.

3.4.3 Signature wrapping attack

Cloud network applies XML signature for confirming service honesty. The intruder changes the snooped texts without signature authorization. Besides, an intruder may implement random guidelines and process like an authentic employer.

3.4.4 Web browser attack

The browser engines are enabled to complete instructions and guidelines on a remote server, like authentication as well as authorization mechanism. But, the web-browser may not yield scrambled XML tokens. Intruders take advantage over this weakness for improving admittance without verification. The cloud-web services may create some meta-data that can include a massive cloud content and implementation code. When the attackers attain this additional information, the intruders might have a chance to do cloud attack.

3.4.5 SQL injection attack

By incorporating SQL queries with IoT-data, a weekly intended programmable code can be vulnerable towards SQL Injection Attack. Intruders utilize these SQL queries for accessing, inserting, and removing actions. This attack cannot individually obtain private data; besides, it aims to attack the whole network's database. After the Web applications attack via SQL injection, existing webpage displays dissimilar results related to the original content.

3.5 Attacks—Application layer

The application layer threats primarily aim to take out user's authentication in an indirect way. Intruders characteristically reach the code and service susceptibilities such as code injection, memory-buffer overflow, and malicious access to outbreak the authenticity. Counterfeiting identity is one of the techniques for an unofficial agent to attain identical sanction (or) approval like an authenticated user. Apart from these threats, this layer is also endangered through system malware like Worms, Trojans and viruses. It can also be demoralized by other vulnerable codes like cross-site scripting, spyware, adware, etc.

3.5.1 Code injection

In this attack, the injection of vulnerable programs is fed into the system through manipulating the uniqueness of the code. It can be applied to snip the sensitive data, gain node access, and spread the viruses. The shell and HTML script injections are the most common vulnerabilities of these attacks. This attack gains the system's access control and the attackers take advantage of the authenticated user's confidentiality, and even make a whole network completely shut down.

3.5.2 Buffer overflow

Buffer Overflow is one of the malicious threats that destroys the program limit or memory buffer through abusing code susceptibilities. Most of the system programs executed in the primary memory, comprising code with fragments of data. This kind of attacker inscribes a lengthy program to a predefined area which is already accessed by the programmer. The resultant action may be the variation of other data, implementation of malicious code, and demolition of the program control flow. Additionally, this type of attack gives the affirmation to an illegal agent by expanding the system-administrator rights and implements the malicious programs.

3.5.3 Sensitive data permission and manipulation

The vulnerable control also destroys the user's sensitive information and their authenticity. This threat usually takes advantage over the designed errors in the consensus model for regulating the IoT smart services. The events can be used to direct the sensitive information of an IoT-Device to SmartApp; these events can be used by the SmartApp to screen the IoT-Device's data. But, the lack of protection in the events can leak out sensitive information and generate serious issues.

3.5.4 Phishing attack

There are more susceptibilities occurring in this attack. Here, the intruder plays like an actual user or faithful organization in order to attain the user's data like login-passwords and pin number details. An E-mail is the most common public channel attack. Here, the user's complex data is manipulated through the intruder while valid workers try to access the mail.

3.5.5 Authentication and authorization

Authentication and Authorization procedures act as a significant portion in the guard of IoT sensitive information. The current verification methods are not sufficient to offer a full-fledged authentication. Additionally, the susceptibilities also occur in the authorization network model. Here, the trick is taking advantage of the user's privacy that permits the intruder to retrieve data by ignoring mandatory details. Additionally, because of the lack of perfect verification of algorithm in the IoT services, the intruder performs malicious access.

4. IoT—Edge computing

Being an innovative architecture, IoT-Edge computing contains edge server characteristics. They are placed on the web edge, nearer by the IoT-sensors, edge workers, and the mobile devices. They evolve IoT terms such as "Cloudlets, micro data centers, and fog," which are used in day-to-day life. They are meant by the minor types of edge-installed computational techniques. It can be characterizing the contrasts of merging with enormous cloud computing data centers.

IoT-Edge computing denotes the empowering computational technique which permits the data to be executed at the network edge. The term "downstream data" is to be used for cloud applications and "upstream data" term can be used for IoT applications [6]. "Edge" is a well-defined term and is a continuation, by means of all computations emerging with network properties on the pathway amid primary-data and massive cloud storage [7]. An internet edge is an exclusive room which is positioned generally prior to the linked IoT-end devices. This technique suggested perfect location for minimum latency offload structure towards the sustenance evolving services like AR (Augmented-Reality), public-protection, transportation, industries, trade and health care.

When cloud computing diffuses with edge computing, innovative tasks and openings are increasing.

4.1 Functions

The edge computing consists of Bi-directional computational streams: one is data upstream in which the sensitive data moves from the IoT-devices to the cloud. Another one is defined as data extracted from cloud to IoT-devices. This is termed as data downstream which is described in Fig. 3. IoT-end devices in the edge computing model are considered as consumers as well as producers of data.

The main goal of IoT is not only intended for cloud services, but also, should they require accomplishing computational performance from the edge.

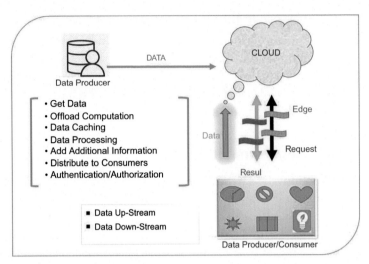

Fig. 3 IoT-edge computing.

Edge plays several roles like offload computing, massive data preservation, and caching along with execution. Additionally, it can dispense the requirements and distribute the applications through the cloud to the clients [8]. Besides these functionalities, the edge encounters the following prerequisites proficiently like perseverance, privacy, and data authentication.

4.2 Three-Tier edge computing model

The edge computing properties are needed to be examined; Fig. 4 depicts an intellectual three-tier edge computing model which comprises IoT-tier1, edge-tier 2, and cloud-tire3. In the IoT-tire, there are more drones; sensors are available for health appliances, plans and approaches which are executed for smart homes, and provisions existing for web trading. In order to relate IoT with edge tire, numerous data transmission procedures are organized. Through 4G/LTE, assigning drones with cellular tower and by using home gateway-Wi-Fi, smart home sensors can be linked together.

The cellular-tower, gateway, edge servers with its requests, and massive loading with the execution of cloud capabilities will perform various tasks [9]. There are procedures in edge and the cloud. They help to increase and achieve high-throughput as well as the speed. The transmission protocols between the edge and the cloud enhances the performance, such as ethernet, optical fibers and 5G technology.

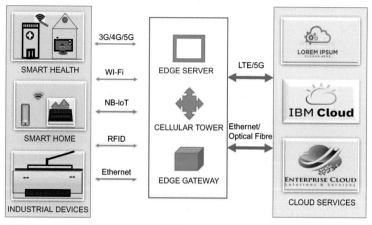

Fig. 4 Three-tier edge computing model.

4.3 Edge vs cloud

Edge computing and cloud computing have reliable connections. The IoT device's pervasiveness and quick progression of current technology together with cloud virtualization have diverted the attention to edge computing [10]. Edge computing requires authoritative power–consumption and also massive storage centers. The cloud data centers require the edge computational technique for progressing the massive secured information. Edge computing significant advantages are classified as follows. Basically, a massive amount of data is not fully sent to the cloud, but most of it is executed at the network edge itself [11]. It significantly diminishes the network bandwidth heaviness as well as resource utilization. Second, the data existing in edge are parallelly executed and they are independent in nature. This feature significantly weakens the latency along with the progression of response competence. Finally, in edge-computing, the authenticated user's sensitive data are stored in the edge-device instead of uploading. The resultant of this feature used to decrease the network hazard of data drip and shelter the data efficiency.

4.4 Attacks on edge nodes

A better picture needed for meeting vulnerabilities so that one can suggest an effective protection. There are two aspects to this: recognizing the ways that an intruder can compromise the node, and understanding the hardness of such action.

Attack Modes: There are four ways to get into edge node:
4.4.1. Network attack
4.4.2. External ports attack
4.4.3. Side-channel attack
4.4.4. Breaking into the device

4.4.1 Network attack

While the network is the greatest protected port, it is only as good as the safety in place. Entirely unprotected nodes can no longer continue by hoping to remain under the radar. For e.g., Web tools can crawl the network by recognizing every vulnerable node. The vulnerabilities may persist due to bugs in the poor random number usage in crypto algorithms, unobserved malware and violent protocol attacks by determining professionals, and even by weaknesses in the protocols themselves. Even with a significantly protected network, an intruder might compromise a poorly fortified edge node by replicating a firmware update and substitute the genuine code with code written by the attacker.

4.4.2 Port attack

The network port may be the only linking available on a small, bare-bones edge node. Sophisticated edge nodes, yet, may have segment ports for working in dissimilar sensors, or they may have USB or other ports for fittings, consumables, or assessment and debug equipment. Each of these ports affords an opportunity to access the edge node. Either attack could originate through an unused port, or an attachment could be detached and substituted by some other hardware intended to execute the attack.

4.4.3 Side-channel attack

Sophisticated attacks can also happen without producing any connection to the edge node. By tapping the power line or measuring emissions or vibrations on an insecure device, it is probable to extract information about the keys. By manipulating undocumented behaviors or faults like issuing a power flow, it may be possible to put a device into an undocumented and leaky state.

4.4.4 Physical attack

Finally, a determined attacker may substantially disassemble the edge node in an effort to inquire the internal circuits either with and without power or even eliminate and reprocess ICs to acquire the contents of embedded memories. Comprehensive security must shield against all of these modes of attack.

5. Requirements for integration of blockchain and edge computing

There are some pre-conditions required to be encountered before the incorporation of blockchain with Edge computing.

5.1 Authentication

As has been discussed earlier, edge computing settings include service providers, infrastructure set-up with innovative services; therefore, it is important to ensure entity verification, which is combined with consistent edge-interfaces for getting smart contracts. The properties and characteristics of these entities are confirmed through blockchains in overall contract process execution. Therefore, the blockchain authentication process is essential to formulate secure transmission medium over edge-ecosystems even though they have different security measures [12].

5.2 Adaptability

In the current situation, IoT-edge devices, their exertion services, and predominant restricted blockchain resources are highly increasing [13]. Consequently, the diffusion of blockchain with edge computing technology must support the extensive number of users with their complicated tasks. In addition to that, the emerging technology must have the capability of adapting fluctuated settings to let the system or thing in the network or to leave from the settings easily.

5.3 Data integrity

It is defined as the preservation of data-honesty throughout its life cycle and gives the assurance of data precision and reliability. When the massive decentralized computing resources are manipulated, the updated edge-server along with the distributed blockchain-secured framework, achieving the authentication services are violated. As a result, they may lead to some changes in the outsourced data and they may create some space for illegal uploading. Hence, additional techniques are required for the authentication process. They are mandatory for the Individual Service Provider and customers.

5.4 Verifiable computation

It allows some untrusted customers to offload their data for computing while doing so, it preserves accurate outcomes [14]. Edge computing allows to execute outsourcing computation. This feature covers various kinds of computations without using blockchain technology, whereas the inspirational and self-sufficient Ethereum blockchain smart contract must guarantee the efficient computation process and the perfect resultant service.

5.5 Low latency

Normally the application latency can be defined as the combination of two terms: (1) Transmission latency and (2) Computational latency. Computational-latency specifies the time period taken together for data execution and blockchain mining. It's function is based on the system computational power. The emerging blockchain technology with edge computing is supposed to regulate its planning among the computational types and its performance, because the volume to afford the high-speed computational process increases from IoT–users to cloud servers which cause to raise the transmission latency.

5.6 Network security

Heterogeneousness devices and threat susceptibility are the two serious issues that are in the edge computing technology. The emergence of blockchain with edge computing features plays a vital role to substitute the substantial significant management over some communication protocols. It offers quick access from the preservation of large decentralized edge servers and creates well-organized monitoring in order to avoid vulnerable activities such as sniffing, and DDoS attack [15].

6. Integration of blockchain and edge computing
6.1 Blockchain role in edge computing

Blockchain relates to its capability for permitting the authenticated–user to record their data in a distributed shared ledger in the well-organized network settings. Additional consideration is rewarded for blockchain essentials like consent protocol, ledger topology, credits and agreement. Fig. 5 depicts that the integrated system extensions can be adopted into various kinds of edge computing and in the already created appropriate groups. The significant blockchain characteristics is the authentication and authorization

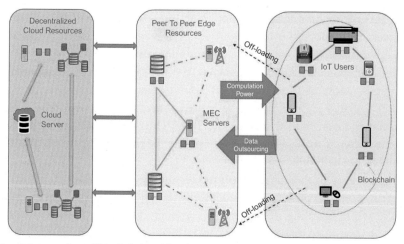

Fig. 5 Integration of blockchain with edge computing model.

which assures the data security enhancement and the necessity for improving scalability measures [16]. Edge computing has the capacity to achieve networking with the IoT-device and to store the massive data and computational capability over the decentralized edge; The subsequent benefit is followed with the application management.

The edge computing dynamic facts consist of attaining distributed scalability. The essential absorption is to afford effective monitoring in a protected way [17]. Consequently, the united architecture and characteristics of blockchain and edge computing- systems are projected at furnishing protected resources for attaining the client request by enchanting with network examination, massive storage with a computational capability that shelters the core blockchain framework and edge computing core competences [18]. An integrated opportunity originates through both the identical distributed network architecture and the same storage volume with computational functionalities; whereas, integration stipulation lies between the blockchain and edge computing disadvantages and it is needed to enhance their corresponding functionalities.

6.2 Mixing—Blockchain and edge computing

6.2.1 Edge computing—Inadequate security

The edge computing, distributed architecture contains several advantages. However, edge computing security remains an important issue. In the future, edge computing is considered a complicated interlacing of several and diverse technological developments such as Peer-to-Peer network, wireless technology, virtualization, etc. The various IoT devices dealings along with massive storage of edge computing servers and the global as well as local message transmission generate the probable causes for vulnerable activities. Most of the threats like jamming attacks, sniffer attacks, etc., can be initiated to break down the entire connection via shut-off the whole network, otherwise to screen the data flow. Thus, disciplinary actions taken by the network managers take responsibility for the authentication process, which is needed to be tested repetitively because of the high-vitality as well as the honesty of the edge computing technology.

It is a tedious progression to detach the data traffic from management traffic while handling the diverse set of edge devices in the network. It gives a chance to the opponents to regulate the entire network easily. Furthermore, distributed authority at the internet edge might be carrying a heavy load to the system administration. In edge computing networks, the data are detached into several fragments, and then, they are kept over diverse data repositories. That makes it even easier for the intruders to drop data fragments; it accumulates the data imperfectly. Therefore, edge computing cannot assure the data authenticity. Similarly, sensitive data drip along with some other confidentiality problems that may happen when the transmitted data from numerous IoT-edge nodes could be altered (or) mistreated through illegal adversaries. One more resistance for data-storage is confirming data consistency; meanwhile, outdated mechanism is used to rectify the data vulnerabilities by applying error-removal mechanisms and network security mechanisms. It produces larger storage over the edge computing structures.

Another significant security challenge in edge computing network is to preserve the security and privacy in uploading computational responsibilities to edge computation nodes. Some computation schemes exist in a place where the computation is outsourced with the computation function or the public key to one or more servers, which return the outcome of the process as well as a proof to verify the execution. Thus, it can be perceived that the security problems such as secure control at the edge, data storage, computation and the protection of network may require new thoughts to adapt to decentralization, synchronization, heterogeneity and mobility of edge computing. In particular, the grouping of scalability with security in such huge joints but avoiding unnecessary encryption outlays.

6.2.2 Challenges and restrictions of blockchain

In spite of blockchain and its excessive performance, it comes across several issues that could confine its extensive operation. The blockchain systems can insist and can have two concerns of the distributed network, scalability and authenticity [19]. Explicitly, decentralization allows the network to be permitted less and transparent. Security refers to the data reliability along with the common attacks, and scalability which refers to the capability to progress the communications, correspondingly. When it comes to current scalability issues, mostly the restrictions of low-throughput, high-latency, and resource draining inhibit the practical blockchain-based resolutions. Blockchain needs to increase its storage space when the number of transactions-records increases.

In the blockchain size of 158 GB in September 2017, reboot time is coarsely taking four days when a new node participates into the network chain. Ethereum framework is hard to undergo similar progression of requests, that is, upgradation of the whole blockchain record-history requires maintenance at each node. However, even now, the internet is enormously large; it leads to a rise in the restrictions in the blockchain distributed technologies. In addition, the blockchain size meets a centralization challenge, if a minor body of broad business environments is proficient to execute all systems which sometimes act like a forger whereas the light nodes do not have any technique of noticing this malicious activity immediately.

Having been regulated by the exact maximum block size (Block size = 1 MB) and the processing block time period, both are the origins to produce the next block. The public blockchains such as Bitcoin and Ethereum framework could manage only an average of 7 to 20 transactions for each second, which is too low for a payment processor, like master credit card, to deal with the average of two thousand transactions per second. Moreover, the transaction fee could differ for the application services, miners and the charge amount for the transaction record confirmation. The hardware charges and power consumption for the transactional mining process are not a negligible measure.

Towards increasing a blockchain's throughput, a simple method of various blockchain frameworks and huge block size have been disapproved due to the security cost and the decentralization challenge. Currently, on-chain network scaling of "sharding" and off-chain network scaling of "state-channels," both are certainly needed; Whereas all these mechanisms are yet to be improved with the combination of other techniques.

Therefore, the foremost research objective is to concentrate on growing transactional throughput, reducing the system bandwidth, storage, and processing consumption, in the meantime nominally losing its authenticity.

7. IoT framework: Secure edge computing with blockchain technology

The secure framework is the concrete outline since blockchain is incorporated with edge computing for IoT computational processing with storage demands. Architecture is comprised of systemized layers for migrating blockchain's comprehensive processes to an isolated layer which can be external to the application layer, holding IoT-devices and possessing restricted resources. The processes hooked on an individual layer of the framework are explained as follows.

7.1 Design overview

The cloud layer, edge layer and the device layer are the three layers composed in the IoT Framework as demonstrated in the Fig. 5 below. This framework consists of layers originating through the edge computing architecture since this framework is enhanced along with a Point-to-Point device connected within the individual layer to offer more storage with computational proficiencies. The device layer establishes Point-to-Point coupled with IoT destination systems which initiate the data as well as they exploit edge network resources. The edge layer contains various servers and storage facilities that are associated with Point-to-Point fashion to make additional storage available. It confirms firmness also reduces the single points of failure hazards. The edge-layer takes responsibility towards the temporary data storage, instantaneous data processing combined with data analysis and controlling data transactions swapped between the diverse set of systems. The cloud layer is inclusive of much authoritative proficiency to afford enduring data analysis, processing with massive storage capability and trade level transactions and reporting. The cloud resources can be systematized by blockchain, which confirms the data authentication.

7.2 Blockchain framework layered architecture

The proposed framework contains a significant blockchain framework layer, specifically, distributed IoT-device layer, Peer-to-Peer network edge servers and decentralized cloud properties which explain the issues and also encounter specific goals.

7.3 Distributed-IoT device layer

This layer contains IoT-devices to be located in Point-To-Point network settings, which are incorporated with the blockchain system for interchanging the sensitive information between the cloud resources and IoT-data. The distributed device-layer holds various IoT edge devices i.e., sensors and actuators for assembling the data and forwarding to other systems in the peer network. Otherwise, it will send the sensitive data to the higher layers. The smart devices use the data transmission mechanisms for accessing either the centralized systems via edge-servers or distributed by a Point-to-Point networking system. A private blockchain can be executed in the centralized message transmission by means of communication between the nodes regulated through the edge server. It is also responsible for appending new systems in the network, block validation and elimination of an active system in the network. The communication among the IoT-peers is simplified by a secret key sharing between the IoT-nodes through the server. Contrarily, IoT-devices and servers together could contribute within the public blockchain by Point-to-Point transmission.

Due to determination of end device resources, their contribution in blockchain is simplified through proficient servers located at the upper layers, edge and cloud servers. The IoT-end devices can complete only the easiest task like forwarding transaction reports with end nodes in the network (or) hardware updating, while the servers complete the excessive processes. The following Fig. 6 explains the process of distributed IoT-device layer. At the request of IoT-devices with constrained properties which are placed in both centralized and distributed network, edge servers, powerfully,

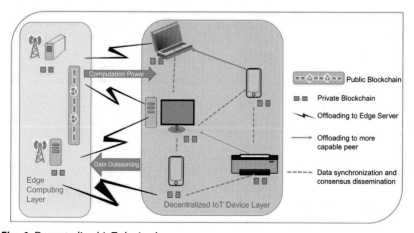

Fig. 6 Decentralized IoT device layer.

offer huge extended massive storage and high computational capabilities. Besides, edge servers are closely located at IoT-end users which lead to quick responses to their IoT-Clients.

The P2P connected IoT-devices can effortlessly offload all their data. Here, the data refer to either stored data or computational data. The data got loaded to an IoT-edge server (or) an adjacent peer for immediate response. In the offloading process, unlike the entire thing only a part of main transactions in the chain is accumulated by the IoT-devices. Here, the absorption is intensive computational management. Additionally, because of the lack of various standards of smart devices among diverse sellers, blockchain permits all these devices to contribute to the same blockchain network.

7.4 Point-to-point edge servers network

The edge layer maximizes the cloud scalability in order to carry the facilities nearer to IoT-intended devices for enhancing the processing performance and decrease the latency. The massive availability of resources to IoT-devices, edge servers can interact themselves to create replicated data repositories and synchronize data processing management. To gain this benefit, blockchain is positioned on the edge servers to launch a decentralized framework which assures the authenticated transmission over the network. During the P2P network interaction, small computational analytics is performed by the edge nodes for peers in the network and for the sake of the network, it is to attain self-systemization by means of adding and removing the edge nodes. In addition, they also take responsibility for data processing and forwarding the sensitive data report either to a decentralized cloud for permanent storage or send back to IoT-end devices based on the request.

7.5 Decentralized resources of cloud

The cloud layer is mainly intended for providing storage and computing services as per the requirement of the client while the blockchain considers each system as an individual node on the blockchain network for proficiently joining in the active mining progression. The cloud layer which has massive data storage with computational capabilities requires a consensus mechanism in the blockchain framework to assure security, low-cost and the enhanced processing capabilities. Properties of nodes in the blockchain architecture might be imposed once it is performed efficiently and they are executed consequently when they move towards the united data authenticity services. The nodes located at the cloud layer are autonomous data and by the implementation of blockchain, record sustainability will be maintained through the complete repetition process.

8. Factors to be addressed in secure edge computing

The addressing factors for securing edge computing execution that have been recognized include:

8.1 Low latency

By its nature, the edge is nearer to the IoT device than the cloud. This means a quicker round-trip for communications to attain local processing power, suggestively speeding up data communications and processing.

8.2 Longer battery life for IoT devices

Being able to expose communication channels for a short span of time due to improved latency means that battery life of battery driven IoT devices could be prolonged.

8.3 More efficient data management

Processing data at the edge makes simple data quality management such as clarifying and prioritization are more competent. Finishing this data management at the edge means cleaner data sets can be represented to cloud-based processing for further analytics.

8.4 Access to data analytics and AI

Edge processing power, information processing and data storage could be all combined processing to permit analytics and AI. They involve very fast response times or require the processing of large "real-time" data sets.

8.5 Resilience

The edge proposes more probable communication paths than a centralized model. This distribution assures that flexibility of data communications is more secure. If there is a failure at the edge, other properties are obtainable to deliver continuous operation.

8.6 Scalability

In decentralized with the edge model, a lesser amount of load should eventually be positioned in the network. This means that scaling IoT devices should have a fewer resources effect on the network, particularly if application and control planes are positioned at the edge together with the data.

9. Advantages—Integration of blockchain and edge computing

Various features of edge computational process, data storage repository, integrated network, required for both the blockchain and the edge computing are identically decentralized mechanisms. These technologies dissimilar in compatible significances and are intended for their combinational working mechanism. There are several benefits that originated through the diffusion of these two technologies. The blockchain and edge computing combination increase the data authorization, confidentiality, and resource utilization.

• There is a possibility for constructing a distributed network. It contains dozens of nodes over the blockchain architecture. Through mining processes and cryptographic methods, it achieves privacy, reliability and data validation with guidelines in a crystal-clear method [20]. Hence, it considers being an operating result in which larger diverse users are divided and located on the edge (or) moving among detached physical edges.

• In edge computing, privacy is a challenge. Here, the data is uploaded either locally or the entire data is divided into fragments between various nodes in the network for exposing the form of synchronizing edge processing. The blockchain technique uses sharing crypto keys among the participants for accessing and regulating the data without a mediator. So, it's significant feature permits a bottom-up management approach without revealing the additional information like source node, destination node and the sensitive data. Therefore, it is also termed as an immutability network.

• The resource interaction between the fog nodes is achieved through active synchronization [21]. Through dynamic execution of on-demand resource allocation and algorithm, the Smart contract integrated with blockchain can utilize the resources based on the service demanded. Here, the service level agreement for both the client and service provider is authenticated by monitoring the resource utilization. Therefore, the integration of these two technologies and edge computing resource management achieves the consistent, self-regulation, powerful distributed network and effective executions suggestively minimize the functioning amount and massive edge storage.

- The edge computing design framework starts through P2P followed by the extension of the other edge devices in the network. It merges with P2P cloud computing across the blockchain technology. The blockchain scale measures are significantly implemented by using this hierarchy level ordered framework. It supports blockchain information and propagation.

- The blockchain technology enhances the offloading authentication quality by the verification of resource-restricted end users of the blockchain network. Currently, Bitcoin is considered as the most immutable technique, on the other hand, it could hardly execute on the IoT-edge devices like sensors, mobile devices, etc. It might be comprehended through implementing Proof-of-Work [PoW] mechanism to the edge users. Similarly, more financial approaches can be executed by the edge with blockchain computational processes.

- The essential requirement of public blockchain is a storage durability and the private chain demands the authenticated and a self-regulating atmosphere. All these provisions are efficiently afforded by an edge server. Blockchain offers only the restricted storage for multimedia applications which requires off-loading storage and it applies some cutting-edge multimedia mechanisms to enhance the edge computing authenticity.

10. Use cases—Blockchain with edge computing

The edge computing technology amalgamated with the blockchain technologies ensure security and data integrity in the following application areas viz. Smart city, Smart transportation, Industrial-IoTs, Smart home and Smart grid.

10.1 Smart city

Deployment of IoT and blockchain technologies is used to build the Smart city not only does it enhance the efficiency of urban management and operation, but also promotes the leap forward development of the city by sensing information, securely communicating information and proficiently processing information.

The blockchain-based decentralized cloud architecture with a Software Defined Networking (SDN) which incorporates blockchain technologies comprises three layers: IoT devices, edge servers and cloud servers. It resolves the demerits of the outdated cloud computing. The inter-domain of these two technologies such as blockchain and edge computing combinedly create a framework which ensures the security and privacy of smart cities [22].

Approximately, all the IoT-devices produce and execute sensitive user's data without authorization. A variant type of blockchain framework evolving with IoT edge computing ensures the authorization of sense data. A micro service which is founded on Blockchain decentralized architecture is employed in a hierarchical blockchain-based edge computing to shield the data execution across various service providers. Such things are located in the smart public safety system.

Blockchain technologies are used to enhance the consistency and durability of user's data. A blockchain database resolves the security issues of the Smart city connected with home devices and sensors.

10.2 Smart transportation

The smart transport system is a fusion of blockchain, whereas edge computing is vital in the field of transportation for future research. For instance, an effective and protecting transportation network system assists various things such as constrained authorization technique, one-to-many adjacency matching, destination co-ordination and data controllability through merging the vehicular fog computing with blockchain framework. Here, the fog servers can match travellers' details with transport drivers details in order to create a private blockchain.

In the blockchain-based MaaS, the smart contracts employ the edge servers, which can link travellers to providers in a more efficient way, and attain many advantages including validation, authorization and development.

10.3 Industrial IoTs

In the progression of industrial IoT, the incorporation of edge computing and blockchain technology offers a protected interface for the current cloud computing technology, data processing and execution, and authorization control, which speed up the delivering processes to the intended edge servers.

The extensible and safe edge application management is achieved through Edge intelligence- Industrial IoT architecture and blockchain security. An inter-domain edge resource scheduling mechanism and a credit differential edge transaction approval mechanism are used to decrease the edge services cost and expand facilities capabilities [23].

In the Industrial IoT, key features of validation, transparency, authority and privacy are achieved through lightweight key agreement protocol built by securing public keys [24]. A novel Blockchain-based Internet-of-Edge model is constructed through the combination of edge computing, blockchain and Industrial IoT. This framework generates a privacy defense mechanism for scalable and controllable IoT systems. For instance, IIoT Bazar is a distributed business edge service market, which generates auditability for all participants through the blockchain secured framework. It is used to deploy monitoring and tracking solutions mounted on IoT-edge devices. Low computational power IoT-Edge devices fused with the IIoT Bazaar by fog computing technology. Augmented Reality (AR) connects the IoT-users with edge devices.

10.4 Smart home

An innovative secured cross domain framework ensures the honesty, privacy and accessibility through the integration of blockchain technology with smart home. For example: A smart home emergency facility system can be employed by the Ethereum blockchain framework. This architecture is used to distribute the trusted IoT-smart applications, for illustration, and the Home Service Providers assures the authorized control to the smart home IoT devices.

The Error Control and the data transmission quality are attained by an innovative architecture that has been built on blockchain. They combined with edge computing technologies. Blockchain architecture permits the healthcare organization for assembling the sensitive health data from the home environment by the IoT-sensors as well as sending this protected data with other peers. Certainly, the life concerned data accumulated by the inter-domain framework which are very important to the IoT-User's treatment is approved by some sensors. These are used to monitor the biological as well as physical features and environmental quality. Therefore, they supply these data to the edge server for further processing.

10.5 Smart grid

A permissioned blockchain with edge model resolves two key problems viz. privacy and security in smart grid. Edge computing is integrated with the blockchain by the common authentication mechanisms and the key agreement protocol for the secured smart grid. The main goal of this key agreement protocol includes restricted anonymity and other shared key management without other complex encryption conditions. The blockchain can confirm secured energy transactions among the grids. The utility functions authenticate the various transactions in the smart grid to approve the node in the network.

11. Further challenges and recommendations

There are more favorable benefits and the bright foreseen future for Blockchain with IoT edge computing. Yet, the substantial tasks in the development and distribution of current and prospective framework will require more analysis:

11.1 Technical threats

The features of blockchain with edge computing are such as efficient reliability, security enhancement and the extension of scalability. But these applications require some advanced technical measures for the improvement of the overall framework. Furthermore, blockchain technologies always have some restrictions when it is incorporated with IoT-devices like transaction volume, implementation in authentication mechanisms (or) some smart contract execution [25]. Additionally, the decentralized mechanisms must be generated to resolve a flexibility issue.

11.2 Interoperability and standardization

When it comes to the participants in blockchain and IoT end users, both are playing a vital role to attain the entire fulfilment of integrated authorities as well as the effective implementation. Encryption standards along with the enhancement of data security and consensus among the users are requirements for the adoption of collaborative technologies and reach the international standards.

11.3 Blockchain framework

There is a need to construct a complete authorized infrastructure which meets all the requirements of inter-domain execution. That can fulfil all the necessities for the use of blockchain in IoT systems. Many advanced mechanisms need to be developed in order to face the issues which arise with the integration of the blockchain with edge computing strategies.

11.4 Administration, authority, controlling and legal characteristics

Addition to technical issues, significant monitoring is another major problem for revealing the probability of Blockchain with IoT. There is a possibility that some intruders counterfeit their blockchain attainment for fascinating their customer through the predictable revenues.

11.5 Rapid field testing

In the future, there is a need for optimization for combining blockchain with several applications when a user needs blockchain security for the IoT applications. The initial stage is to discover the adaptability (i.e.) the user must find out which blockchain framework fits to their IoT-requirement. Consequently, there must be a concrete mechanism needed to examine diverse blockchains. The two core steps of the above concept can be explained as follows: (i) standardization (ii) testing. In standardization, it includes deep consideration of blockchain applications like supply-chains, marketplaces, goods, and security solutions. All services should be reviewed and approved. In testing, more than one type of criteria must be estimated with respect to confidentiality, security, power consumption, throughput, latency and application of blockchain among the users.

12. Conclusion

This chapter mainly focuses on distributed applications and management to meet privacy, execution performance, scalability and inter-domain adaptivity for the future networks and systems. Eclipse attacks, selfish mining attacks and 51% of attacks are some of the attacks that still exist in the blockchain technology, though it provides various productive methods for the edge computing. Here, the main challenges are authorization and secure encryption algorithms. Resource utilization constraints in order to improve the software quality after the incorporation of blockchain with edge computing. In addition, there is a need to concentrate on transactional security and privacy in the IoT-devices.

This chapter explains the inter-domain of two eminent technologies which comprises IoT-framework, possible threats and attacks, edge-devices, and the security measures of collaborative blockchain along with edge computing technologies. At last, the opportunities and challenges are discussed. In addition, this chapter gives an outline about evolution of the current applications such as smart home, smart cities, smart grid, smart transportation and smart industries over the blockchain technologies. Having given the comprehensiveness of research areas, it is also determined that hard investigations are essential with better attention to be focused on transforming well-established fog computing.

References

[1] O. Novo, Blockchain meets IoT: an architecture for scalable access management in IoT, IEEE Internet Things J. 5 (2) (2018) 1184–1195.

[2] J. Pan, J. McElhannon, Future edge cloud and edge computing for internet of things applications, IEEE Internet Things J. 5 (1) (2017) 439–449.

[3] W. Shi, J. Cao, Q. Zhang, Y. Li, L. Xu, Edge computing: vision and challenges, IEEE Internet Things J. 3 (5) (2016) 637–646.

[4] S. Mostafavi, M.A. Dawlatnazar, F. Paydar, Edge computing for IoT: challenges and solutions, J. Commun. Technol. Electron. Comput. Sci. 25 (2019) 5–8.

[5] K. Chen, S. Zhang, Z. Li, Y. Zhang, Q. Deng, S. Ray, Y. Jin, Internet-of-Things security and vulnerabilities: taxonomy, challenges, and practice, J. Hardware Syst. Sec. 2 (2) (2018) 97–110.

[6] B. Varghese, N. Wang, S. Barbhuiya, P. Kilpatrick, D.S. Nikolopoulos, Challenges and opportunities in edge computing, in: 2016 IEEE International Conference on Smart Cloud (SmartCloud), IEEE, 2016, pp. 20–26.

[7] W. Shi, G. Pallis, Z. Xu, Edge computing [scanning the issue], Proc. IEEE 107 (8) (2019) 1474–1481.

[8] K. Cao, Y. Liu, G. Meng, Q. Sun, An overview on edge computing research, IEEE Access 8 (2020) 85714–85728.

[9] M.A. Rahman, M.S. Hossain, G. Loukas, E. Hassanain, S.S. Rahman, M.F. Alhamid, M. Guizani, Blockchain-based mobile edge computing framework for secure therapy applications, IEEE Access 6 (2018) 72469–72478.

[10] F.J. Ferrández-Pastor, H. Mora, A. Jimeno-Morenilla, B. Volckaert, Deployment of IoT edge and fog computing technologies to develop smart building services, Sustainability 10 (11) (2018) 3832.

[11] H. Bangui, S. Rakrak, S. Raghay, B. Buhnova, Moving to the edge-cloud-of-things: recent advances and future research directions, Electronics 7 (11) (2018) 309.

[12] J. Pan, J. Wang, A. Hester, I. Alqerm, Y. Liu, Y. Zhao, EdgeChain: an edge-IoT framework and prototype based on blockchain and smart contracts, IEEE Internet Things J. 6 (3) (2018) 4719–4732.

[13] T.M. Fernández-Caramés, P. Fraga-Lamas, A review on the use of blockchain for the internet of things, IEEE Access 6 (2018) 32979–33001.

[14] S. Naveen, M.R. Kounte, Key technologies and challenges in IoT edge computing, in: 2019 Third International conference on I-SMAC (IoT in Social, Mobile, Analytics and Cloud) (I-SMAC), IEEE, 2019, pp. 61–65.

[15] S. Huh, S. Cho, S. Kim, Managing IoT devices using blockchain platform, in: 2017 19th International Conference on Advanced Communication Technology (ICACT), IEEE, 2017, pp. 464–467.

[16] E.F. Jesus, V.R. Chicarino, C.V. de Albuquerque, A.A.D.A. Rocha, A survey of how to use blockchain to secure internet of things and the stalker attack, Sec. Commun. Netw. 2018 (2018).

[17] Y. Ai, M. Peng, K. Zhang, Edge computing technologies for Internet of Things: a primer, Digit. Commun. Netw. 4 (2) (2018) 77–86.

[18] P. Mendki, Blockchain enabled IoT edge computing, in: Proceedings of the 2019 International Conference on Blockchain Technology, 2019, pp. 66–69.

[19] M.A. Khan, K. Salah, IoT security: review, blockchain solutions, and open challenges, Fut. Gen. Comput. Syst. 82 (2018) 395–411.

[20] H.N. Dai, Z. Zheng, Y. Zhang, Blockchain for Internet of Things: a survey, IEEE Int. Things J. 6 (5) (2019) 8076–8094.

[21] P.J. Escamilla-Ambrosio, A. Rodríguez-Mota, E. Aguirre-Anaya, R. Acosta-Bermejo, M. Salinas-Rosales, Distributing computing in the internet of things: cloud, fog and edge computing overview, in: NEO 2016, Springer, 2018, pp. 87–115.

[22] M.A. Rahman, M.M. Rashid, M.S. Hossain, E. Hassanain, M.F. Alhamid, M. Guizani, Blockchain and IoT-based cognitive edge framework for sharing economy services in a smart city, IEEE Access 7 (2019) 18611–18621.

[23] I. Sittón-Candanedo, R.S. Alonso, S. Rodríguez-González, J.A.G. Coria, F. De La Prieta, Edge computing architectures in industry 4.0: a general survey and comparison, in: International Workshop on Soft Computing Models in Industrial and Environmental Applications, Springer, Cham, 2019, pp. 121–131.

[24] C. Wang, G. Yang, G. Papanastasiou, H. Zhang, J. Rodrigues, V. Albuquerque, Industrial cyber-physical systems-based cloud IoT edge for federated heterogeneous distillation, IEEE Trans. Ind. Inform. (2020).

[25] R. Yang, F.R. Yu, P. Si, Z. Yang, Y. Zhang, Integrated blockchain and edge computing systems: A survey, some research issues and challenges, IEEE Commun. Surv. Tutor. 21 (2) (2019) 1508–1532.

About the authors

Dr. M.P. Anuradha, Assistant Professor in the Department of Computer Science, Bishop Heber College, Tiruchirappalli, Tamil Nadu, India, has been working in the area of Wireless Sensor Networks and Internet of Things, Block Chain, Secure Supply chain management, Network Security. She received her M.C.A., and M.Phil, degrees in Computer Science from Bharathidasan University, Tiruchirappalli, India in 2002 and 2005 respectively. She has obtained Ph.D. in "Establishment of Reliable, Credible Routing Protocol with Maximum Connectivity for Wireless Sensor Networks through Secured and Aggregated Data Approach" from Bharathidasan University, Tiruchirappalli, India in 2014. She is guiding number of research works and published significant number of research papers in Scopus, Web of Science journals. She has presented more than 10 papers in national, international conferences and also acted as reviewer of the national and international peer reviewed journals.

She has a significant contribution in carrying out research projects on the design and development of IoT based secure supply chain management.

K. Lino Fathima Chinna Rani is working as an Assistant Professor in the Department of Computer Applications, Bishop Heber College, Tiruchirappalli, Tamil Nadu, India. She post graduated in Master of Information Technology from Bharathidasan University in 2008. She had completed her Master of Philosophy in Computer Science in September 2009 at Bharathidasan University, Tiruchirappalli. Currently, she is pursuing doctorate of philosophy in computer science under the broad topic, "An Enhanced Approach to Improve the Performance Efficiency of IoT with Block Chain". Her area of interest is Network Security; with a specific focus on Block chain Decentralized Technology. She is passionate to explore on implementation of Block chain and distributed technology to enhance the quality in IoT sector. She has published research papers in international journals.

CHAPTER FIFTEEN

EDGE/FOG computing paradigm: Concept, platforms and toolchains

N. Krishnaraj[a], A. Daniel[b], Kavita Saini[b], and Kiranmai Bellam[c]
[a]Department of Networking and Communications, School of Computing, SRM Institute of Science and Technology, Kattankulathur, Tamil Nadu, India
[b]School of Computing Science and Engineering (SCSE), Galgotias University, Delhi, Uttar Pradesh, India
[c]Department of Computer Science, A & M University, Prairie View, TX, United States

Contents

Abstract

At the moment, Web applications operating on smart phones create huge amounts of data that may be processed on the Cloud. Yet, one of a Cloud's core limitations is its connection to endpoint devices. Through the use of distributed compute, communication, and storage services along the Cloud to Things (C2T) continuum, fog computing overcomes this constraint and empowers new application possibilities such as smart cities, augmented reality (AR), and virtual reality (VR) (VR). Furthermore, the use of Fog-based computing resources and its incorporation with the Cloud brings new resource management issues, necessitating the development of new techniques to ensure application quality of service (QoS) compliance. In this setting, a critical challenge

Advances in Computers, Volume 127
ISSN 0065-2458
https://doi.org/10.1016/bs.adcom.2022.02.012

is how to link application QoS needs to Fog and Cloud resources. One possibility is to classify the applications that arrive at the Fog into Classes of Service (CoS). Thus, this article provides a set of CoS for fog applications that incorporates the QoS criteria that most accurately describe these fog applications. Additionally, this article suggests the use of a standard machine learning classification approach to differentiate Fog computing applications based on their QoS needs. Additionally, this technique is demonstrated through the evaluation of classifiers' efficiency, accuracy, and robustness to noise. Adopting a technique for machine learning-based categorization is a first step in defining methods for providing QoS in fog computing. Additionally, categorizing Fog computing applications can aid the Fog scheduler's decision-making process.

1. Introduction

Cloud computing provides widespread access to the available reservoirs of programmable resources and services through the Internet, which may be deployed fast and with little administrative work. However, as the Internet of Things (IoT), mobile, and multimedia applications gain traction, the transmission traffic jams between the Cloud and an end device have indeed been deemed excessive and unsuitable for latency-sensitive applications, posing the primary constraint on the Cloud's use for latency-sensitive and mobile applications. Cloud computing has enabled the provision of several services, including platform, software, and infrastructure, as services, with the possibility of presenting anything as a service. Nevertheless, it is not always practical to send sensor-generated large data to the cloud for processing and storage. Additionally, some IoT applications need quicker processing, which only modern cloud infrastructure is capable of fulfilling [1]. The problem is addressed through the use of FC, which utilizes the computing capacity of devices located near a user to assist with data storage and processing. FC's many objectives include increased efficiency and a decrease in the quantity of data that must be sent to the cloud for data processing, analysis, and storage. Fig. 1 The primary reason for this is frequently performance, although security and privacy can also be factors. Recently, AI algorithms have been used to the processing of IoT data. End–user devices at the network's lowest layer have a variety of undesirable characteristics, including insufficient memory, insufficient connection bandwidth, limited computing power, and heterogeneous hardware that differs from cloud infrastructure [2].

Throughout the previous decade, computing technology have progressed in a variety of areas, including AI, GPU computing, cloud computing, and other hardware advancements. Machine learning is the most frequently

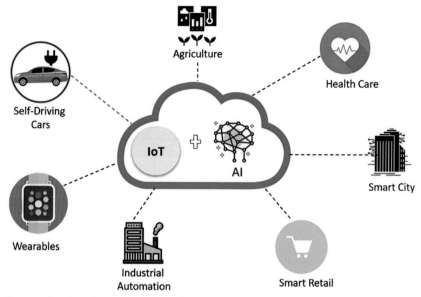

Agriculture

Health Care

Self-Driving
Cars

IoT

AI

Smart City

Wearables

Industrial
Automation

Smart Retail

Fig. 1 Role of machine learning in IoT devices.

used AI algorithm in a variety of disciplines. Several previous research have utilized machine learning to solve networking challenges, including network routing, security, traffic engineering, and resource allocation. ML is critical in establishing a smart/intelligent environment capable of autonomous administration and operation. The importance of machine learning is extended to IoT, as without it, IoT cannot fulfill functional, monitoring, or preprocessing duties. Additionally, satisfying the varied QoS requirements of the IoT remains a difficult challenge due to the resource-constrained nature of IoT devices. Thus, it is important to describe machine learning in terms of fog, cloud, and edge computing for IoT implementation. Using existing technologies such as Cisco's IOS XR, the ability to examine data on network equipment such as routers and switches is simple to deploy. Typically, machine learning research involves actuators, sensors, and low-level fog nodes [3,4]. However, fog nodes may manage frameworks such as Weka and Scikit-learn to create a variety of AI applications. Machine learning is used to automate, optimize, allocate, and monitor operational activities such as clustering, routing, duty-cycle management, data aggregation, and medium access control [2,5–7]. It is challenging to control the processes that occur in fog nodes due to their dynamic, complex, and diverse nature.

2. Machine learning (ML) in FC

Applying AI to fog and IoT enable the provision of services and applications, as well as the optimization of operation of the system and network management. FC is about redistributing part of the cloud's computational capacity to the network's edge, which is often occupied by IoT devices and human users. Thus, two strategies exist for incorporating intelligence into FC [8]. The first is device-driven intelligence, which occurs when IoT devices and fog layers get smarter with the addition of sensors, increased computing power, storage, and communication capabilities. Local server and base stations, IoT gateways, and nodes for mobile data gathering carried by a human user are all examples of devices with these characteristics. The research intended to increase these devices' intelligence capabilities, for example, by adding intelligent data processing and networking services to an e-health gateway or a gateway capable of doing machine learning [9]. Additionally, by monitoring wireless channel characteristics using a neural network model, efficient coverage and connection may be achieved. With these capabilities, edge devices may gather data with unprecedented granularity, enabling the network to be context-aware, make choices, and manage local resources. The second method is intelligence that is driven by humans. Human users are critical since they molded the IoT landscape's architecture. Humans are often the data sources in IoT networks. As a result, their behavioral patterns are critical in teaching the network to become smarter [5]. Numerous academic works are devoted to meeting human requirements, such as the e-butler for managing household appliances and the cognitive IoT-based smart home for enhancing living quality. The goal is to create a system that uses user circumstances while serving people and learns how to do network-specific activities effectively and with an eye toward power usage. In this way, user intelligence converts individual domain knowledge into beneficial network domain decision-making [8]. Thus, device and human-driven intelligence can be regarded as a viable option when designing an FC system that satisfies the QoS requirements of an IoT application.

3. Classes of service for fog applications

Fog computing allows innovative solutions, particularly those requiring low latency and mobility. Such new applications can have diverse QoS needs, necessitating the use of fog management methods to deal with this heterogeneity efficiently.

Therefore, resource management is extremely hard in fog computing, necessitating the development of integrated systems capable of constantly adjusting the resource allocation. The initial stage in managing resources is to classify incoming requests into Classes of Service (CoS) based on their Quality of Service (QoS) needs. Bandwidth. Certain applications require a minimum guarantee of throughput, referred to as a Guaranteed Bit Rate (GBR). Although some multimedia applications utilize adaptive coding techniques to encode digitized speech or video at a rate that suits the presently offered bandwidth, multimedia applications are bandwidth-sensitive Fig. 2. Sensitivity to delay. Certain applications, particularly real-time applications, have a specified latency threshold below which latency must be guaranteed. The loss sensitivity parameter specifies the percentage of packets that do not reach their intended destination. Reliability refers to the capacity of Fog components to perform the required operation in the face of a variety of failure modes. Certain applications require the rapid re-establishment of failed Fog components in order to complete activities within specified latency constraints. Availability is a metric that indicates how frequently the Fog's resources are accessible to end users. Applications and services that must run continuously, such as mission-critical applications, require high availability [5].

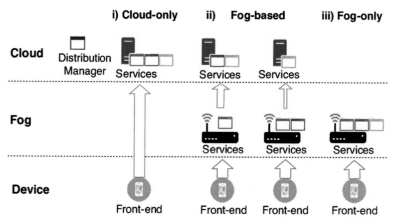

Fig. 2 Services of Fog computing toward high availability devices.

4. Clusters for lightweight edge clouds

Security is the process of developing and implementing authentication and authorization mechanisms to safeguard personally identifiable and sensitive information generated by end users. The data location specifies the location of the application's data. Locally, at the end device; remotely, at a Fog node; or in a distant repository, in the Cloud. The data placement requirements for an application are determined by a variety of variables, including reaction time limitations, the computing capability of each Fog layer, and the available capacity on network lines. Numerous edge devices are intrinsically mobile. Continuity of supplied services should be guaranteed, even for end customers who are very mobile. Continuous connection is required to do the necessary processing Fig. 3. Scalability refers to an application's capacity to run efficiently even when confronted with a rising number of requests from end users. The number of users in a Fog might vary depending on their movement and the activation of applications or sensors. In large data processing, streams of data may need to be processed within a certain time limit. Demand for Fog nodes is subject to fluctuation, and resource flexibility must be given to meet these demands [10].

Because of the widespread deployment of artificial intelligence, we delved deeper into the AI department [5]. The categorization in this part is based on where applications or services are placed in Fog, task scheduling, job offloading, and resource allocation. As a result, the authors chose the

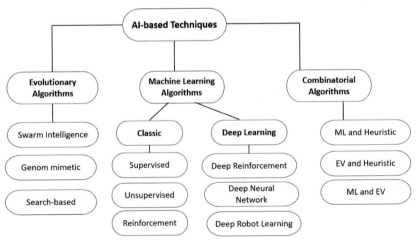

Fig. 3 AI-based fog computing application placement strategies.

algorithms stated in each category since they were their favorites, and they did not include any other algorithms in this area.

4.1 FAP's machine learning algorithms

Classic learning algorithms and deep learning algorithms are two subsets of machine learning algorithms. supervised learning, unsupervised learning, and reinforcement learning are three types of traditional learning algorithms (RL). Learning that is supervised. The supervised learning algorithm is a type of machine learning (ML) algorithm that uses a labeled data set as input. The basic goal of supervised learning is to build a model of input-output communication and anticipate the intended outcomes. Regression and classification are two types of supervised learning. Decision Tree (DT), ANN, Random Decision Forest (RF), K-Nearest Neighbor (K-NN), Support Vector Machine (SVM), Apriori, and Logistic Regression are among the classifications (LR). Regression entails the use of a regression tree.

Regression is a statistical technique used to model the relationship between dependent variables. Additionally, it is utilized to establish a link between independent and dependent variables. The advantages of linear regression include the following: it performs well regardless of the volume of data and when the information is separated linearly; it provides information about the significance of features. Although linear regression is prone to overfitting, this can be minimized by utilizing dimensionality reduction techniques including such regularization and cross-validation. Linear regression's drawbacks include the assumption of linearity, susceptibility to noise and overfitting, susceptibility to outliers, and susceptibility to multicollinearity. Polynomial regression has the following advantages: it is applicable to data sets of any size and performs exceptionally well on non-linear issues. Disadvantages of polynomial regression include the necessity to select the appropriate polynomial degree for optimal bias, a variance trade-off, and being overly sensitive to outliers. The module placement strategy in mobile fog computing is based on classification and regression trees to choose the optimal fog device. In comparison to the First Fit algorithm, the suggested approach consumed less power, responded faster, and performed better [4].

Linear regression is a type of mathematical modeling that falls under the classification category and is a generalized form of linear regression. The advantages of logistic regression include the following: it performs well when the dataset is linearly separable; it is less prone to overfitting than linear regression; nonetheless, it can overfit in high-dimensional datasets [11]. The disadvantages of logistic regression include the following: In contrast to real-world data, which is typically separable linearly, this approach assumes a linear relationship between the dependent and independent variables. Whereas if number of observations is lower than the number of features, Logistic Regression should not be used. In this case, the likelihood of overfitting increases. Only discrete functions may be predicted using Logistic Regression. Bashir et al. suggested a dynamic resource management approach for fog computing and used logistic regression to determine the load on each Fog node and its impact on subsequent decisions. The proposed algorithm enhanced performance by 98.25% Fig. 4.

KNN is among the most straightforward data mining and classification techniques available. This algorithm performs simple categorization operations and gives accurate results in the form of predictions. The term "closest" serves as the foundation for K's classification process. Each new sample is compared to all previous samples, and the closest previous samples are allocated to those samples Fig. 5. One of the benefits of this algorithm is that it requires no training, which means that fresh data may be introduced effortlessly without affecting the algorithm's accuracy. This approach is simple to construct and has applications in both classification and regression. The downsides of this technique include its inability to handle huge datasets, its inability to handle high-dimensional data, its requirement for feature scaling, and its sensitivity to noisy data, missing values, and outliers. Pastor et al. recommended the use of IoT Edge and Fog computing technologies, as well

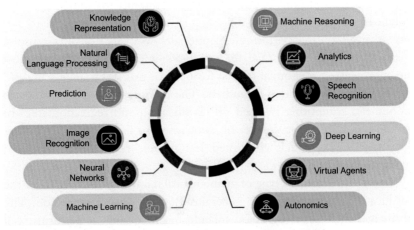

Fig. 4 Machine learning framework for different data flow variables.

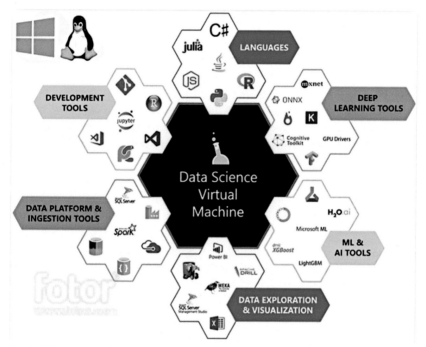

Fig. 5 Different development tool sets for machine learning algorithm.

as the integration of Edge and Fog paradigms, to construct smart building services. KNN and decision tree algorithms are being utilized to regulate energy use and generation from renewable sources [9,12].

A support vector machine (SVM) is a non-linear machine learning algorithm that is used for classification and regression. SVM's goal is to generate a hyperplane that separates the classes (classifies the data), such that this hyperplane is as far away from the samples in the classes as possible. The potential benefits of the SVM algorithm also included the following: it is appropriate for situations in which we have no clue about the data, it is best suited for unorganized and semi-structured data, it is capable of solving the most complex problems, SVM is not optimized for local optima, SVM models exhibit generalization in practice, and the risk of overfitting is reduced with SVM. The disadvantages of this technique are the difficulty of selecting the optimal kernel and the lengthy training time for huge datasets. He et al. suggested a multi-tier fog computing model for smart city applications that includes large-scale data analytics services [10]. Two classification algorithms are used to present the experiments: logistic regression and SVM.

ANN is the abbreviation for artificial neural network. An ANN can establish a link between the input and the desired output. The Multi Layer Perceptron (MLP) is a prominent neural network model that resembles the human brain's transfer function. The following are some critical characteristics concerning neural networks: first, neural networks are very dependent on data sets. Second, while neural networks are slow during the learning phase, they are lightning quick during execution. The advantages of ANNs include the capacity to store data over the entire network, to be fault tolerant, to have a distributed memory, to tolerate gradual corruption, and to perform parallel processing. Limitations of ANNs include hardware dependence, inexplicable network behavior, the inability to design an ideal network structure, the network's longevity is unpredictable, and neural networks become stuck in the local optimal. Deep learning frameworks such as TensorFlow and Keras, as well as Google Cloud Platform and Deep Neural Networks on Fog servers could be used for object detection [13].

4.2 Machine learning for the protection of security and privacy

The studies in this part are focused on enhancing the security of FCs or networks in general. Additionally, some authors sought to safeguard the privacy of user data [5]. It identifies two primary security concerns: recognizing zero-day attacks on IoT protocols and identifying cyberattacks launched from IoT networks. Additionally, they stated that network-based intrusion detection systems (NIDS) use storage for attack signatures and are unable to detect fresh attacks in future network traffic. Additionally, host-based IDS (HIDS), such as anti-virus software, is incompatible with low-resource IoT devices. They then recommended utilizing a Random Forest machine learning algorithm implemented in the fog layer to detect dangers in IoT systems. The system is designed around a master security fog node that monitors network traffic, detects intrusions, and sends cloud-based notifications. The RF model is trained using the UNSW-NB15 cyberattack data set. Python was used for both implementation and training. The classification accuracy was 99.34%, with a 0.02% false-positive rate. However, the confusion matrix revealed that the precision for attack detection is only 0.79% and the recall is 0.97%. They addressed how a centralized threat detection system would be ineffective in IoT applications due to issues of scalability, distribution, resource constraints, and latency.

Additionally, the majority of existing solutions rely on supervised machine learning algorithms which needs a lot of labeled data for training.

Additionally, accessible data sets (such as KDDCup99) are out of date, resulting in less reliable conclusions. As a result, the authors provided an intrusion detection and prevention framework that is based on the FC paradigm and employs a semi-supervised Fuzzy C-means (ESFCM) algorithm that is ELM-based. While connected to the IoT devices through a base station, the fog node detects the TCP/IP layers. We utilized SFCM and ELM to train the model on labeled data in order to label the unlabeled data set. This work makes use of the KDDTest-21 and KDDTest+ data sets. The centralized technique and other supervised algorithms were used as reference points for evaluation (SVM, KNN, Logistic Regression, Random Forest, and Bayesian Network). According to the results, the framework required less detection time than the centralized framework and beat other standard classifiers for both data sets. Additionally, such encryption algorithms are not meant to work on low-power IoT devices. Thus, the authors presented a technique for multifunctional data aggregation methods that preserves privacy (such as additive and non-additive aggregation). The technique utilizes a trained machine learning model in the fog layer to predicate the results of the aggregation query, which are then delivered to the cloud via the fog. In terms of privacy, the sensor IoT separates the original data and sends it to distinct fog nodes individually. Three entities comprise the proposed system model: IoT sensors, fog nodes, fog centers, and the cloud. In this case, the cloud sends the aggregation functions (minimum, maximum, median, -percentile, mean, and summing) to the fog center, which generates the appropriate queries for each sensor. After the fog nodes collect all of the sensors' reported data, a training data set is formed for training a basic linear regression algorithm. The regression model is then used to forecast the query set supplied by the fog center, that can store sensor data in order to respond to the fog center's aggregate queries. The authors demonstrated rationally that constructing the training data set satisfies the concept of differential privacy, which is accomplished by introducing Laplace noise to the entire training set [14]. We simulated and trained the machine learning-based Laplace differential privacy technique (MLDP) using two real-world data sets: the Mobile Health Data Set (MHEALTH) and the Reference Energy Disaggregation Data Set (REDD). To determine the suggested model's accuracy, the mean absolute error (MAE) was determined by comparison to the classic Laplace differential privacy approach (LapDP). The results indicate that LapDP outperforms MLDP with a small enough query set. Therefore, as the size of the query set expanded, MLDP outperformed LapDP Fig. 6.

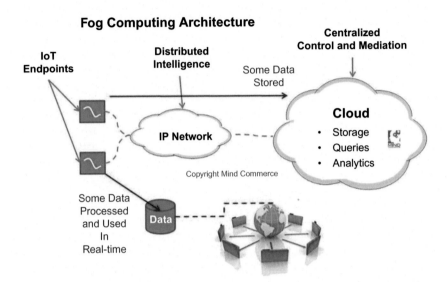

Fig. 6 Fog computing architecture for storage of cloud data.

Machine learning as a practical application The subsequent experiments used machine learning with reinforcement learning to solve a problem and create an application. Three modules comprise the system (FC, server computing, and back-end interactive communication modules). We trained and tested the Lie Group classifier model using photos of a mobile circuit board. The photos were carefully labeled with 12 distinct fault categories. The taught machine learning model is distributed to local fog nodes, which receive images from camera sensors suspended above assembly lines. The cloud server stores the classification results. The experiments were carried out using MATLAB simulations. The system's performance was evaluated using the Receiver Operating Characteristic (ROC) curve, which has been compared to the contour detection method, the pixel-based method, and the K-means algorithm. Additionally, efficient runs for a variety of items on the assembly line eliminated the time and delay associated with fog processing. In comparison to directly transmitting photos to the cloud for fault identification, the solution increased running speed by 53%. Additionally, it reduced time by 42% and boosted accuracy by 28% when compared to previous categorization approaches [7].

To solve the proof-of-work (PoW) challenge, blockchain applications need enormous amounts of computer power. Fuzzy services should be incentivized to sell fog resources while still guaranteeing QoS. They devised an optimum auction for fog resource allocation based on deep learning to mine blockchain data. There were two FFN networks utilized to address the assignment's optimization problem: one for payments, and the other for assignment submissions using a single FFN network. When creating the first FFN, the hidden layer employed the sigmoid activation function, and the output layer used the SoftMax function to generate the auction's winners. Since the payments in the second FFN are not negative, the output layer used the rectifier activation function. In objective unsupervised learning, the loss function included weighted restrictions, and the Augmented Lagrangian approach was utilized. In order to identify weights which minimizes the loss function, TensorFlow was used to execute deep learning and then the model was uploaded in the cloud. There are 5000 application evaluation profiles for miners in the data sets for both FFN. Fast and smooth convergence are the goals of the Adam optimizer during training. As a comparison, we used the greedy algorithm, which selects the top bidders in order to maximize income. Individualized rationality and incentive compatibility (IC) violations were compared against multiple rounds of Lagrange multiplier values in order to see which was worst. In comparison to the greedy strategy, the recommended auction quickly converged on a higher revenue value [8].

5. IoT Application with fog real time application

There has been a noticeable improvement in the overall quality of life as the Internet of Things (IoT) has progressed. To support real-time applications, high-end processing and storage units are necessary. In order to compute and store these enormous amounts of raw data, cloud computing is critical. However, the cloud-only set-up is not an energy-efficient and delay-aware solution for handling such a vast volume of data. Fog and edge computing have been invented to address this issue. On the other hand, \seamless communication owing to the mobility of IoT devices is a critical feature to process the data in the faraway cloud servers. For time-critical applications such as health care, connectivity interruption and therefore the increase in delay in providing the processed information, result in poor Quality of Service (QoS). Delivering processed data/information becomes difficult when the device becomes disconnected because to mobility. This needs a hierarchical\ infrastructure, where each tier (IoT, edge, fog or cloud) either gathers, maintains and processes information for minimizing the latency. That work proposes a collaborative Cloud-Fog-Edge system for

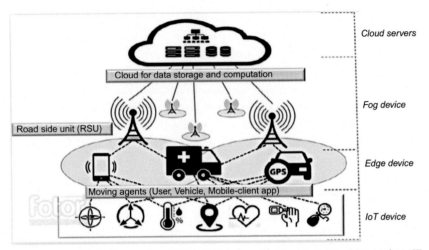

Fig. 7 Hierarchical placement of IoT, edge, fog devices and cloud in Mobi-IoST framework.

processing IoT data and presenting the selection based on mobility analysis to overcome the aforementioned issues. We have considered a hierarchical mobility-based infrastructure made of four layers: IoT layer, edge layer, fog layer, and cloud layer. Nowadays smart phone has become a popular medium for ubiquitous Internet access and various user-specific IoT apps are accessible through smart phones. It is possible that these smart phones might move about a lot because they are used as edge devices [15]. The users of our system utilize these time-critical IoT applications while commuting throughout. The edge layer is connected to the fog layer via the IoT layer and collects the raw data generated there. The cloud layer connects the fog layer, that are used for high-end processing and mobility analysis Fig. 7.

6. Safeguarding data consistency at the edge

Clients located near the network's edge, commonly known as the Internet of Things, can access a variety of services provided by several cloud-deployed apps (IoT). Applications which require minimal latency, such as augmented reality or online games, are frequently run using edge devices. Much of these apps can only be used if they have a response time of less than 6–40 ms. By using fog nodes to be doing portion of the computation that otherwise would have been performed on a cloud server, edge computing saves network traffic while also producing fast results. Fog nodes more anticipated to be administered by a number of different local providers and to be located in more vulnerable physical locations. As a result, fog nodes are extremely vulnerable, and developers of edge computing applications and middleware must priorities security over all other considerations. Offering safe middleware elements, like as secure storage services, that can protect the applications against fog node weaknesses is a viable way to make app development easier in this environment [9,12].

6.1 Data embedding on computing device with IoT on fog environment

Connecting connected systems to an existing Internet infrastructure and creating a computing environment is what is referred to it as the Internet of Things (IoT). The cloud servers process the raw data that IoT devices acquire. In contrast, storing and processing large amounts of raw data in a remote cloud adds delay and consumes more energy. Fog computing has been offered as a solution to this problem. To save time and energy, IoT devices' raw data is then processed locally on the fog device rather than in the faraway cloud. Connection interruptions during data processing become more difficult if the user is using a mobile device. For example, there is an Internet of Multimedia Things (IoMT) and an Internet of Health Things (IoHT) comment thread, as well as an Internet of Vehicles (IoV) sub-domain. IoST, or Internet of Geographical Things, is a new branch of the Internet of Things that focuses on managing spatial data. Geospatial interoperability standards across networks are commonly represented as numerical values about physical objects that may be transmitted and received by ubiquitous and embedded computing devices, which are referred to as IoST. Fog-based IoT uses switches, routers, and other IoT devices as fog devices to analyze raw data more quickly. Edge devices, such as smartphones and tablets, act as a bridge between the Internet of Things (IoT) devices and the network. However, these mobile gadgets face significant resource constraints. As a result, mobile devices must save their data on cloud servers. Additionally, mobile devices can offload computationally intensive tasks to the cloud to conserve battery life when resources are limited. Even for tiny amounts of work, using a distant cloud adds delay and increases the mobile device's battery usage. Numerous current techniques to offloading have already been targeted towards reducing energy and latency [5,7].

7. Cloud-fog-edge-IoT collaborative framework

Organizations and individuals are increasingly relying on smart gadgets and computers in the Digital Age. Data is generated by apps and sensors, which use electronic devices. This means that a lot of businesses are responsible for storing a lot of data. Businesses need a dynamic IT infrastructure now because of the shift to cloud computing, which offers benefits such as scalability, access to resources on demand, and pay as you go pricing models. Platform as a Service (PaaS), Software as a Service (SaaS), and Infrastructure as a Service (IaaS) are all moving toward "Anything" as a Service thanks to cloud technology. Unfortunately, some of the large amounts of data collected by sensors cannot be sent to the cloud for processing. Many Internet of Things (IoT) applications which require faster processing, however current cloud capacity will not be able to handle them. Fog computing is a way to solve this issue since it uses the processing capability of nearby devices (i.e., idle computer power) to help with storage, networking, and processing [16]. Fog computing can be used to achieve a wide range of objectives, including improving efficiency and reducing the amount of data that must be transmitted to the cloud for several reasons, including data processing, analysis, and storage. Most of the time, this is done for reasons of efficiency, but it can also be done for security and regulatory compliance considerations. In IoT data analyzing techniques, AI algorithms have already been deployed. In both industry and academics, the phrases "fog computing" and "edge computing" are interchangeable. In this study, we make a distinction between the two words that have been mentioned. Fog computing and edge computing have the same goal, which is to reduce network congestion and end-to-end delay, but they differ in how data are processed and managed, as well as where processing and cognitive power is located. Fog computing, on the other hand, differs significantly from traditional computing in that it is decentralized (i.e. It does not involve centralized computing). Basically, data processing and storage are performed between both the source as well as cloud infrastructure in a decentralized computing architecture. With edge computing, a compute facility is moved closer to data sources like mobile devices, sensors, and actuators by "pushing" the computation facility. Edge elements handle data locally instead of transferring it to the cloud, and each one plays a different role. A fog node uses its resources to decide whether it should analyze data from numerous sources or to transfer it to the cloud. This determination is taken by the fog node. In addition, edge computing doesn't really support several virtualized services such as SaaS, IaaS, PaaS, and others [8].

8. Edge computing with machine learning

There are sensors, hardware, and edge connections everywhere in IoT systems. Many Internets of Things (IoT) applications must meet bandwidth, latency, and security requirements. Despite this, cloud computing falls short of meeting these standards. Edge computing is a currently available technology that could meet these needs. In that other example, data commutation by automobiles can be performed using edge networks, and vehicles here on road move in a coordinated manner to improve consumer efficiency Fig. 8. For example: virtual reality and augmented reality apps that require high bandwidth can obtain material from an edge network. Depicts a model of the edge computing problem in IoT networks. The model makes it possible

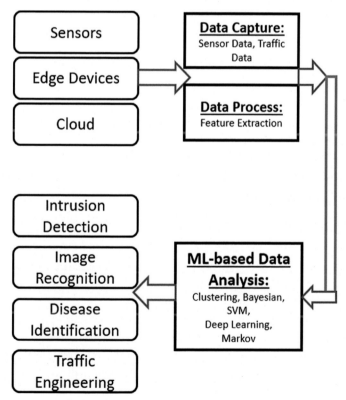

Fig. 8 Connection of edge devices with machine learning based data analysis.

to analyze information from sensors and traffic. Classification of data using machine learning to identify patterns in the attributes gleaned from various data sources. In intrusion detection, disease diagnosis, imaging recognition, and traffic engineering, the use of outcomes can be controlled [9].

8.1 Resource management in fog computing

Our algorithm's main purpose is to estimate the amount of fog device resources that will be accessible. The software also has other capabilities, such as ensuring the accuracy of rendered results and determining how many jobs should be repeated. The authors of this study, on the other hand, concentrated on proactive network association and open-loop wireless communication allowed by ML [17]. There was also a study done to forecast the overall length of time required from processing to data transmission, as well as link utilization for different pre—processing workloads through end to end delay Table 1.The authors used the GENI infrastructure's image processing ensemble services to create a realistic fog computing environment. This standard IoT network protocol, Wireless Application Protocol, was proposed by the authors in order to overcome the energy and communication challenges connecting end resource—constrained devices. Four different ML algorithms were used to predict real sensor readings. Researchers have also shown that implementing several ML approaches in layers reduces the amount of energy consumed. ML approaches used in ubiquitous computing applications entail several phases, and these processes can be performed at different levels of abstraction [2,8].

In most cases, coordinating data processing there at network's edge is a difficult problem for applications around the world (fog). Despite the fact that fog computing has solved many processing issues by offering preprocessing and moving computation to the network's edge rather than relying on centralized processing, fog computing still has a number of issues that need to be addressed. The use of machine learning (ML) has made a significant impact on the way data processing is managed. Fog computing can benefit greatly from the application of machine learning (ML) in the right way. After that, we'll talk about fog computing's processing and computation management concerns and unresolved topics. Fog design should really be scalable and adaptable enough to handle networked applications' unpredictable workloads. Increasing sensor and information quantity demanded connectivity and energy limits, resulting in new issues requiring effective IoT cloud architecture. Existing DL systems are constrained by their

Table 1 Comparison statement of various techniques and its application usage of Machine learning Algorithms.

Problem	Technique	Data	Application
Edge device communication	Supervised (linear regression)	Real-world seismic data traced from Parkfield, California	Seismic imaging
Continuous and real-time patient monitoring	Supervised (classification—SVM)	Data provided in Physiobank as "The Long-term ST Database" and ECG with arrhythmia in the middle of a normal ECG signal	Healthcare–patient monitoring
Automatically generates musical score from a huge amount of music data in an IoT network	Unsupervised (clustering—hidden Markov model)	Audio files or record audio signals in real time	Music cognition
Large amount of IoT sensor data adopted in industrial productions	DL	A total of 10 categories, each of which has 200 images for the training process and 50 test images for network testing	Smart industry
Centralized data processing	Unsupervised (density estimation—CRBMs)	One week of FCD generated in Barcelona City	Traffic modeling
Centralized data processing	Unsupervised (clustering—PCA)	MIRS dataset	Smart dairy farming
Energy efficiency and latency requirements for time-critical IoT	Supervised (classification—SVM, decision tree, and Gaussian naïve Bayes)	Sensor data from real human subjects	Time-critical IoT applications
Accuracy and adaptability of data analytics on the edge of a network	Supervised (classification—SVM)	"Long-term ST Database"	Health monitoring systems

computational speed, therefore creates issues with transmission and processing as the volume of in-the-wild data grows. When processing is done at the sensor level, energy consumption at the edge and end-to-end delay are minimized. To enhance information processing outcomes, a self-learning algorithm can be used, and this can minimize the overall data processing/communication volume needed in the complete IoT network. Existing potential for the creation of new and creative [4].

9. Security challenges in fog computing

Devices for fog computing being typically used without being closely monitored or protected, leaving them wide open to a wide range of security risks and vulnerabilities. As a result, building trust in the fog is the most difficult challenge to overcome. A public key encryption is one method for addressing some of the problems mentioned above Fig. 9. Fog computing is vulnerable to several hostile assaults, and as a result, a network's capabilities may be seriously weakened if no practical security measures are in place. Malicious attacks, such as denial-of-service (DoS) attacks, are common. DoS attacks are simple to initiate because most devices connected to the network aren't mutually authenticated. Sending bogus processing/storage requests from several devices can also launch a Denial-of-Service attack. Fog computing has an open network, making existing defensive

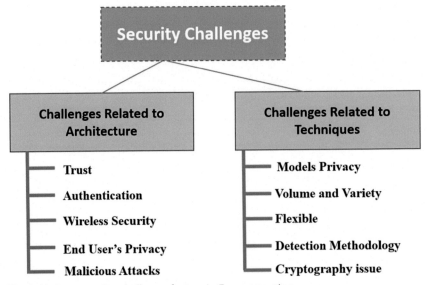

Fig. 9 Various security challenge factors in Fog computing.

techniques for other kinds of networks ineffective. A large network poses a significant problem. Fog/cloud services could be used by hundreds of thousands of IoT nodes to solve computation and storage constraints and improve performance [18].

Fog computing protection has a severe challenge with authentication while front fog nodes provide services to large numbers of end users [19]. And using fog network services, a device must first join the network by logging in to the fog network using its own credentials [20]. These step is critical in preventing unauthorized computers from entering the network. Furthermore, it's a significant challenge due to the obvious numerous constraints placed on the equipment connected to the network [8,9].

10. Conclusion

The purpose of this article was to investigate the role of ML in the FC. To narrow the study options, researchers used the Google Scholar search engine. The initial selection of research was made using abstracts and keywords from the corresponding webpages as a second constraint. There's a lot of potential for machine learning to replace humans as the go-to technology in many fields. As a result, it can be an effective analytical tool for FC. An application solution, security and privacy preservation, and paradigm enhancement were the three FC perspectives offered in this paper. The papers examined present exemplary FC-environment-based intelligent solutions. FC's varied properties, lack of testbeds, and other challenges hampered real-world evaluations. Statistics and supervised learning algorithms seem to be the most common ML approaches, as seen in the tables in Section 4. Unsupervised learning ML methods in an SDN-enabled FC context should be included in future study. In addition, the IDS should be implemented on actual hardware rather than just through simulation. Finally, the examined research addressed a wide range of concerns and obstacles, as well as verifying the pliability and efficacy of ML implementation in the FC setting.

Reference

[1] L. Hernandez, H. Cao, M. Wachowicz, Implementing an edge-fog-cloud architecture for stream data management, in: 2017 IEEE Fog World Congress, FWC, 2017, 2018, pp. 1–6, https://doi.org/10.1109/FWC.2017.8368538.
[2] S.R. Jena, R. Shanmugam, K. Saini, S. Kumar, Cloud computing tools: Inside views and analysis, in: International Conference on Smart Sustainable Intelligent Computing and Applications under ICITETM2020, Elsevier, 2020.

[3] D. Belli, et al., A capacity-aware user recruitment framework for fog-based mobile crowd-sensing platforms, in: 2019 IEEE Symposium on Computers and Communications (ISCC) IEEE, 2019. Available at: https://ieeexplore.ieee.org/document/8969754/ (Accessed: 6 October 2021).

[4] D. Borthakur, et al., Smart fog: fog computing framework for unsupervised clustering analytics in wearable Internet of Things, in: 2017 IEEE Global Conference on Signal and Information Processing (GlobalSIP). IEEE, 2017. Available at: https://ieeexplore. ieee.org/document/8308687/ (Accessed: 6 October 2021).

[5] S.R. Jena, R. Shanmugam, R. Dhanaraj, K. Saini, Recent advances and future research directions in edge cloud framework, Int. J. Eng. Adv. Technol. 9 (2) (2019) 2249–8958. https://doi.org/10.35940/ijeat.B3090.129219.

[6] B.B. Ma, S. Fong, R. Millham, Data stream mining in fog computing environment with feature selection using ensemble of swarm search algorithms, in: 2018 Conference on Information Communications Technology and Society (ICTAS). IEEE, 2018. Available at: https://ieeexplore.ieee.org/document/8368770/ (Accessed: 6 October 2021).

[7] F. Mehdipour, B. Javadi, A. Mahanti, FOG-engine: towards big data analytics in the fog, in: 2016 IEEE 14th Intl Conf on Dependable, Autonomic and Secure Computing, 14th Intl Conf on Pervasive Intelligence and Computing, 2nd Intl Conf on Big Data Intelligence and Computing and Cyber Science and Technology Congress(DASC/PiCom/DataCom/CyberSciTech). IEEE, 2016. Available at: https://ieeexplore.ieee.org/document/7588914/ (Accessed: 6 October 2021).

[8] J. Wang, et al., Maximum data-resolution efficiency for fog-computing supported spatial big data processing in disaster scenarios, IEEE Trans. Parallel Distrib. Syst. 30 (8) (2019). Available at: https://ieeexplore.ieee.org/document/8630038/ (Accessed: 6 October 2021).

[9] K. Saini, V. Agarwal, A. Varshney, A. Gupta, E2EE for data security for hybrid cloud services: a novel approach, in: IEEE International Conference on Advances in Computing, Communication Control and Networking (IEEE ICACCCN 2018) organized by Galgotias College of Engineering & Technology Greater Noida, 12–13 October, 2018, https://doi.org/10.1109/ICACCCN.2018.8748782.

[10] L. Hernandez, H. Cao, M. Wachowicz, Implementing an edge-fog-cloud architecture for stream data management, in: 2017 IEEE Fog World Congress (FWC). IEEE, 2017. Available at: https://ieeexplore.ieee.org/document/8368538/ (Accessed: 6 October 2021).

[11] R. Iqbal, et al., Context-aware data-driven intelligent framework for fog infrastructures in internet of vehicles, IEEE Acc. 6 (2018). Available at: https://ieeexplore.ieee.org/document/8488344/ (Accessed: 6 October 2021).

[12] D.N. Le, R. Kumar, B.K. Mishra, J.M. Chatterjee, M. Khari, (Eds.), Cyber Security in Parallel and Distributed Computing: Concepts, Techniques, Applications and Case Studies, John Wiley & Sons, 2019.

[13] A. Khochare, et al., Dynamic scaling of video analytics for wide-area tracking in urban spaces, in: 2019 19th IEEE/ACM International Symposium on Cluster, Cloud and Grid Computing (CCGRID). IEEE, 2019. Available at: https://ieeexplore.ieee.org/document/8752708/ (Accessed: 6 October 2021).

[14] R. Jaiswal, A. Chakravorty, C. Rong, Distributed fog computing architecture for real-time anomaly detection in smart meter data, in: 2020 IEEE Sixth International Conference on Big Data Computing Service and Applications (BigDataService). IEEE, 2020. Available at: https://ieeexplore.ieee.org/document/9179551/ (Accessed: 6 October 2021).

[15] K. Ahmad, A. Kamal, K.A.B. Ahmad, M. Khari, R.G. Crespo, Fast hybrid-MixNet for security and privacy using NTRU algorithm, J. Inf. Sec. Appl. 60 (2021), 102872.

[16] S. Nguyen et al., A low-cost two-tier fog computing testbed for streaming IoT-based applications, IEEE Int. Things J. 8 (8) (2021) Available at: https://ieeexplore.ieee.org/document/9249378/ (Accessed: 6 October 2021).

[17] S. Ghosh, et al., Mobi-IoST: Mobility-aware cloud-fog-edge-IoT collaborative framework for time-critical applications, IEEE Trans. Network Sci. Eng. 7 (4) (2020) 2271–2285, https://doi.org/10.1109/TNSE.2019.2941754.

[18] F. Faticanti, et al., Cutting throughput with the edge: app-aware placement in fog computing, in: Proceedings—6th IEEE International Conference on Cyber Security and Cloud Computing, CSCloud 2019 and 5th IEEE International Conference on Edge Computing and Scalable Cloud, EdgeCom 2019, 2019, pp. 196–203, https://doi.org/10.1109/CSCloud/EdgeCom.2019.00026.

[19] K.H. Abdulkareem, et al., A review of fog computing and machine learning: concepts, applications, challenges, and open issues, IEEE Acc 7 (2019) 153123–153140, https://doi.org/10.1109/ACCESS.2019.2947542.

[20] J. Clemente, et al., Fog computing middleware for distributed cooperative data analytics, 2017 IEEE Fog World Congress (FWC) IEEE, 2017 Available at: https://ieeexplore.ieee.org/document/8368520/ (Accessed: 6 October 2021).

About the authors

Dr. A. Daniel currently working as an Associate Professor in School of Computing Science and Engineering in Galgotias University, Greater Noida, Uttar Pradesh. He completed his B.E and M.E both in Anna University. He has completed his Ph. D. in Computer Science and Engineering at Shri Venkateshwara University, Uttar Pradesh. His research interests are Cloud Computing, Networking etc. He has published several article in reputed international journals. He has membership in IEEE, ACM, IFERP, IAENG and CSTA.

Dr. N. Krishnaraj is working as an Associate Professor, School of Computing, SRM Institute of Science and Technology, Kattankulathur, Tamilnadu, India. He is having 13 years of experience in teaching and research, his research areas are Biometrics, Wireless sensor networks, Internet of Things, Medical image processing. He has completed one funded research project supported by DST, India. He is Cisco certified Routing and switching professional. He has published more than 65 articles in reputed international journals and 15 articles presented in international conferences. He is being serving as an Editorial board member of MAT Journal,

IRED and Allied Academic Sciences. He has delivered several special lectures in workshops and seminars. He is a professional society member for ISTE, IEI and IAENG.

Kavita Saini is presently working as Professor, School of Computing Science and Engineering, Galgotias University, Delhi NCR, India. She received her Ph.D. degree from Banasthali Vidyapeeth, Vanasthali. She has 18 years of teaching and research experience supervising Masters and Ph.D. scholars in emerging technologies.

She has published more than 40 research papers in national and international journals and conferences. She has published 17 authored books for UG and PG courses for a number of universities including MD University, Rothak, and Punjab Technical University, Jallandhar with National Publishers. Kavita Saini has edited many books with International Publishers including IGI Global, CRC Press, IET Publisher Elsevier and published 15 book chapters with International publishers. Under her guidance many M.Tech and Ph.D. scholars are carrying out research work.

She has also published various patents. Kavita Saini has also delivered technical talks on Blockchain: An Emerging Technology, Web to Deep Web and other emerging Areas and Handled many Special Sessions in International Conferences and Special Issues in International Journals. Her research interests include Web-Based Instructional Systems (WBIS), Blockchain Technology, Industry 4.O, and Cloud Computing.

Kiranmai Bellam now at Professor in the Computer Science Department at Prairie View A&M University, she completed his B.S from Madras University, Chennai and M.S from New Mexico Tech, United States. He has completed his Ph.D. in Computer Science from Auburn University. Her research interests are Cloud Computing, Networking etc.

CHAPTER SIXTEEN

Artificial intelligence in edge devices

Anubhav Singh[a], Kavita Saini[b], Varad Nagar[c], Vinay Aseri[c], Mahipal Singh Sankhla[c], Pritam P. Pandit[c], and Rushikesh L. Chopade[c]
[a]School of Forensic Science and Risk Management, Rashtriya Raksha University, Lavad, India
[b]School of Computing Science and Engineering (SCSE), Galgotias University, Delhi, Uttar Pradesh, India
[c]Department of Forensic Science, Vivekananda Global University, Jaipur, Rajasthan, India

Contents

Advances in Computers, Volume 127
ISSN 0065-2458
https://doi.org/10.1016/bs.adcom.2022.02.013

Abstract

In the current era, advancements in deep learning have seen the services and uses of artificial intelligence (AI) flourish. From individual support and commendation structures to video/audio monitoring, there is something for every person. With the development of mobile computing and the Internet of Things (IoT), billions of mobile and IoT devices are now connected to the Internet, generating tons of data at the edge of the network. As a result of this inclination, there is a demanding need to push AI beyond its limits to the networks edge in order to fully understand the capability of big data. Edge computing, a developing model that encourages computing everyday jobs as well as network services, which are basic to the edge of the network, has long been touted as a hopeful explanation for meeting this demand. The resulting new interdisciplinary field known as edge AI or edge intelligence (EI) has been attracting an incredible amount of attention. Although EI experimentation is still in its early stages, both the computer system and AI societies would benefit from a committed forum for exchanging recent EI accomplishments. At present, we are conducting a detailed survey of recent EI research events. We go over the context and enthusiasm along with inspiration for AI running first at the edge of the network.

1. Introduction

We live in the age of AI that has never before seen such rapid growth. Deep learning has been propelled forward by recent advances in algorithms, processing power, and large amounts of datasets [1]. Computer vision, speech recognition, and language processing, as well as chess (for example, AlphaGo) and robotics, have all seen major advancements owing to AI's remarkable reach [2]. A plethora of intelligent applications, such as intelligent personal assistants, personalized purchasing recommendations, video monitoring, and smart home appliances, have sprung up as a result of these advancements. These intelligent applications are well known for greatly enhancing people's lifestyles, increasing human work rate, and improving social productivity. Big data have lately seen a drastic change in the knowledge source from giant-scale cloud data centers to more ubiquitous end devices, such as mobile devices and Internet of Things (IoT) devices, as a significant driver that boosts AI development. Big data, such as online shopping records, social media–related items, and business information systems, have traditionally been created and mainly stored in large data centers. Cisco predicts that by 2021, people, machines, and things at the network edge will have created nearly 850 ZB [3]. Global data center traffic, on the other hand, will only reach 20.6 ZB by 2021. However, due to concerns about efficiency, cost, and privacy, driving the AI frontier to the sting ecology that lives at the bottom of the web is not trivial [4]. When transmitting a large amount of data across a wide area network (WAN), the pecuniary cost as well as the transmission time may be excessively high, making privacy leaks a major concern [5]. Another option is on-device analytics, which uses AI programs to process IoT data directly on the device, but it could suffer from poor performance and energy efficiency [6]. Recently, moving cloud services from the networks core to its edge, i.e., closer to IoT devices and data sources, has been proposed. A helper node can be any adjacent terminal devices capable of communicating with each other via device-to-device (D2D) communication [7], servers connected to access points (e.g., WLANs, routers, and base stations), network gateways, or even microdata gateways that can be accessed by neighboring devices. While edge nodes can vary in size from a credit card-sized computer to a microdata center with multiple server racks, the most important feature highlighted by edge computing is physical closeness to the information-generating sources. In essence, the physical proximity of information-processing and information-generating sources promises many advantages over traditional cloud-based computing paradigms, including low latency, energy efficiency, data protection, low bandwidth consumption, local awareness, and context [6,8]. In fact, the combination of edge computing and AI has generated a new field of study known as edge intelligence (EI) or edge AI [9,10]. Rather than relying solely on the cloud, EI leverages the best of widely available edge resources to provide AI insights. EI has gained a lot of interest from both industry and academia. For example, EI has been included in the well-known Gartner's advertising cycle as a new technology that could reach productivity levels in the next 5–10 years [11]. Pilot projects have been proposed by major corporations such as Google, Microsoft, Intel, and IBM to illustrate the benefits of edge computing in paving the way for AI. This initiative has enhanced various AI applications, from real-time video analytics to machine learning [12], reasoning support [13] to correct farming, smart homes [14], and industrial Internet of Things (IIoT) [15]. Notably, studies on and practice of this new interdisciplinary field of EI are still in their infancy stages. In both industry and academia, there is a widespread lack of a platform dedicated to reviewing, debating, and sharing the current advancements in EI. To close this gap, we perform a comprehensive and detailed assessment of the current EI research efforts in this study. First, we will go through the history of artificial intelligence. Next, the rationale, definition, and grading of EI will be discussed. Following that, we will categorize and discuss the evolving computer architectures and supporting technologies for EI model formation and inference. Finally, we will discuss about some of the open research problems and possibilities for EI. The following is a breakdown of this chapters structure.

(1) Section 2 provides an outline of AI's core ideas, focusing on deep learning—chosen AI's field.

(2) The purpose, definition, and rating of EI are discussed in Section 3.

(3) The architectures, enabling methodologies, systems, and frameworks for training EI models are discussed in Section 4.

(4) Section 5 discusses EI model inference structures, enabling methodologies, systems, and frameworks.

(5) EI's future directions and difficulties are discussed in Section 6. We believe that by conducting this poll, we will be able to evoke more interest, spark constructive debates, and inspire new research ideas on EI.

2. Primer on artificial intelligence

In this section, we go through AI ideas, reports, and approaches, focusing on deep learning, which is one of the most prominent branches of AI+.

2.1 Artificial intelligence

Although artificial intelligence (AI) has received a lot of press recently, it is not a new idea; it was originally coined in 1956. Simply put, AI is a technique used for building intelligent machines that can perform tasks as well as humans. This is an extremely broad term, and it may apply to everything from Apples Siri to Googles AlphaGo as well as to too strong technologies that are yet to be developed. In order to simulate human intelligence, AI systems must exhibit at least a few of the following behaviors: planning, learning, thinking, problem-solving, knowledge representation, perception, motion, and manipulation, as well as social intelligence and creativity, to a lesser extent. AI has risen, fallen, and risen again throughout the course of the last 60 years. Deep learning, a technology that has achieved human-level precision in a variety of intriguing domains, has been the driving force behind AI's rapid growth after the 2010s.

2.2 Deep learning and deep neural networks

Machine learning (ML) is a strong tool for AI to achieve its objectives. Many machine learning (ML) approaches have been developed to train machines to categorize and make predictions using data from the 000 world, including decision trees, K-means clustering, and Bayesian networks. One of the current machine learning techniques is deep learning, which uses artificial neural networks (ANNs) [16]. To tell the truth, deep representation of information has produced excellent outcomes in a range of tasks, such as image classification and face recognition. The model is dubbed as a deep neural network because deep learning models often utilize an ANN with several layers (DNN). Each layer of a DNN is made up of neurons that can produce nonlinear outputs based on the information provided by the neurons input. The information is received by the input layer neurons, which then pass it on to the neurons in the central layer that is also known as the hidden layer. The weighted sums of the computer file are created by the middle layer neurons, which are then produced using certain activation functions, and the outputs are then sent to the output layer. The final results are displayed on the output layer. Because DNNs contain more complex and abstract layers than does a traditional model, they are equipped with high-level learning abilities that allow them to make highly accurate inferences on tasks. There are three in-demand DNN architectures, namely, multilayer perceptrons (MLPs), convolutional neural networks (CNNs), and recurrent neural networks (RNNs) [17]. Convolutional layers in CNN models, unlike fully connected layers in MLPs, extract basic characteristics from input by conducting convolutional operations. CNN models, which use several convolutional filters to capture the high-level representation of a computer file, are the most popular choice for computer vision applications like photo categorization (e.g., AlexNet [18], VGGNet [19], ResNet [20], and MobileNet [21]) and artifact acknowledgment (e.g., Fast R-CNN [22], YOLO [23], and SSD [24]). RNN models, which employ consecutive information nourishing, are another type of DNN. The fundamental element of an RNN is called a cell, and each cell consists of layers. A sequence of cells allows RNN models to be processed sequentially. In the task of tongue processing, RNN models are extensively used and when it comes to tongue processing, RNN models are commonly used e.g., dialectal modeling, computational phonology, enquiry responding, and document organization. Deep learning is the most sophisticated AI approach as well as a resource-intensive task that is best suited to edge computing.

In the rest of this chapter, we will focus on the interaction between deep learning and edge computing for space reasons. We believe that the concepts provided may be applied to a variety of AI models and processes. For example, stochastic gradient descent (SGD) is the most favored training technique for a variety of AI/ML systems (e.g., K-means, support vector machine, and lasso regression) [25]. As a result, the SGD training optimization methods described in this chapter may be carried out using different AI means and methodologies as well as the training process of other AI models.

2.3 From deep learning to model training and inference

For each neuron in a DNN layer, there is a weighting vector that is proportional to the size of the layer's computer file. Of course, the weights of a deep learning model must be refined through a training process. Weight values are generally set at random during the training phase of a deep learning model. A loss function is used to calculate the mean squared error of the error rate between the results and the actual label, and the precision of the results is determined by calculating the mean error of the error rate between the results and the actual label. The neural network returns the error rate using the backpropagation method [26,27].

The weights are set to accommodate both the gradient and the learning rate. The DNN model is deduced after training. In a photo classification task, for example, the DNN is trained to figure out how to recognize a picture by providing it with an excessive number of coaching examples, and, then, using real-world images as inputs, inference makes fast predictions/classifications. The training technique utilizes both the feedforward and backpropagation procedures. It is worth mentioning that the inference is solely dependent on the feedforward process, which means that the whole neural network is aware of the input from the 000 world, and, therefore, the model generates the prediction.

2.4 Popular deep learning models

For a more complete understanding of deep learning and its applications, this section provides an outline of several typical deep learning models.

2.4.1 Convolutional neural networks

In 2012, AlexNet became the first CNN team to win the ImageNet Challenge for image classification [18]. In this picture, there are five convolutional layers and three totally connected layers. AlexNet requires

61 million weights and 724 million MAC to categorize a picture with a resolution of 227×227 pixels (multiply-add computation). In order to achieve greater accuracy, VGG-16 [19] is trained on a deeper structure of 16 layers, comprising 13 convolutional layers and 3 completely connected layers. To categorize an image with a size of 224×224, it took 138 million weights and 15.5G MAC [28]. GoogLeNet provides an inception module consisting of various sized filters to improve accuracy while decreasing DNN inference computation. GoogLeNet outperforms VGG-16 in terms of accuracy while using just 7 million weights and 1.43G MAC to analyze a picture of the same size. The "shortcut" structure is utilized to attain human-level accuracy with a top-five error rate of less than 5% in the state-of-the-art endeavor, ResNet [20], to increase accuracy while decreasing expenses. The "shortcut" module is used to address the problem of disappearing gradients during the training phase, allowing a DNN model with a deeper structure to be coached. In computer vision, CNNs are often used. The AI system learns to automatically extract the properties of these inputs to complete a job, such as image classification, face identification, or image semantic segmentation, given a set of photographs or videos from across the world.

2.4.2 Recurrent neural networks

RNNs were designed to solve the problem of time series in sequential computer data. An RNN's input is made up of this and the preceding samples. In an RNN, each neuron has its own internal memory that holds the computation data from previous samples. Backpropagation through time (BPTT) is used to train RNNs [29]. RNNs with long short-form memory (LSTM) [30] are a form of RNNs that is more sophisticated. In LSTM, the gate represents the fundamental unit of a neuron. Each memory cell in LSTM includes a multiplicative forgetting gate, an entry gate, and an exit gate. These gates are used to regulate access to memory cells and prevent them from being confused with unwanted entries. Information can be added or removed from the memory cell through the door. The information that may be stored on a memory cell is controlled by several neural networks called gates. The forget gate may identify whether or not knowledge is remembered throughout training. Since it is often used to handle data with a variable input length, an RNN is extensively used in natural language processing. Language modeling, word embedding, and computational linguistics are some of the AI approaches that might be utilized to build a system that understands languages that people speak.

2.4.3 Generative adversarial networks

GANs (generative adversarial networks) [31] are made up of two main processes, the generator and discriminator networks. After cramming the information dissemination of a training dataset with actual data, the generator takes the responsibility of creating new data. The discriminator is responsible for distinguishing the 000 data from the bogus generator data. Imaging, image transformation, image synthesis, image super-resolution, and other applications often use GANs.

2.4.4 Deep reinforcement learning

Deep reinforcement learning (DRL) is made up of DNNs with RL. DRL's sixth goal is to develop a genius agent that can utilize effective techniques to maximize the advantages of long-term goals while taking up activities that are under their power. DRL is frequently used to solve scheduling difficulties such as game selection concerns, video transmission rate selection, and so on. In the DRL method, the DNN is in charge of representing a large number of states and estimating the action values to evaluate the quality of the action within the provided states. The RL is responsible for finding the best policy of action toward states in the environment, whereas the DNN is responsible for representing a large number of states and approximate action values to assess the quality of the measures within these states. The reward might be a function that indicates the difference between a predefined requirement and, as a consequence, action performance. The DRL model's agent may be utilized for a variety of activities, including gaming [32], owing to continual learning.

3. Edge intelligence

EI is the result of the union of edge computing with AI. The rationale, advantages, and definition of EI are discussed in this section.

3.1 Motivation and benefits of edge intelligence

Because AI and edge computing have a lot in common, it makes sense to merge the two. Edge computing, in particular, aims to coordinate a large number of cooperating edge devices and servers in order to analyze the produced data in close proximity, whereas AI aims to imitate genius human behavior in devices/machines through data learning. Putting AI into the mix benefits each other in the following ways, in addition to enjoying the benefits of edge computing (low latency and reduced bandwidth use).

AI is necessary to fully realize the potential of data created at the networks edge: As the number and type of mobile and IoT devices increase, significant amounts of multimodal data (for example, audios, images, and videos) from the physical environment are constantly being recognized on the device side. Due to its ability to quickly evaluate large amounts of data and obtain high-quality decision-making information, AI is functionally essential in this environment. As proven by population distribution, deep learning, the most widely used AI technique, has the ability to automatically detect patterns and anomalies in the data captured by the edge device. Traffic flow, humidity, temperature, pressure, and air quality are factors that must be taken into account. The insights acquired from sensed data are integrated into real-time predictive decision-making (e.g., public transportation planning, control, and driver alert) in response to quickly changing surroundings and increasing operational efficiency. Gartner [33] predicted that by 2022, more than 80% of business IoT projects would use AI, up from 10% at the time of this study.

Edge computing is ready-to-prosper AI with more data and application scenarios: It is well known that the four factors that have contributed to the current surge in deep learning are algorithms, hardware, data, and application scenarios. Although the influence of algorithms and technologies on the advancement of deep learning is obvious, the importance of data and application situations has been largely neglected. To improve the performance of a deep learning system, the most commonly used method is to modify the DNN by adding additional layers of neurons. As a result, we seek to learn more about the parameters of the DNN, and the amount of data necessary for training will rise. This illustrates the importance of data in the case of AI. After recognizing the value of information, the next issue is determining where the information comes from. Traditionally, data have been created and stored mostly in large-scale data centers. Despite this, the tendency is currently reversing due to the rapid development of IoT. According to Cisco's research [3], significant IoT data will be created at the edge in the near future. If AI algorithms are used to analyze these data in the cloud data center, it will require a lot of bandwidth and put a lot of strain on the cloud data center. To solve these issues, edge computing is offered as a way to accomplish low-latency processing by sinking computing capacity from the cloud data center to the edge side, i.e., the data creation source, thus allowing for high-performance AI processing. Although edge computing and AI complement one another in terms of technology, their application and adoption are intertwined.

AI democratization requires edge computing as a key infrastructure: Many digital goods or services in our daily lives, such as online shopping, service suggestions, video surveillance, smart home devices, etc., have achieved outstanding success owing to artificial intelligence technology. Self-driving cars, intelligent finance, disease diagnosis, and drug development are just a few of the emerging creative areas where AI is the driving force. Beyond the examples provided above, AI democratization or ubiquitous AI might be utilized to enable a broader range of applications and push the limits of what is possible [34]. AI has been declared as "for each person and each organization everywhere" by major IT companies, with the objective of "creating AI for each person and each organization everywhere." For the time being, AI should come "closer" to people, data, and end devices. In this regard, edge computing is definitely superior than cloud computing. Because of these advantages, edge computing is a natural facilitator for ubiquitous AI.

Edge computing is frequently used to popularize AI applications: In the cloud computing industry, throughout the early development of edge computing, there has always been an emphasis on which high-demand applications could take edge computing to the next level that cloud computing could not and what killer applications for edge computing are. Microsoft has been investigating what sort of data should be transferred from the cloud to the premises since 2009 to eliminate any concerns [35]. Voice command recognition, augmented reality/virtual reality, and interactive cloud gaming are just a few examples of [36] real-time video analytics. On the other hand, real-time video analytics is predicted to be a game changer for edge computing [12,37,38]. Real-time video analytics is a new technique based on computer vision that captures high-definition footage from security cameras in real time and analyzes it with high processing speed, high bandwidth, high privacy, and low latency. The only viable option that can meet these stringent requirements is state-of-the-art computing. In light of the aforementioned rise of edge computing, new AI applications coming from industries such as IIoT, intelligent robotics, smart cities, and smart homes will likely play a crucial role in popularizing edge computing. This could be because many mobile and IoT-related AI applications fall into the category of practical applications that are energy- and computation-intensive, sensitive to data protection and lag, and thus are, of course, compatible with edge computing. Due to its attractiveness and need of running AI applications on the Internet, edge AI has recently garnered a lot of attention. A report titled *A Berkeley View on Systems Challenges for AI* was published in December 2017 [39]. According to a study published by the

University of California at Berkeley, the AI system at the edge of the cloud is seen as a critical research path to achieve the goal of business-critical and tailored artificial intelligence. In August 2018, the Gartner Hype Cycle included edge AI for the first time [40]. According to Gartner, edge AI is still in its innovation activation phase and will reach a productivity plateau in the next 5–10 years. Throughout the industry, many pilot programs aimed toward edge AI have been distributed. Traditional cloud providers like Google, Amazon, and Microsoft have built service platforms to deliver intelligence to the sting by allowing end devices to conduct machine learning inferences using pretrained models locally on the sting AI service platform. Several high-end processors specialized for running ML models are commercially available, as demonstrated by Googles Edge TPU, Intels Nervana NNP, and Huaweis Ascend 910 and Ascend 310. B. Scope and Rating of Edge Intelligence. Although the term "edge AI" or "EI" is relatively new, research and activity in this field have long existed. As mentioned above, Microsoft built an edge-based prototype in 2009 to demonstrate the benefits of edge computing by supporting mobile voice command recognition, an artificial intelligence application. Despite the fact that this is the first survey, there is currently no acknowledged description for EI. Most administrations [41,42] and journalists [43] consult with EI. At this point, the standard has been set for operating AI algorithms directly on a device, generating data (sensor data or signals) on the device. Although this is the most common approach to EI in the world today (e.g., using high-end AI processors), it is important to note that this definition severely limits the scope of the EI solution. Consecutive computationally intensive processes, such as DNN models, are resource-intensive and require the use of high-end processors in the device. Such a strict requirement not only increases the cost of EI but also renders legacy end devices with poor computing capabilities incompatible and unfriendly. We propose in this chapter that EI's scope should not be limited to AI models that operate just-on-edge servers or devices. In fact, a dozen recent studies have found that using edge–cloud synergy to execute DNN models may reduce both end-to-end latency and power consumption compared to running locally. We believe that a collaborative hierarchy like this should be integrated into the design of effective EI systems because of these benefits. Furthermore, the current EI concepts are mostly focused on the inference phase (i.e., executing the AI model), assuming that the AI model is learned in high-performance cloud data centers, where the training phase uses significantly more resources than does the inference phase. However, because of concerns about data privacy, a large amount of training data must be transferred from devices or edges to the cloud, resulting in prohibitive communication costs. Instead, it is believed that EI would be the paradigm that makes full use of the available data and resources in the hierarchy of end points, edge nodes, and cloud data centers to improve the efficiency of training and inference of a DNN model. This suggests that EI does not require that the DNN model be fully trained or derived at the time but that data offload can serve as a mechanism to coordinate cloud edge devices. Based on the number and length of information dumping pathways, we categorize EI into six types. The numerous EI levels and their meanings are shown below.

1. *Cloud intelligence*: The DNN model is entirely trained and inferred using cloud computing.
2. *Cloud–edge coinference and cloud training*: The DNN model is trained but inferred through edge–cloud collaboration. Data are partially offloaded to the cloud as part of the edge–cloud cooperation.
3. Level 2: Cloud training and in-edge coinference: The DNN model is trained in the cloud, but it is inferred on the edge. Model inference at the network's edge is referred to as "in-edge," and can be done in whole or in part by downloading data to nearby nodes or devices (via D2D communication).
4. Level 3: Cloud training and on-device inference: The DNN model is trained in the cloud, but the DNN model on-device is inferred in a highly localized manner. The phrase "on-device" refers to the fact that no data will be sent off of the device.
5. Level 4: Cloud–edge cotraining and inference: The DNN model is trained and inferred via the edge–cloud collaboration.
6. Level 5: All-in-edge: The DNN model is completely trained and inferred on the in-edge.
7. Level 6: On-device training and inference: The DNN model is completely trained and inferred on the device.

As EI rises, the number and duration of information dumping routes decrease. As a result, information offloading transmission latency is reduced, data privacy is enhanced, and WAN bandwidth costs are reduced. However, this leads to an increase in processing time and energy consumption. This paradox suggests that there is no universal "best-level" EI; rather, the "best-level" EI is not independent of application and should be determined by taking into account a variety of variables such as latency, energy efficiency, privacy, and WAN bandwidth cost. We will look at the current empowerment methods and solutions for different levels of EI in the following sections.

4. Edge intelligence model training

As a result of the growth of mobile and IoT devices, data are being created at the networks edge, which is important for AI model training. The architectures, essential performance measurements, enabling methodologies, and existing systems and frameworks for distributed DNN training at the sting are all covered in this section.

4.1 Architectures

For DNN training distributed at the edge, there are three popular designs: centralized, decentralized, and hybrid (cloud–edge device). End devices, which are data sources, are represented by cell phones, cars, and security cameras, whereas the cloud refers to the central data center. We utilize base stations for the edge server because of the methodology.

(1) Centralized: In a centralized DNN training, the DNN model is trained within the cloud data center. Training data are generated and collected from a variety of distributed end devices, including mobile phones, cars, and security cameras. When the data are received, they will be used by the cloud data center to perform DNN training. The system that supports the centralized design may be classified as cloud intelligence level 1 or level 2 or a level depending on the particular inference technique used by the system.

(2) Decentralized: In the decentralized mode, each computer node trains its own DNN model locally with local data, keeping private information private. The nodes in the network will communicate with each other to exchange local model updates to preserve the global DNN model by combining local training enhancements.

(3) Hybrid: This style combines the centralized and decentralized approaches. The sting servers, as the system's hub, could train the DNN model either decentralized with each other or centralized with the cloud data center. Therefore, the hybrid architecture covers levels 4 and 5. Hybrid architecture is also known as cloud edge device training because of the responsibilities involved.

4.2 Key performance indicators

There are six important performance metrics to consider when evaluating a distributed training approach.

(1) Training loss: The DNN training technique fundamentally solves an optimization problem by reducing training loss. Since it records the difference between the learned (for example, expected) value and the labeled data, the training loss shows how well the trained DNN model matches the training data. As a result, the training loss is expected to be modest. Drill forfeiture is mostly due to drill samples and drill methods causing anguish.

(2) Convergence is the decentralized technique's particular convergence indicator. If the distributed training processes converge in a consensus, the result of the strategy formation, then this indicates that a decentralized method seems to work. "Convergence" is a term that explains whether a decentralized strategy converges to such a consensus and how quickly it does so. In the decentralized training mode, the convergence value is defined by how the gradient is synchronized and updated.

(3) When training a DNN model with data from a large number of end devices, the information or intermediate data should be sent out of the top devices. In this circumstance, catering to privacy concerns is unavoidable. To safeguard privacy, less privacy-sensitive data are likely to be sent out of end devices. Whether data are sent to the sting affects whether or not privacy is safeguarded.

(4) Communication costs: Training the DNN model is data-intensive because data or intermediate data must be sent between the nodes. On the surface, this communication overhead looks to increase training latency, energy, and bandwidth use. The transmission overhead is controlled by the size of the original input file, the mode of transmission, and the bandwidth available.

(5) Latency is one of the most significant performance variables for distributed DNN model training since it influences when the trained model is available for use. The latency of a distributed training process is made up of computation delay and communication latency. The potential of sting nodes has a significant impact on calculation time. Communication latency may be affected by the size of the raw or intermediate data as well as the network connection's capacity.

(6) By training the DNN model in a decentralized manner, both computing and communication activities consume a lot of energy. However, most end devices are power-limited. Consequently, energy efficiency is highly desirable in training DNN models. Energy efficiency is strongly influenced by the size of the target training model and the device resources used. It should be noted that training loss and convergence are standard performance measures; thus, some DNN training literature might not explicitly claim them.

4.3 Enabling technologies

When training the EI model, we consider the underlying technologies to improve one or more of the abovementioned performance metrics.

Federated learning: Federated learning is dedicated to optimizing privacy issues within the top performance metrics listed above. It is a promising new technique for maintaining privacy while using input from multiple clients to train the DNN model. In federated learning, rather than collecting and sending data to a centralized data center for training [44], clients (e.g., mobile devices) are left with the information while the server trains a shared model by adding locally calculated updates. Optimization and communication are the most difficult aspects of federated learning. The present task is to improve the gradient of a shared model through distributed gradient updates on mobile devices for the optimization problem. SGD is used in federated learning on this topic. SGD is a simple but widely used gradient descent method that updates the gradient in extremely small subsets (mini batches) of the dataset. Shmatikov and Shokri [45] are developing an SGD (SSGD) protocol that allows clients to train their own datasets while selectively sharing small sections of their model's important parameters with the centralized aggregator. Due to the ease with which SGD may be parallelized and run asynchronously, SSGD addresses both privacy and training loss. When compared to training only on their own inputs, the loss of training will be minimized by sharing the models across clients while maintaining client privacy. One shortcoming [45] is that unbalanced and non-dependent identically distributed data (non-IID) are not taken into account. As an extension, McMahan et al. [44] advocate a decentralized approach known as federated learning and present a FedAvg method for associative learning with an average of iterative models supported by a DNN. Iterative version averaging means that the customers replace the version regionally with the use of a one-step SGD, and the server then averages the consequent fashions the use of weights. Because dispersed data might originate from a variety of sources, the optimization in the study by McMahan et al. [44] highlights the features of unbalanced and non-IID. The issue of communication efficiency is posed by the unstable and unpredictable network in the case of the communication problem. In federated learning, each client transmits to the server a whole model or a full model update in a standard round. Because of the unstable network connections, this phase will most likely be the bottleneck for large models. McMahan et al. [44] propose extending the calculation of client-side local updates. It is, however, impracticable when the customers' computation resources are severely limited. In answer to this problem, Konečný et al. [46] offer two novel update techniques, structured update and sketched update, to reduce communication costs. In an extremely structured update, the model learns an update directly from a limited space that is parameterized with a reduced number of variables, such as B. Random or low-rank masks. When using a sketched update, the model first learns a full model update and then compresses it with a mixture of quantization, random rotations, and subsampling before sending it to the server. Despite the fact that the federated learning method makes use of a fresh new decentralized deep learning architecture, it relies on a central server to aggregate local updates. Lalitha et al. consider the case of teaching a DNN model across a totally decentralized network, that is, a network without a central server [47]. They present a Bayesian-distributed method in which each device changes its belief by merging data from its neighbors in a single hop to form a model best suited for network observations. Furthermore, using the newly developed blockchain technology, Kim et al. [48] offer blockchain federated learning (BlockFL), which uses blockchain to share and verify device model updates. BlockFL can also be used in a completely decentralized network, where the machine learning model can be trained without central coordination, although some devices do not have their own training dataset.

Aggregation rate control: This approach focuses on reducing communication overhead while training a DNN model. A popular approach (for example, federated learning) to train deep learning models in an edge computing context is to first teach locally distributed models and then centrally add changes. The frequency of update addition has a major impact on the communication overhead in this situation. As a result, the aggregation process, including aggregation content and frequency, should be closely monitored. Hsieh et al., based on the foregoing knowledge [49] for geo-distributed DNN model training, build the Gaia system as well as the approximate synchronous parallel (ASP) model. Gaia's core concept is to separate communication within a data center from communication between data centers and to allow alternative models of communication and consistency for each. To do this, the ASP model was created to dynamically eliminate unimportant communication between data centers, with the aggregation frequency governed by a significance threshold. Gaia, on the other hand, concentrates on capacity-unlimited geo-distributed data centers, rendering it inapplicable to edge computing nodes with restricted capacity Wang et al. [50] present a sway method that calculates the most effective trade-off between the local update and aggregation of global parameters within a given resource budget, taking into account the capacity constraint of edge nodes. The technique is based on distributed gradient descent convergence analysis and can be used for federated learning in edge computing with proven convergence. Nishio and Yonetani [51] explore the problem of selecting a resource-constrained client to perform federated learning in a high-capacity, resource-constrained computing environment. Particularly, FedCS is an update aggregation protocol that allows the centralized server to collect as many client updates as feasible in order to improve ML model performance.

Gradient compression: Another logical technique for compressing the model update is gradient compression, which reduces the communication costs imposed by decentralized training (i.e., gradient information). Gradient quantification and gradient preservation have been proposed as solutions to this problem. Gradient quantization achieves, in particular, lossy compression of gradient vectors by quantizing each of their components to a low precision finite bit value. Sending only a portion of the gradient vectors saves the gradient, thus minimizing communication costs [52] Gradient compressibility is demonstrated from the fact that 99.9% of the distributed SGD (DSGD) swap is redundant. Lin et al. proposed deep gradient compression (DGC), which compresses gradients of 270–600 for various CNNs and RNNs. DGC uses four approaches to maintain accuracy during compression: momentum correction, local gradient cutting, momentum factor masking, and warm-up training. Motivated by Lin et al. [52], Tao and Li [53] suggested the advantages of SGD (ESGD), a series of dispersed methods that ensure both convergence and practical performance. ESGD comprises two techniques to improve gradient-based first-order optimization of stochastic objective functions in edge computing: (1) identifying which gradient coordinates are relevant and communicating only these and (2) constructing accumulated residual momentum to track obsolete residual gradient coordinates to avoid poor convergence rates caused by infrequent updates. Stich et al. [54] provided a brief convergence study of the scattered SGD, where the SGD is assessed using k-sparsification or compression (e.g., top-k or random-k). When equipped with error correction, this technique converges at the same rate as vanilla SGD according to the study (keeping track of accumulated errors in memory). To put it in another way, communication is decreased by a factor of the problem's complexity (often much more) while staying convergent at the same pace and reducing the transmission bandwidth by quantizing the gradients to values with low precision. During this time, Tang et al. [55] created a framework for decentralized compression training and introduced two alternative methods for compression by extrapolation and differential compression. Examination of the two methods shows that they both converge to $O\ (1/nT)$, where n is the number of customers and T is the number of iterations. This is the convergence rate for precision centralized training. Amiri and Gunduz [56] used a remote parameter server to implement DSGD at the wireless edge and construct DSGD in both digital and analog formats. With each iteration of the DSGD algorithm, the digital DSGD (DDSGD) assumes that customers are concerned about the limit of the multiple access channel (MAC)'s capacity range, and, to communicate its estimations of the gradient within the permitted bits budget power, it employs gradient quantization and error accumulation distribution. DSGD has an analog format (A-DSGD). The clients first use error accumulation to sparsify their gradient estimations; then, the available channel bandwidth dictates that they be projected to a lower dimensional space. Without using any digital coding, these projections are sent straight across to the MAC.

DNN splitting: DNN splitting is used to protect privacy. By providing partially processed data instead of raw data, DNN splitting protects the privacy of users. It takes place between the top devices and the edge server to enable edge-based data protection and preservation training of DNN models. Mao et al. [57] used a differentially private technique to reduce the price of mobile devices by partitioning a DNN after the first convolutional layer. The proof in their study [57] ensures that training jobs may be outsourced to untrusted edge servers via a differentially private method on activation. Wang et al. [58] looked at the topic from the perspective of mobile devices and cloud data centers. To take advantage of the computing power of cloud data centers and avoid privacy concerns, Wang et al. [58] Arden (private inference framework that supports deep neural networks) is a framework that divides the DNN model using a lightweight privacy approach. Arden achieves data protection through random data cancellation and random noise addition. Taking into account the detrimental impact of personal disturbance on the first data, Wang et al. [58] enhanced the cloud-side network's resilience to disturbed data by utilizing a noisy training technique. With regard to the privacy problem, Osia et al. [59] presented a hybrid user-cloud architecture that uses a private feature extractor as a core component and breaks down large and complicated depth models for collaborative analysis and privacy protection. The feature extraction module in this framework is built correctly to dump the private feature that is restricted to retain the initial information while rejecting all other sensitive statistics. To make unexpected sensitive measurements, three distinct approaches are used: dimensionality reduction, noise addition, and Siamese fine-tuning. It is astonishing to observe how effectively this system handles a DNN's huge computation when employing DNN splitting for privacy preservation. Taking advantage of the fact that edge computing generally entails handling DNN processing on a large number of devices, parallelization methods are commonly used. Data parallelism and model parallelism are two types of parallelisms used in DNN training. Data parallelism might result in significant communication overhead, whereas model parallelism frequently results in significant underutilization of computational resources. To deal with these issues, Harlap et al. [60] Pipeline parallelism is a variation on model parallelism in which many micro-batches are injected directly into the system to ensure efficient and concurrent computing resource utilization. Pipedream is a technology developed by Harlap et al. [60] that allows pipeline training and automatically identifies how a particular models work is systematically divided among the available computational nodes based on pipe parallelism. Pipedream demonstrates the benefits of minimizing communication overhead and effectively utilizing computational.

Knowledge transfer learning: The DNN splitting approach is strongly linked to knowledge transfer learning or transfer learning for short. To reduce the energy costs of training the DNN model on shaking devices, we first train a basic network (teacher network) on a basic dataset and then use the newly learned functions, i.e., transfer learning. To be trained on a target dataset, they should be moved to the second target network (student network). This procedure will attempt to determine whether the characteristics are universal (i.e., applicable to both base and target tasks) rather than task-specific. An interaction from consensus to particularity is engaged with the change. The transfer learning approach seems promising for erratic device learning due to extremely low resource requirements, but an intensive study of its efficacy is lacking. To fill this gap, Sharma et al. [61] and Chen et al. [62] carried out significant research on the performance of transfer learning (in terms of accuracy and speed of convergence), taking into account the different topologies of student networks and the different strategies for transferring information from the teacher to the student.

Techniques fail miserably, and a few even have a detrimental influence on performance. The flat layers of a previously trained DNN in a dataset are treated in the transfer learning approach as an overall component extractor that can be applied to various objective assignments or informational collections. Movement in learning is utilized in various exploration projects and inspires the improvement of different systems in light of this property. Osia et al. [59] used this technique to figure out how to decide the level of oversimplification and distinction of an individual property, as described in Section IVC4. Arden, as suggested in Wang et al. [58], shares a DNN between the mobile device and thus the cloud data center, with the flat sections of the DNN on the mobile device side transforming the information. As mentioned in Wang et al. [58], the planning of the DNN division in Arden is inspired by transfer learning.

Gossip training: Gossip training, which is based on randomized gossip algorithms and aims to reduce training latency, might be a new decentralized training technique. Gossip averaging [63] is an early study on random gossip algorithms that aims to quickly come to an agreement with hubs by sharing the distributed data. Because they do not require centralized nodes or variables, gossip-distributed algorithms offer full asynchronization and ultimate decentralization. Gossip SGD (GoSGD) [64] is a proposed method for decentralized and asynchronous training of DNN models. GoSGD is a system that maintains a collection of autonomous hubs, every one of which contains a DNN model and is rehashed in two stages: angle update and mix update. In the gradient update phase, each node internally changes its hosted DNN model and, in the combined update step, shares its knowledge with another randomly selected node. The processes are repeated until all of the DNNs have reached an agreement. The goal of GoSGD is to solve the problem of convolutional network training speeding up. Instead, gossiping SGD [65], a gossip-based algorithm, is designed to preserve the benefits of both synchronous and asynchronous SGD techniques. Gossiping SGD achieves asynchronous training by replacing the all-diminish aggregate activity of coordinated preparing with a tattle total technique [64,65] When gossip algorithms are applied to SGD updates, none of them experience a significant performance degradation. Daily et al. [66] show that enormous-scope tattle-based paltry calculations lead to correspondence irregularity, helpless combination, and high correspondence overhead when implemented in large systems. To address these difficulties, Daily et al. [66] offer GossipGraD, an SGD calculation dependent on the tattle correspondence convention, which is valuable for scaling profound learning calculations for enormous-scope organizations. GossipGraD diminishes the general correspondence intricacy from (log (p)) to 0 (1) and takes diffusion into account so that computing nodes indirectly share their updates (gradients) after each log (p) step. It also takes into account the rotation of communication partners to support direct gradient diffusion and asynchronous distributed sample shuffling during the forward phase to avoid overfitting.

4.4 Summary of the existing systems and frameworks

The systems and methodologies of distributed EI model training on the Internet are summarized in this section, including EI levels, aims, technology used, and efficacy of the aforementioned current systems and frameworks. The problem of data privacy is a major difficulty for distributed EI model training in general organizations. For users, they are sensitive to their own private data and do not allow any disclosure of private information. For businesses, they should consider data protection regulations to avoid subpoenas and extrajudicial surveillance. Therefore, the emergence of distributed training systems should carefully consider the protection of privacy. Systems that address privacy issues in a decentralized architecture are of course data protection-friendly, so, systems based on a decentralized architecture, such as BlockFL and GossipGraD, generally protect privacy better. On the other hand, a centralized architecture incorporates a unified information assortment activity, and, along these lines, the crossover design requires an information move activity. FedAvg, BlockFL, GossipGraD, and other such systems all consider privacy issues. The centralized design, on the other hand, necessitates a centralized data collecting activity, whereas the hybrid architecture necessitates a data transfer operation. As a result, systems based on these two designs would go to great lengths to ensure privacy. Compared to DNN training in a cloud-based framework, DNN training in an edge-based framework is more about protecting user privacy and training an already accessible deep learning model. Furthermore, edge-based training is highly desired in some circumstances, such as military and catastrophe applications, when connection to the cloud center is difficult. On the other hand, a cloud data center can collect a larger amount of data and form an artificial intelligence model with stronger resources. Therefore, cloud intelligence has the advantage of being able to create a much larger and more accurate AI model.

5. Edge intelligence model interface

The rapid implementation of model inference at the edge will be crucial to enable the delivery of high-quality EI services after distributed training of deep learning models. This section covers the architectures, key performance indicators, enabling approaches, and current systems and frameworks for DNN model inference in sting.

5.1 Architectures

We describe different edge-centered inference architectures and classify them into four DNN model inference modes, namely, edge-based, device-based, edge–device, and edge–cloud. The following is how each mode's core process is described.

Edge-based mode: Device A is in the edge-based mode, which means that the device will receive the input file and subsequently send it to the sting server. The prediction results are sent to the device when the DNN model inference is completed on the sting server. Because the DNN model resides on the sting server in this inference mode, it is simple to build the application on many mobile platforms. The biggest drawback is that the induction depends on the capacity of the network between the gadget and, subsequently, the edge server.

Device-based mode: Device B is in the device-based mode. The DNN model is downloaded from the sting server to the mobile device, which then performs model inference locally. The versatile gadget does not interface with the sting server amid the deduction preparation. Hence, the deduction is solid but requires a part of assets on the versatile gadget, e.g., CPU, GPU, and RAM.

Edge–device mode: Device C is in the edge–device mode. The device divides the DNN model in the edge–device mode into different sections according to the parameters of the current system environment, such as network bandwidth, device resources, and edge server workload. The gadget at that point runs the DNN model up to a certain layer some time recently sending the middle information to the edge server. The taking after layers will be executed by the sting server, which can at that point provide the expectation comes about to the gadget. The edge–device mode is more dependable and versatile than are the edge-based and device-based modes. Since the convolution layers at the starting of a DNN demonstrated are computationally serious, it can too require a part of assets on the portable device.

Edge–cloud mode: Device D is in the edge–cloud mode. It is similar to the edge–device mode and is appropriate when the device's resources are limited. The device is responsible for computer file gathering in this mode, and the DNN model is implemented via edge–cloud synergy. The quality of the network connection has a significant impact on the model's performance. It is worth noting that the four edge-centric inference modes listed above can be used in the same system to perform complicated AI model inference tasks (z, edge nodes, and clouds).

5.2 Key performance indicators (KPIs)

The following six metrics are used to describe the QoS of the EI model inference.

(1) Latency: The time spent all through the derivation interaction, including pre-handling, model surmising, information moves, and post-handling, is called idleness. Some ongoing savvy versatile applications (like AR/VR portable games and keen mechanical technology) have severe cutoff times, such as an inactivity of 100 ms. The assets of anxious gadgets, the technique for information move, and the gratitude to execute the DNN model are largely factors that impact inactivity.

(2) Accuracy: The proportion between the quantity and the complete number of information tests, which procure the right forecasts from deduction, mirrors the exhibition of DNN models. Ultrahigh accuracy on the DNN model deduction for a couple of portable applications that require an undeniable degree of unwavering quality, like self-driving vehicles and facial confirmation, is sought. Because of the restricted assets of sting gadgets, certain information tests might be avoided in a video investigation application with a high rate of care, influencing a lower precision.

(3) In contrast to the edge waiter, and in this way the cloud server farm, the tip gadgets are for the most part battery-restricted with regard to executing a DNN model. DNN model deduction burns through a ton of energy due to the calculation and correspondence overheads. Energy effectiveness is basic for an EI application, and it is affected by the DNN model's measurements just as the assets of jittery gadgets.

(4) IoT and cell phones create a huge volume of information that can be security-sensitive. During the induction period of an EI application model, the protection and security of the information near the data source should be ensured. The manner in which the first information is handled decides the degree of security insurance.

(5) Aside from the gadget-based strategy, correspondence overhead essentially affects the induction of different modalities. In an EI application, the overhead of the DNN model surmising should be decreased, particularly the exorbitant WAN data transfer capacity use for the cloud. The correspondence overhead here is for the most part controlled by the DNN derivation mode just as the accessible data transmission.

(6) Memory footprint lessens the memory impression of a DNN model surmising on cell phones. From one viewpoint, a high-accuracy DNN model generally contains an enormous number of boundaries and requires a great deal of assets on cell phones. Then again, dissimilar to elite discrete GPUs in cloud server farms, portable GPUs on cell phones do not have committed high-transmission capacity memory [67]. Moreover, versatile computer chips and GPUs are often in a contest for shared and restricted memory transmission capacities. The memory impression might be an unimportant sign for improving DNN deduction on the sting side. The underlying DNN model's size, just as the technique for stacking the huge DNN boundaries, leaves an enormous memory impression.

5.3 Enabling technologies

This part checks out the basic advances to work on at least one of the above basic presentation measurements for EI model surmising.

Model compression: To facilitate the strain between asset-hungry DNNs and asset-helpless end gadgets, DNN pressure is generally used to limit model intricacy and asset necessities and to empower neighborhood surmising on the gadget, thus lessening reaction inertness and security concerns. The model pressure strategy, as such, advances the four markers referenced above: idleness, energy, protection, and memory impression. Weight pruning, information quantization, and smaller compositional plan are among the DNN pressure approaches recommended. Weight pruning is, by a wide margin, the most normally utilized model pressure approach. This methodology eliminates pointless loads (that is, joints between neurons) from a prepared DNN. To do this, it first scores the neurons in the DNN as per their commitment and afterward eliminates the lower-positioned neurons to decrease the size of the model. Since eliminating neurons diminishes the accuracy of the DNN, the fundamental trouble is discovering a procedure to decrease the organization while retaining the accuracy. A pilot study in enormous-scope DNNs has been directed toward current huge-scope

DNNs. Han et al. [68] utilized an extent-based weight pruning method to deal with this issue. The essential idea driving this methodology is to eliminate smaller loads with sizes under a specific limit (e.g., 0.001) and afterward refine the model to reestablish precision. This method can diminish the quantity of AlexNet and VGG-16 loads by 9 and 13, respectively, without losing ImageNet accuracy. The subsequent works profound pressure [69], The pressure proportion is additionally moved to 35–49 by consolidating the advantages of pruning, weight sharing, and Huffman coding to pack DNNs. Notwithstanding this, the earlier craft of size-based weight managing may not be straightforwardly pertinent to control-restricted gadgets, as the real information shows that lessening the quantity of loads does not generally bring about critical energy reserve funds [70]. This is on the grounds that the energy of the folding layers incorporates the whole energy cost in the DNN like AlexNet; however, the sum inside the completely associated layers makes up the vast majority of the absolute number of loads in the DNN. This suggests that the measure of loads may not be a decent pointer of force and that the weight trim for the end gadgets should be unequivocally energy-cognizant. MIT has fostered a web-based DNN energy assessment apparatus (https://energyestimation.mit.edu/) to empower speedy and simple DNN energy assessment. The energy to move data from various layers of the capacity chain of command, the quantity of Macintoshes, and the scarcity of information in the granularity of the DNN layers are all illustrated using this fine-grained instrument. An energy-mindful pruning approach known as EAP was utilized to help this energy-assessing apparatus [71]. Maybe of utilizing the 32-digit skimming direct arrangement toward depict layer data sources, loads, or both, this methodology utilizes a more minimized construction. Information quantization expands generally speaking registering and energy proficiency by encoding a number with less pieces, which diminishes memory impression and velocities up handling. Most past quantization techniques tuned the piece width just for a decent number sort impromptu, which may bring about a helpless result. Late exertion was performed to resolve this issue [72]. To decide the best piece width for the sanctioned configuration dependent on the IEEE 754 norm, the analysts analyzed the ideal number of portrayals for layer granularity. Because of the combinatorial multiplication of conceivable number structures, this issue is hard to address. Thus, the creators created a versatile programming interface known as number dynamic information type (ADT). It permits clients to characterize the data to be quantized in a layer as a number sort (e.g., sources of information, loads, or both). ADT typifies the inward portrayal of variety along these lines, hence confining the significance of building an effective DNN from the significance of expanding the measure of portrayal at the touch level. While most current endeavors use a solitary pressure approach, this will not be sufficient to meet the assorted necessities and limitations enforced by different IoT gadgets as far as exactness, inactivity, stockpiling, and energy is concerned. Late exploration has shown how numerous pressure approaches might be consolidated to pack DNN models to their most extreme limit (e.g., precision, dormancy, and energy) and, in this way, the distinctive accessibility of assets between stages (for instance, stockpiling and handling limit). To this end, the programmed enhancement structure proposed by adadeep [73] efficiently details accuracy, inactivity, stockpiling, and energy targets and imperatives in a solitary streamlining issue and uses DRL to effectively find a fair mix of pressure calculations.

Model partition: As delineated, to ease the weight of the EI application on end gadgets 13, one clear idea is that by apportioning the model and reappropriating the computationally exorbitant parts to the sting server or nearby cell phones, model derivation might be improved. Model division is for the most part disturbed by certain issues. Inactivity, energy, and protection are factors to largely consider. The model segment is divided into two kinds: server-to-gadget segment and gadget-to-gadget segment. Neurosurgeon is a model partitioned between the server and the gadget. Kang et al. [74] represents an incredible exertion. The DNN model in Neurosurgeon is parted between the gadget and along these lines the waiter, and the primary issue is tracking down a satisfactory parcel point that outcome in the best deduction execution of the model. The creators offer a relapse-based strategy to evaluate the dormancy of each layer inside the DNN to demonstrate and produce an ideal split point that will bring about the model's derivation, thus meeting the idleness or force necessities. Ko et al. [75] present a model apportioning approach that joins lossy element encoding with edge-have parceling. That is, after the model is parceled, the middle information is packed by lossy element encoding the preceding transmission. Moreover, through the mix of model dividing and lossy element coding, the structure forms model parceling as an essential issue in applied science (ILP). Upgrading the model parcel to limit inactivity utilizing a coordinated noncyclic diagram (DAG) rather than a series has been demonstrated to be NP-hard before. Subsequently, Hu et al. [76] present a most pessimistic scenario execution ensured estimation strategy dependent on the chart min–cut technique. Each of the structures mentioned above expect that the server has the EI application's DNN model. For EI applications,

IONN [77] proposes a gradual offloading approach. IONN partitions the DNN layers and loads them into stages so that cell phones and edge servers can cooperate on the surmising of the DNN model. IONN fundamentally expands inquiry execution and force utilization when stacking DNN models contrasted with the technique that heaps the full model. The segment between gadgets is one more kind of model segment. MoDNN [78] uses Wi-Fi Direct to deal with setting up a microscale processing group in the WLAN with a few supported Wi-Fi-empowered cell phones for divided DNN model surmising because of the spearheading work of model parcel across gadgets. The gathering proprietor is the cell phone that plays out the DNN work, whereas the others fill in as specialist hubs. In MoDNN, two segment techniques are proposed to accelerate DNN layer execution. MoDNN speeds up DNN model deduction by 2.17–4.28 for two to four laborer hubs, as indicated by the test. In a subsequent study, MeDNN [79], an eager 2D parcel, was introduced for versatile dividing of the DNN model on different cell phones and to deal with the pressure of the DNN model utilizing an organized dispersion pruning approach. With two to four specialist hubs, MeDNN expands the DNN model surmising by 1.86–2.44 and saves an extra of 26.5% calculation time and an extra of 14.2% correspondence time. In MoDNN and MeDNN, DNN layers are parceled evenly, yet, in profound settings, DNN layers are apportioned upward [80]. To decrease the memory impression, it utilizes a combined tile parceling approach that partitions the DNN layers upward. DeepX [81] additionally attempts to isolate the DNN models; yet, it basically divides them into submodels and disperses them to the nearby processors. DeepX proposes two methodologies: RLC (Runtime Layer Pressure) and Father (Profound Engineering Decay) (Father). After pressure, the layer is executed by neighborhood processors (computer chips, GPUs, and DSPs). Something else to remember is that once we have many model parcel occupations, we would prefer to enhance the scheduler. LEO [82] is another detecting calculation scheduler that allotments detecting calculation execution and conveys occupations across the computer chip, co-processor, GPU, and cloud to upgrade execution for a long-time versatile sensor application.

Models early exit: A high-accuracy DNN model generally comprises a deep structure. Executing a DNN model on a tip device requires a significant number of resources. The models early exit technique uses the output data of the early layer to induce the classification result, implying that the inference process is complete when utilizing the partial DNN model; the optimization objective of the models early departure is latency.

BranchyNet [83] may be a programming framework that implements the early exit mechanism in the model. BranchyNet [83] modifies the quality of the DNN models structure. Exit branches can be added to specified layer locations. Each exit branch is an exit point, and the normal DNN model shares certain DNN layers with it. A CNN model with a score of 14. Three out of five points. At these many early exit points, the computer file is categorized. A framework called DDNN [84] for distributed DNN spanning the cloud, edge, and devices has been suggested based on BranchyNet [83]. The device layer, edge server layer, and cloud layer form the three-layer framework of DDNNs. Each level indicates an output from the system BranchyNet [83]. Maximum pooling (MP), average pooling (AP), and concatenation (CC) are three suggested aggregation methods. When multiple mobile devices stream intermediate data to a foothold server or when multiple edge servers stream intermediate data to the cloud data center, aggregation techniques work. By taking the maximum of each component, MP adds the information vectors. By choosing the common of all components, AP aggregates the information vectors. The information vectors are simply concatenated in concert vectors by CC. In addition, Edgent was constructed on top of BranchyNet [10]. When using model early exit and model partition together, it is recommended to traverse the accuracy latency trade-off. The core concept of Edgent is to use a regression-based layer latency prediction model to maximize accuracy under a given latency constraint. There are a variety of techniques for implementing model early departure in addition to BranchyNet, e.g., the cascading network [85]. The MP layer and the fully bonded layer are simply added to the quality DNN model, resulting in 20% acceleration. DeepIns [15] introduces an intelligent industrial manufacturing inspection system based on the DNNs early exit model. Edge devices are in charge of data gathering, the sting server is the first exit point, and the cloud data center is the second departure point in DeepIns. Then, Lo et al. [86] propose that the basic BranchyNet model be supplemented with an authentic operation (AO) unit. By setting different threshold criteria of a confidence level for different output classes of the DNN model, the AO unit assesses whether an input needs to be sent to the edge server or cloud data center for further execution [87]. By adding regularization to the DNN model's evaluation latency, one can train a guide to decide whether the current samples should move on to the next layer.

Edge caching: Edge reserving may be a novel, sensible methodology for accelerating DNN model surmising, i.e., storing the DNN deduction results to lessen inertness. Edge reserving's primary objective is to lessen EI applications, questioning inertness by putting away and reusing position results like picture characterization forecast at the organization edge. The fundamental course of the semantic reserve procedure is that if a solicitation from a cell phone discovers the reserved outcomes put away on the edge server, the edge server returns the outcome; in any case, the solicitation is shipped off the cloud server farm for derivation utilizing the full accuracy model. Impression [88] may very well be a creative endeavor to utilize reserving procedures for DNN deduction. Impression recommends that for an item acknowledgment application, the obsolete acknowledgment result is reused to perceive the article at the current edges. Impression removes a portion of these reserved outcomes and computes the optical progression of provisions between the handled edges and, furthermore, the current casing utilizing the consequences of the distinguished item from the out-of-date outlines put away on cell phones. The bouncing box will be moved to an ideal spot in the current edge utilizing the optical stream registering discoveries. Impression speeds up at a pace of 1.6–5.5. However, on the grounds that storing discoveries locally does not scale past several photographs, Cachier [89] has been proposed to acknowledge great many item–recognizable proofs. Cachier reserves the after effects of EI applications on the edge server, saving the information highlights (e.g., picture) just as the work results. The least sum generally used (LFS) is then utilized as the reserve renewal method by Cachier. The sting server communicates the section to the cloud server farm if the passage cannot be reserved. Cachier can expand responsiveness by a factor of at least three. The Cachier augmentation is Precog [90]. The reserved information in Precog is not just put away on the edge server yet in addition on the cell phone. Precog prefetch information onto the cell phone utilizing Markov chain forecasts, bringing about a 5 × speedup. Precog likewise suggests that the stored include extraction model on the cell phone be progressively changed in light of the climate data. Cachier has been upgraded again in Shadow Manikins. Hazy Reserve [91] is proposed to lessen repetitive estimations in the application situation where a similar application runs on various gadgets in closeness and the DNN model frequently parses comparative information records. Hazy Reserve has two difficulties: one is that the design of the information document is shaky, so the trouble is ordering the PC record with consistent hunt quality, and the other is communicating the comparability of the PC document. To reconcile these two issues, versatile LSH (A-LSH) and homogenized KNN ($H-$KNN) techniques are proposed by Hazy store. Hazy store cuts handling deferral and energy use by a factor of 3–10.

Input filtering: Information separation is a valuable methodology for accelerating DNN model surmising, especially in video examination. The main idea behind information sifting is to avoid nontarget outlines from the information, eliminate repetitive computations from DNN model deduction, and, accordingly, increase induction accuracy, diminish derivation inactivity, and decrease power costs. Noscope [92] has been proposed to accelerate video examination by skipping outlines with minor adjustments. For this reason, Noscope contains a distinction finder that features the fleeting contrasts between outlines. For instance, the indicator watches edges to check whether vehicles show up in them, and just those edges containing vehicles are handled in the DNN model derivation. Lightweight twofold classifiers are utilized to recognize the distinction. On account of a multitude of robots sending videos progressively, Wang et al. [93] advance for DNN surmising's essential bounce remote transfer speed. To decrease transmission by and large, four strategies are proposed: early disposing, just-in-time learning (JITL), Reachback, and Setting Mindful FFS-VA [94]. It is conceivable that a pipeline framework for multistage video examination might arise. The FFS-VA sifting framework is developed in three stages. The essential identifier may be a stream-explicit distinction indicator (SDD), which is utilized to dispose of edges with simply a background. A stream-specific organization model (SNM) for target object outline identification may be the subsequent choice. The third choice is to utilize a Tiny-YOLO-Voc (TYOLO) model to avoid outlines with focal points that are under a specific limit. For video investigation, Canel et al. [95] offered a two-stage separating method. It removes the semantic measurements of the edges for sending out DNN's middle-of-the road information, after which yield capacities are amassed in an incredible body cushion. The support is viewed as a DAG, and the separating calculation decides the top-k fascinating casings utilizing Euclidean distance as a similitude metric. Derivation of the DNN model for cross-chamber investigation is sped up. Rex-cam channels video outlines utilizing a learned spatiotemporal model. Rex-cam cuts computation time by 4.6% and lifts DNN model surmising precision by 27%.

Model determination: A model choice methodology is introduced to work on the idleness, accuracy, and force of DNN derivation. The essential thought behind model choice is that we can prepare various DNN models for an equivalent disconnected undertaking with various model sizes and afterward pick the best model for online derivation. The principal contrast is that the leave point imparts a portion of the DNN layers to the model with more branches, while the models are isolated inside the model determination process. Park et al. [96] provide a structure to the choice of huge/little small DNN models. To put it in another way, a sensitive model is utilized for arranging the PC record, and the enormous model is possibly used when the minute model's self-importance is not exactly a foreordained edge. Taylor et al. [97] show that for different pictures, distinctive DNN models (e.g., versatile net, ResNet, and Origin) accomplish the least derivation inactivity or best exactness on different assessment rules (top-1 or top-5). They then, at that point, provide an approach to figure out which DNN is best as far as dormancy and precision is concerned. A model selector is prepared in this setting to choose the best DNN for different information pictures. Similarly, for IF-CNN [98], the prevalence indicator (RP), a model selector, is additionally prepared to alter the model utilized in the undertaking. RP may be a DNN model that performs various tasks, which implies that it has many yields. The likelihood of every potential DNN model's best 1 name is addressed by the yield of RP. The image is the contribution of RP, and if the yield of RP surpasses a foreordained limit, then coordination with DNN model is picked. Beside the energy-saving improvement for DNN model induction delay, Stamoulis et al. [99] Considering the exactness and correspondence restrictions of the gadgets, we can transform the versatile DNN model choice issue into a hyperparameter improvement issue. Then, at that point, to resolve the issue, a Bayesian streamlining (BO) is utilized; this prompts an improvement of up to 6 as far as least energy per picture dependent on accuracy prerequisites is concerned.

Multitenancy support: By and by, an end or edge gadget frequently runs a few DNN applications simultaneously. For instance, in Web vehicles, the high-level driver-help framework (ADAS) executes DNN calculations for vehicle distinguishing proof, passer-by location, traffic sign acknowledgment, and path line recognition simultaneously. Numerous DNN applications would vie for restricted assets in the present circumstances. Generally, effectiveness would be lost without satisfactory help for a long time, that is, both asset distribution and assignment booking for these simultaneous projects were seriously influenced. Multi-occupant support centers around decreasing energy and capacity prerequisites. Considering the elements of the assets at runtime [100], for each DNN model, home DNN has been proposed to provide customizable asset exactness compromises. Home DNN utilizes a model substitution pruning and recuperation methodology to transform the DNN model into a minimal variety model comprised of an assortment of descendent models. Each relative model has a special asset exactness compromise. For every relative model that is simultaneously dynamic, home DNN fosters an asset-precision runtime scheduler to shape the ideal compromise for each simultaneous descendent model by encoding its exactness and idleness into a worth capacity. Essentially, Standard [101] utilizes the famous exchange-learning DNN preparing system to show different DNN models with fluctuating levels of accuracy and an insatiable way to deal with picking the best developer for the worth spending plan. On a solitary gadget, a few DNN models might be executed Hivemind [102]. For these simultaneous errands, HiveMind has been prescribed to expand GPU utilization. A HiveMind compiler and a runtime module are the two principal parts. The compiler enhances information move, preprocessing, and calculation among jobs, and the runtime motor believers the improved models into a runtime DAG. This will be run on the GPU fully intent on getting as much simultaneousness as practical. At a point when granularity is better, profound eye [103] has been proposed to plan the executions of heterogeneous DNN layers to improve the performance of various task derivations on cell phones. Profound eye divides all DNN layers into two classifications: convolutional layers and completely connected layers. For convolutional layers, a line-based first in (endash), first out (FIFO) execution strategy is utilized. Expected for the totally related stores, profound eye to capitalize on memory, it adopts an avaricious strategy to putting away the boundaries of the completely connected layers.

While the enhancement approaches referenced above are by and large pertinent to EI applications, application-explicit streamlining can be utilized to work on the exhibition of EI applications, like exactness, inertness, power utilization, and memory space. For video-based applications, for instance, two handles, for example to decrease asset utilization, outline rate and assurance and are much of the time deftly altered. Be that as it may, on the grounds that asset-sensitive handles debase derivation precision, they unavoidably bring about an expense exactness compromise. When setting the video outline rate and assurance, we wanted to track down a decent harmony between the expense of assets and the accuracy of the deduction. While heading to the ethereal objective, Chameleon [104] changes the controls for every video investigation task by moving top-k's best powerful settings starting with one assignment and then onto the next. Video errands in Chameleon are positioned by their spatial connection, and, then, at that point, the gathering chief looks for the best Top-k blends and offers them with the devotees. Profound Choice [105]. The knob-tuning issue is reformulated as a different decision multi-imperative rucksack program that is addressed using a superior beast power scan approach.

Additionally, it is worth noting that equipment speed increase for proficient DNN surmising has been a hot issue in the PC design field and has drawn a ton of consideration. Endeavors to do explore. Peruses who are intrigued can counsel the most recent monograph [106] for a more inside and out outline of current advancements in DNN handling equipment speed increase.

5.4 Summary of the existing systems and frameworks

Important frameworks and systems to show the use of the above empowering methods for IE model deduction, including viewpoints of target applications, design and IE level, targets, and streamlining procedures. Utilized, and adequacy. Obviously, particular subsets of empowering techniques have been embraced by current frameworks and structures, each adjusted for exceptional EI applications and necessities. Far-reaching empowering approaches and different enhancement methodologies should cooperate for broad plan adaptability to boost the general exhibition of a nonexclusive EI framework. In any case, we might confront a high-dimensional setup challenge where an inordinate number of execution basic arrangement boundaries are not set in stone continuously. High-dimensional arrangement boundaries for video investigation can incorporate, for instance, video outline rate, goal, model determination, and expected model yield. The high-dimensional arrangement issue, because of its combinatorial nature, has an enormous quest space for boundaries, which makes it extremely hard to address.

6. Future research directions

We are presently recognizing the critical open inquiries and future exploration ways for EI dependent on the phenomenal input above on current drives.

6.1 Programming and software platforms

Numerous associations across the world are presently zeroing in on man-made intelligent distributed computing. Amazon's Greengrass, Microsoft's Sky Blue IoT Edge, and Google's Cloud IoT Edge are among the significant organizations dispatching programming/programming stages to offer edge registering administrations. In any case, the greater part of those frameworks presently serves essentially as transfers for connecting to incredible cloud server farms. EI as a service platform (EiaaS) could turn into an inescapable worldview as increasingly more simulated intelligence-controlled and registered, escalated, and portable IoT applications hit the market and EI stages with powerful edge artificial intelligence highlights are developed and carried out. This is often unmistakable from the ML as a help (MLAAs) presented by open mists. MLAA is a sort of cloud knowledge that focuses on picking the right server setup and AI structure to cost-effectively show models in the cloud. Although EiaaS is essentially focused on a strategy for model preparation and surmising in asset-compelled and protection-delicate edge processing settings, there is an unmistakable distinction. There are various significant hindrances to tackle to completely satisfy the guarantee of EI administrations. The EI stages ought to, above all else, be heterogeneously viable. Since there will be various EI specialist organizations/suppliers later on, a solitary open standard ought to be set up so that purchasers can encounter consistent administrations on heterogeneous EI stages. Second, there are a few simulated intelligence programming systems to browse (for instance, TensorFlow, Lite, and Caffe). The future ought to take into account the versatility of stings simulated intelligence models that were created by different programming systems across heterogeneously conveyed edge hubs. Third, a few programming systems (like TensorFlow, Lite, Caffe2, Coral, and Maxent) are developed especially for edge gadgets; in any case, experimental measures fail to [107] show that there is no champion that can outperform different structures in all measurements; later on, we might expect a structure that performs well on a more extensive scope of measures. To wrap things up, lightweight registering and virtualization approaches, for example, compartment and capacity processing, should be additionally investigated to empower compelling position and relocation of EI administrations in asset-obliged web-based business conditions.

6.2 Resource-friendly edge AI model design

Numerous contemporary man-made intelligence models, like CNN and LSTM, have been created with PC vision and discourse correspondence handling as a main priority. Most profound learning-based artificial intelligence models are asset-concentrated, requiring vigorous figuring power upheld by broad equipment assets (for example, GPU, FPGA, and TPU) to work on the exhibition of these models. Man-made brainpower. Therefore, as recently observed, a few researches have been conducted to scale down simulated intelligence models utilizing model pressure draws (e.g., weight pruning) to make them more asset-agreeable for edge sending. We will showcase an interesting line, notwithstanding a particular line. Edge computer-based intelligence model plan that is asset mindful. Maybe of depending on asset serious simulated intelligence models that as of now exist, we will impact the Auto ML information [108] NAS (neural engineering search) strategies are likewise utilized [109] to plan asset effective edge artificial intelligence models that oblige the equipment asset limitations of the fundamental edge servers and gadgets. Techniques like RL, hereditary calculations, and BO can be utilized, for instance, to proficiently put requirements on execution measurements, like execution inertness and energy use, through the KI model plan boundary space (d-work memory).

6.3 Computation-aware networking techniques

EI, by and large, runs computationally escalated, computer- and intelligence-based applications in a circulated edge-figuring climate. Hence, refined and PC-agreeable organization arrangements are incredibly attractive, as computation results and information can be successfully traded across numerous edge hubs. Ultra-Reliable Low-Latency Communication (URLLC) is intended for basic business applications that require low deferral and high dependability in the longest 5G organization of things to come. Subsequently, joining the 5G URLLC's best-in-class figuring abilities to provide incredibly solid Low Dormancy EI (URLLEI) administrations will show guarantee. What is more, 5G will utilize complex advancements, for example, programming-characterized systems administration and organization work virtualization. These methodologies take into account more adaptable organization asset the board and empower on-request associations across various edge hubs for handling concentrated simulated intelligence applications. Then again, the plan of an independent organization component is fundamental to the proficient arrangement of EI benefits in a powerful heterogeneous organizational concurrence (e.g., LTE/5G/Wi-Fi/lora), permitting recently acquainted edge hubs and gadgets with self-design in a fit-and-play style. What is more, PC helps specialized techniques, for example, angle coding [110], to decrease the defer impact on appropriated learning and over-the-air calculation for DSGD [111], which could be worthwhile in accelerating model preparation. Edge computer-based intelligence, gain in significance.

6.4 Trade-off design with various DNN performance metrics

There are, for the most part, various DNN model applicants fit for finishing the work for an EI application with a specific target. Nonetheless, on the grounds that quality exhibition measurements, for example, top-k precision or mean exactness, do not address the runtime execution of DNN model induction on fringe gadgets, it is hard for programming engineers to pick a model. DNN fitting for EI application. For a model, during the organization period of an EI application, derivation speed and exactness are significant contemplations. The utilization of assets is additionally a significant measurement. We need to dissect the compromises between these actions and the variables that influence them. For item acknowledgment, Huang et al. [112] explored the impact of different boundaries on the time and accuracy of the derivation, like the quantity of clues, the size of the information picture, and the choice of the element extractor. In light of the aftereffects of their trials, they found that a new blend of these components outflanks the best-in-class approach. Accordingly, it's basic to examine the compromises between different markers to work on the productivity of sending EI applications.

6.5 Resource management and smart services

Edge hubs and gadgets that endow EI with usefulness are appropriated across numerous geospatial areas and locales because of the disseminated idea of edge registering. Distinctive computer-based intelligence models and concentrated simulated intelligence errands can be run on various gadgets and edge hubs. Way. Separating muddled edge computer-based intelligence models into little suberrands and viably reappropriating these assignments to edge hubs and gadgets for synergistic executions are likewise important to completely use the dispersive assets at edge hubs and gadgets. Since help conditions in some EI application situations (e.g., shrewd urban areas) are extremely powerful, it is hard to appropriately expect future events. Along these lines, outstanding Internet-based edge asset provisioning and arrangement capacities would be needed to reliably deal with enormous EI responsibilities. The joint ongoing improvement of heterogeneous process, correspondence and reserve asset assignments, just as the arrangement of high-dimensional framework boundaries (for instance, the determination of suitable model preparing and induction draws near), which are customized to countless undertaking prerequisites, they are imperative. To adapt to the intricacy of calculation plan, another space of exploration is utilizing artificial intelligence methods, for example, DRL to change the ideal asset designation approaches in a self-learning based way in information. To manage the intricacy of the calculation plan, another examination course is to utilize man-made brainpower strategies like DRL to tweak proficient asset designation strategy in an exceptionally information-driven and self-adapting manner.

6.6 Security and privacy issues

As a result of the open idea of edge processing, decentralized trust is important to guarantee that EI administrations provided by many organizations are solid [113]. To provide client validation and access control, information and model honesty, and cross-stage confirmation for EI, circulated and lightweight security systems are significant. While considering the living together of dependable edge hubs with malignant ones, look at new safe steering strategies and trust network geographies for EI administration conveyance. Then again, the principal clients and gadgets would produce a lot of information at the edge of the organization that would be exceptionally dangerous to security, as it could incorporate client area information, well-being or movement records, or assembling data, in addition to other factors. Straightforwardly trading the principal informational indexes among various edge hubs may have the critical danger of security spillage when dependent upon security insurance necessities, like the EU's Overall Information Assurance Guideline (GDPR). Subsequently, combined learning is a practical worldview for protection well-disposed disseminated information preparing when unique informational indexes are put away on your delivered gadgets/hubs and Edge computer based intelligence model boundaries are shared. To further develop information protection, specialists are progressively going to innovations like differential security and homomorphic encryption, and secure multiparty figuring to make protection safeguarding man-made intelligence model boundary trade plans [114].

6.7 The EI ecosystem is a large open collaboration that focuses on incentives and business models

Stage suppliers (for instance, Amazon), artificial intelligence programming suppliers (for instance, SenseTime), edge gadget suppliers (for instance, Hikvision), network administrators (for instance, AT&T), information generators (for example IoT and cell phone proprietors), and fix clients all have a place with EI specialist organizations and clients (i.e., EI clients) [115].

For instance, for embracing improved asset sharing and simple assistance move, the profoundly proficient activity of EI administrations might require close coordination and incorporation between numerous specialist organizations. Advanced, successful, and effective coordinated efforts between all individuals from the EI environment, a suitable motivation component, and a monetary model are needed. In this case, a stand-out savvy estimating structure is important to represent the client's administration utilization just as the worth of their information input. Blockchain with a sensible agreement may be executed into the EI administration by working on decentralized edge servers as a way for decentralized participation. It is helpful to concentrate on the best way to effectively ascertain the cost and divide the pay between the individuals from the EI environment as indicated by their work records. It would likewise be ideal to develop an asset-cordial, lightweight blockchain agreement framework for EI.

7. Conclusions

With artificial intelligence and IoT on the ascent, there is an earnest need to move the computer-based intelligence boondocks from the cloud to the edge of the web. Edge registering has been broadly perceived as a suitable choice to help process serious artificial intelligence applications in asset-obliged settings to meet this pattern. The association between state-of-the-art processing and simulated intelligence provides a new worldview of EI. In this part, we have conducted an exhaustive survey of the most recent examinations on EI. In particular, we initially analyzed the foundation and inspiration of man-made intelligence at the edge of the network. We then, at that point, provided an outline of the arising key designs, structures, and advances for the profound learning model for preparing and surmising at the edge of the organization. Finally, we discussed EI's open issues and potential exploration pathways. We trust that this survey will produce more interest, cultivate useful conversations, and create new exploration thoughts on EI.

References

[1] Y. LeCun, Y. Bengio, G. Hinton, Deep learning, Nature 521 (7553) (2015) 436.
[2] L. Deng, D. Yu, Deep learning: methods and applications, Found. Trends Signal Process. 7 (3) (2014) 197–387.
[3] Cisco Global Cloud Index, Forecast and Methodology [online], 2016–2021, Available https://www.cisco.com/c/en/us/solutions/collateral/service-provider/global-cloud-index-gci/white-paper-c11-738085.html.
[4] B. Heintz, A. Chandra, R.K. Sitaraman, Optimizing grouped aggregation in geo-distributed streaming analytics, in: Proc. ACM HPDC, 2015, pp. 133–144.

[5] Q. Pu, et al., Low latency geo-distributed data analytics, in: Proc. ACM SIGCOMM, 2015, pp. 421–434.
[6] W. Shi, J. Cao, Q. Zhang, Y. Li, L. Xu, Edge computing: vision and challenges, IEEE Internet Things J. 3 (5) (2016) 637–646.
[7] X. Chen, L. Pu, L. Gao, W. Wu, D. Wu, Exploiting massive D2D collaboration for energy-efficient mobile edge computing, IEEE Wirel. Commun. 24 (4) (2017) 64–71.
[8] Y. Mao, C. You, J. Zhang, K. Huang, K.B. Letaief, A survey on mobile edge computing: the communication perspective, IEEE Commun. Surveys Tuts. 19 (4) (2017) 2322–2358. 4th Quart.
[9] X. Wang, Y. Han, C. Wang, Q. Zhao, X. Chen, M. Chen, In-Edge AI: Intelligentizing Mobile Edge Computing Caching and Communication by Federated Learning [online], 2018, arXiv:1809.07857. Available https://arxiv.org/abs/1809.07857.
[10] E. Li, Z. Zhou, X. Chen, Edge intelligence: on-demand deep learning model co-inference with device-edge synergy, in: Proc. Workshop Mobile Edge Commun. MECOMM, 2018, pp. 31–36.
[11] Trends Emerge in the Gartner Hype Cycle for Emerging Technologies [online], 2018. Available https://www.gartner.com/smarterwithgartner/5-trends-emerge-in-gartner-hype-cycle-for-emerging-technologies-2018/.
[12] G. Ananthanarayanan, et al., Real-time video analytics: the killer app for edge computing, Computer 50 (10) (2017) 58–67.
[13] K. Ha, Z. Chen, W. Hu, W. Richter, P. Pillai, M. Satyanarayanan, Towards wearable cognitive assistance, in: Proc. ACM Mobisys, 2014, pp. 68–81.
[14] C. Jie, L. Xu, R. Abdallah, W. Shi, EdgeOS_h: a home operating system for internet of everything, in: Proc. IEEE ICDCS, 2017, pp. 1756–1764.
[15] L. Li, K. Ota, M. Dong, Deep learning for smart industry: efficient manufacture inspection system with fog computing, IEEE Trans. Ind. Informat. 14 (10) (2018) 4665–4673.
[16] D. Svozil, V. Kvasnicka, J. Pospichal, Introduction to multi-layer feed-forward neural networks, Chemom. Intell. Lab. Syst. 39 (1) (1997) 43–62.
[17] R. Collobert, J. Weston, L. Bottou, M. Karlen, K. Kavukcuoglu, P. Kuksa, Natural language processing (almost) from scratch, J. Mach. Learn. Res. 12 (2011) 2493–2537.
[18] A. Krizhevsky, I. Sutskever, G.E. Hinton, Imagenet classification with deep convolutional neural networks, in: Proc. NIPS, 2012, pp. 1097–1105.
[19] K. Simonyan, A. Zisserman, Very Deep Convolutional Networks for Large-Scale Image Recognition [online], 2014, arXiv:1409.1556. Available https://arxiv.org/abs/1409.1556.
[20] K. He, X. Zhang, S. Ren, J. Sun, Deep residual learning for image recognition, in: Proc. IEEE CVPR, 2016, pp. 770–778.
[21] A.G. Howard, et al., MobileNets: Efficient Convolutional Neural Networks for Mobile Vision Applications, 2017, arXiv:1704.04861 [online] Available: https://arxiv.org/abs/1704.04861.
[22] H. Mao, S. Yao, T. Tang, B. Li, J. Yao, Y. Wang, Towards real-time object detection on embedded systems, IEEE Trans. Emerging Topics Comput. 6 (3) (2018) 417–431.
[23] J. Redmon, S. Divvala, R. Girshick, A. Farhadi, You only look once: unified real-time object detection, in: Proc. IEEE Conf. Comput. Vis. Pattern Recognit, 2016, pp. 779–788.
[24] W. Liu, et al., SSD: single shot multibox detector, in: Proc. Eur. Conf. Comput. Vis, 2016, pp. 21–37.
[25] L. Bottou, Large-scale machine learning with stochastic gradient descent, in: Proc. COMPSTAT, 2010, pp. 177–186.
[26] D.E. Rumelhart, G.E. Hinton, R.J. Williams, Learning representations by back-propagating errors, Nature 323 (6088) (1986) 533.

[27] Y. Chauvin, D.E. Rumelhart, Backpropagation: Theory Architectures and Applications, Psychology Press, 2013.

[28] C. Szegedy, et al., Going deeper with convolutions, in: Proc. IEEE Conf. Comput. Vis. Pattern Recognit, 2015, pp. 1–9.

[29] P.J. Werbos, Backpropagation through time: what it does and how to do it, Proc. IEEE 78 (10) (1990) 1550–1560.

[30] S. Hochreiter, J. Schmidhuber, Long short-term memory, Neural Comput. 9 (8) (1997) 1735–1780.

[31] I. Goodfellow, et al., Generative adversarial nets, in: Proc. Adv. Neural Inf. Process. Syst, 2014, pp. 2672–2680.

[32] V. Mnih, et al., Human-level control through deep reinforcement learning, Nature 518 (7540) (2015) 529.

[33] AI Trends for Enterprise Computing, 2017. [online] Available https://www.gartner.com/smarterwithgartner/3-ai-trends-for-enterprise-computing/.

[34] Democratizing AI, 2016. [online] Available https://news.microsoft.com/features/democratizing-ai/.

[35] M. Satyanarayanan, P. Bahl, R. Caceres, N. Davies, The case for VM-based cloudlets in mobile computing, IEEE Pervasive Comput. 4 (2009) 14–23.

[36] Microsoft Interactive Cloud Gaming, 2022. [online] Available https://azure.microsoft.com/en-us/solutions/gaming/.

[37] H. Zhang, G. Ananthanarayanan, P. Bodik, M. Philipose, P. Bahl, M.J. Freedman, Live video analytics at scale with approximation and delay-tolerance, in: Proc. USENIX NSDI, 2017, pp. 377–392.

[38] C.-C. Hung, et al., VideoEdge: Processing camera streams using hierarchical clusters, in: Proc. IEEE/ACM Symp. Edge Comput. (SEC), 2018, pp. 115–131.

[39] I. Stoica, et al., A Berkeley View of Systems Challenges for AI, 2017, arXiv:1712.05855 [online] Available https://arxiv.org/abs/1712.05855.

[40] Trends Emerge in the Gartner Hype Cycle for Emerging Technologies, 2018. [online] Available: https://www.gartner.com/smarterwithgartner/5-trends-emerge-in-gartner-hype-cycle-for-emerging-technologies-2018/.

[41] IEC White Paper Edge Intelligence [online], 2017. Available https://www.iec.ch/whitepaper/edgeintelligence/.

[42] Accelerating AI on the Intelligent Edge [online], 2022. Available: https://azure.microsoft.com/en-us/blog/accelerating-ai-on-the-intelligent-edge-microsoft-and-qualcomm-create-vision-ai-developer-kit/.

[43] Edge Intelligence for Industrial Internet of Things [online], 2021. Available https://www.comsoc.org/publications/magazines/ieee-network/cfp/edge-intelligence-industrial-internet-things.

[44] H.B. McMahan, E. Moore, D. Ramage, S. Hampson, B.A.Y. Arcas, Communication-Efficient Learning of Deep Networks from Decentralized Data [online], 2016, arXiv:1602.05629, Available https://arxiv.org/abs/1602.05629.

[45] R. Shokri, V. Shmatikov, Privacy-preserving deep learning, in: Proc. 22nd ACM SIGSAC Conf. Comput. Commun. Secur, 2015, pp. 1310–1321.

[46] J. Konečný, H.B. McMahan, F.X. Yu, P. Richtárik, A.T. Suresh, D. Bacon, Federated Learning: Strategies for Improving Communication Efficiency, arXiv:1610.05492, 2016.

[47] A. Lalitha, S. Shekhar, T. Javidi, F. Koushanfar, Peer-to-Peer Federated Learning on Graphs [online], 2019, arXiv:1901.11173. Available https://arxiv.org/abs/1901.11173.

[48] H. Kim, J. Park, M. Bennis, S.-L. Kim, On-Device Federated Learning Via Blockchain and Its Latency Analysis [online], 2018, arXiv:1808.03949. Available https://arxiv.org/abs/1808.03949.

[49] K. Hsieh, A. Harlap, N. Vijaykumar, D. Konomis, G.R. Ganger, P.B. Gibbons, Gaia: geo-distributed machine learning approaching LAN speeds, in: Proc. NSDI, 2017, pp. 629–647.

[50] S. Wang, et al., Adaptive federated learning in resource constrained edge computing systems, IEEE J. Sel. Areas Commun. 37 (3) (2019) 1205–1221.

[51] T. Nishio, R. Yonetani, Client Selection for Federated Learning with Heterogeneous Resources in Mobile Edge [online], 2018, arXiv:1804.08333. Available https://arxiv.org/abs/1804.08333.

[52] Y. Lin, S. Han, H. Mao, Y. Wang, W.J. Dally, Deep Gradient Compression: Reducing the Communication Bandwidth for Distributed Training [online], 2017, arXiv:1712.01887. Available https://arxiv.org/abs/1712.01887.

[53] Z. Tao, Q. Li, eSGD: communication efficient distributed deep learning on the edge, in: Proc. USENIX Workshop Hot Topics Edge Comput. (HotEdge), 2018.

[54] S.U. Stich, J.-B. Cordonnier, M. Jaggi, Sparsified sgd with memory, in: Proc. Adv. Neural Inf. Process. Syst, 2018, pp. 4452–4463.

[55] H. Tang, S. Gan, C. Zhang, T. Zhang, J. Liu, Communication compression for decentralized training, in: Proc. Adv. Neural Inf. Process. Syst, 2018, pp. 7663–7673.

[56] M.M. Amiri, D. Gunduz, Machine Learning at the Wireless Edge: Distributed Stochastic Gradient Descent over-the-Air [online], 2019, arXiv:1901.00844. Available https://arxiv.org/abs/1901.00844.

[57] Y. Mao, S. Yi, Q. Li, J. Feng, F. Xu, S. Zhong, A privacy-preserving deep learning approach for face recognition with edge computing, in: Proc. USENIX, 2018.

[58] J. Wang, J. Zhang, W. Bao, X. Zhu, B. Cao, P.S. Yu, Not just privacy: improving performance of private deep learning in mobile cloud, in: Proc. 24th ACM SIGKDD Int. Conf. Knowl. Discovery Data Mining, 2018, pp. 2407–2416.

[59] S.A. Osia, et al., A Hybrid Deep Learning Architecture for Privacy-Preserving Mobile Analytics [online], 2017, arXiv:1703.02952. Available https://arxiv.org/abs/1703.02952.

[60] A. Harlap, D. Narayanan, A. Phanishayee, V. Seshadri, G.R. Ganger, P.B. Gibbons, PipeDream: Fast and Efficient Pipeline Parallel DNN Training [online], 2018, arXiv:1806.03377. Available https://arxiv.org/abs/1806.03377.

[61] R. Sharma, S. Biookaghazadeh, B. Li, M. Zhao, Are existing knowledge transfer techniques effective for deep learning with edge devices? in: Proc. IEEE Int. Conf. Edge Comput. (EDGE), 2018, pp. 42–49.

[62] Q. Chen, Z. Zheng, C. Hu, D. Wang, F. Liu, Data-driven task allocation for multi-task transfer learning on the edge, in: Proc. IEEE 39th Int. Conf. Distrib. Comput. Syst. (ICDCS), 2019.

[63] S. Boyd, A. Ghosh, B. Prabhakar, D. Shah, Randomized gossip algorithms, IEEE Trans. Inf. Theory 52 (6) (2006) 2508–2530.

[64] M. Blot, D. Picard, M. Cord, N. Thome, Gossip Training for Deep Learning [online], 2016, arXiv:1611.09726. Available https://arxiv.org/abs/1611.09726.

[65] P.H. Jin, Q. Yuan, F. Iandola, K. Keutzer, How to Scale Distributed Deep Learning? [online], 2016, arXiv:1611.04581. Available https://arxiv.org/abs/1611.04581.

[66] J. Daily, A. Vishnu, C. Siegel, T. Warfel, V. Amatya, GossipGraD: Scalable Deep Learning Using Gossip Communication Based Asynchronous Gradient Descent [online], 2018, arXiv:1803.05880. Available: https://arxiv.org/abs/1803.05880.

[67] C.-J. Wu, et al., Machine learning at facebook: understanding inference at the edge, in: Proc. IEEE Int. Symp. High Perform. Comput. Archit. (HPCA), 2019, pp. 331–344.

[68] S. Han, J. Pool, J. Tran, W. Dally, Learning both weights and connections for efficient neural network, in: Proc. Adv. Neural Inf. Process. Syst, 2015, pp. 1135–1143.

[69] S. Han, H. Mao, W.J. Dally, Deep Compression: Compressing Deep Neural Networks with Pruning Trained Quantization and Huffman Coding [online], 2015, arXiv:1510.00149. Available https://arxiv.org/abs/1510.00149.

[70] Y.-H. Chen, J. Emer, V. Sze, Eyeriss: a spatial architecture for energy-efficient dataflow for convolutional neural networks, ACM SIGARCH Comput. Archit. News 44 (3) (2016) 367–379.

[71] T.-J. Yang, Y.-H. Chen, V. Sze, Designing energy-efficient convolutional neural networks using energy-aware pruning, in: Proc. IEEE Conf. Comput. Vis. Pattern Recognit. (CVPR), 2017, pp. 5687–5695.

[72] Y.H. Oh, et al., A portable automatic data quantizer for deep neural networks, in: Proc. ACM PACT, 2018, p. 17.

[73] S. Liu, Y. Lin, Z. Zhou, K. Nan, H. Liu, J. Du, On-demand deep model compression for mobile devices: a usage-driven model selection framework, in: Proc. 16th Annu. Int. Conf. Mobile Syst. Appl. Services, 2018, pp. 389–400.

[74] Y. Kang, et al., Neurosurgeon: collaborative intelligence between the cloud and mobile edge, ACM SIGPLAN Not. 52 (4) (2017) 615–629.

[75] J.H. Ko, T. Na, M.F. Amir, S. Mukhopadhyay, Edge-Host Partitioning of Deep Neural Networks with Feature Space Encoding for Resource-Constrained Internet-of-Things Platforms [online], 2018, arXiv:1802.03835. Available https://arxiv.org/abs/1802.03835.

[76] C. Hu, W. Bao, D. Wang, F. Liu, Dynamic adaptive DNN surgery for inference acceleration on the edge, in: Proc. IEEE INFOCOM, 2019.

[77] H.-J. Jeong, H.-J. Lee, C.H. Shin, S.-M. Moon, IONN: Incremental offloading of neural network computations from mobile devices to edge servers, in: Proc. ACM Symp. Cloud Comput, 2018, pp. 401–411.

[78] J. Mao, X. Chen, K.W. Nixon, C. Krieger, Y. Chen, MoDNN: local distributed mobile computing system for deep neural network, in: Proc. Design Autom. Test Eur. Conf. Exhib. (DATE), 2017, pp. 1396–1401.

[79] J. Mao, et al., MeDNN: a distributed mobile system with enhanced partition and deployment for large-scale DNNs, in: Proc. 36th Int. Conf. Comput.-Aided Design, 2017, pp. 751–756.

[80] Z. Zhao, K.M. Barijough, A. Gerstlauer, DeepThings: distributed adaptive deep learning inference on resource-constrained IoT edge clusters, IEEE Trans. Comput.-Aided Design Integr. Circuits Syst. 37 (11) (2018) 2348–2359.

[81] N.D. Lane, et al., DeepX: a software accelerator for low-power deep learning inference on mobile devices, in: Proc. 15th Int. Conf. Inf. Process. Sensor Netw, 2016, p. 23.

[82] P. Georgiev, N.D. Lane, K.K. Rachuri, C. Mascolo, Leo: scheduling sensor inference algorithms across heterogeneous mobile processors and network resources, in: Proc. 22nd Annu. Int. Conf. Mobile Computing Netw, 2016, pp. 320–333.

[83] S. Teerapittayanon, B. McDanel, H. Kung, BranchyNet: fast inference via early exiting from deep neural networks, in: Proc. 23rd Int. Conf. Pattern Recognit. (ICPR), 2016, pp. 2464–2469.

[84] S. Teerapittayanon, B. McDanel, H.-T. Kung, Distributed deep neural networks over the cloud the edge and end devices, in: Proc. IEEE 37th Int. Conf. Distrib. Comput. Syst. (ICDCS), 2017, pp. 328–339.

[85] S. Leroux, et al., The cascading neural network: building the internet of smart things, Knowl. Inf. Syst. 52 (3) (2017) 791–814.

[86] C. Lo, Y.-Y. Su, C.-Y. Lee, S.-C. Chang, A dynamic deep neural network design for efficient workload allocation in edge computing, in: Proc. IEEE Int. Conf. Comput. Design (ICCD), 2017, pp. 273–280.

[87] T. Bolukbasi, J. Wang, O. Dekel, V. Saligrama, Adaptive Neural Networks for Efficient Inference [online], 2017, arXiv:1702.07811. Available https://arxiv.org/abs/1702.07811.

[88] T.Y.-H. Chen, L. Ravindranath, S. Deng, P. Bahl, H. Balakrishnan, Glimpse: continuous real-time object recognition on mobile devices, in: Proc. ACM Sensys, 2015.

[89] U. Drolia, K. Guo, J. Tan, R. Gandhi, P. Narasimhan, Cachier: edge-caching for recognition applications, in: Proc. IEEE ICDCS, 2017, pp. 276–286.

[90] U. Drolia, K. Guo, P. Narasimhan, Precog: prefetching for image recognition applications at the edge, in: Proc. ACM/IEEE Symp. Edge Comput. (SEC), 2017, p. 17.

[91] P. Guo, B. Hu, R. Li, W. Hu, FoggyCache: cross-device approximate computation reuse, in: Proc. ACM Mobicom, 2018, pp. 19–34.

[92] D. Kang, J. Emmons, F. Abuzaid, P. Bailis, M. Zaharia, Noscope: optimizing neural network queries over video at scale, Proc. VLDB Endowment 10 (11) (2017) 1586–1597.

[93] J. Wang, et al., Bandwidth-efficient live video analytics for drones via edge computing, in: Proc. IEEE/ACM Symp. Edge Comput. (SEC), 2018, pp. 159–173.

[94] C. Zhang, Q. Cao, H. Jiang, W. Zhang, J. Li, J. Yao, in: FFS-VA: A Fast Filtering System for Large-Scale Video Analytics, Proc. ACM ICPP, 2018, p. 85.

[95] C. Canel, et al., Picking interesting frames in streaming video, in: SysML'18, February 15–16, 2018.

[96] E. Park, et al., Big/little deep neural network for ultra low power inference, in: Proc. 10th Int. Conf. Hardw./Softw. Codesign Syst. Synth, 2015, pp. 124–132.

[97] B. Taylor, V.S. Marco, W. Wolff, Y. Elkhatib, Z. Wang, Adaptive deep learning model selection on embedded systems, in: Proc. ACM LCTES, 2018, pp. 31–43.

[98] G. Shu, W. Liu, X. Zheng, J. Li, IF-CNN: image-aware inference framework for CNN with the collaboration of mobile devices and cloud, IEEE Access 6 (2018) 621–633.

[99] D. Stamoulis, et al., Designing adaptive neural networks for energy-constrained image classification, in: Proc. ACM ICCAD, 2018.

[100] B. Fang, X. Zeng, M. Zhang, NestDNN: resource-aware multi-tenant on-device deep learning for continuous mobile vision, in: Proc. ACM Mobicom, 2018, pp. 115–127.

[101] A.H. Jiang, et al., Mainstream: dynamic stem-sharing for multi-tenant video processing, in: Proc. USENIX ATC, 2018, pp. 29–42.

[102] D. Narayanan, K. Santhanam, A. Phanishayee, M. Zaharia, Accelerating deep learning workloads through efficient multi-model execution, in: Proc. NIPS Workshop Syst. Mach. Learn, 2018.

[103] A. Mathur, N.D. Lane, S. Bhattacharya, A. Boran, C. Forlivesi, F. Kawsar, DeepEye: resource efficient local execution of multiple deep vision models using wearable commodity hardware, in: Proc. ACM Mobisys, 2017, pp. 68–81.

[104] J. Jiang, G. Ananthanarayanan, P. Bodik, S. Sen, I. Stoica, Chameleon: scalable adaptation of video analytics, in: Proc. ACM SIGCOMM, 2018, pp. 253–266.

[105] X. Ran, H. Chen, X. Zhu, Z. Liu, J. Chen, Deepdecision: a mobile deep learning framework for edge video analytics, in: Proc. IEEE INFOCOM, 2018, pp. 1421–1429.

[106] V. Sze, Y.-H. Chen, T.-J. Yang, J.S. Emer, Efficient processing of deep neural networks: a tutorial and survey, Proc. IEEE 105 (12) (2017) 2295–2329.

[107] X. Zhang, Y. Wang, W. Shi, pCAMP: performance comparison of machine learning packages on the edges, in: Proc. USENIX Workshop Hot Topics Edge Comput. (HotEdge), 2018.

[108] Y. He, J. Lin, Z. Liu, H. Wang, L.-J. Li, S. Han, AMC: Automl for model compression and acceleration on mobile devices, in: Proc. Eur. Conf. Comput. Vis, 2018, pp. 815–832.

[109] B. Zoph, Q.V. Le, Neural Architecture Search with Reinforcement Learning [online], 2016, arXiv:1611.01578. Available https://arxiv.org/abs/1611.01578.

[110] R. Tandon, Q. Lei, A.G. Dimakis, N. Karampatziakis, Gradient coding: avoiding stragglers in distributed learning, in: Proc. Int. Conf. Mach. Learn, 2017, pp. 3368–3376.

[111] G. Zhu, Y. Wang, K. Huang, Low-Latency Broadband Analog Aggregation for Federated Edge Learning [online], 2018, arXiv:1812.11494. Available https://arxiv.org/abs/1812.11494.

[112] J. Huang, et al., Speed/accuracy trade-offs for modern convolutional object detectors, in: Proc. IEEE Conf. Comput. Vis. Pattern Recognit. (CVPR), 2017, pp. 7310–7311.

[113] D. Li, Z. Zhang, W. Liao, Z. Xu, KLRA: a kernel level resource auditing tool for IoT operating system security, in: Proc. IEEE/ACM Symp. Edge Comput. (SEC), 2018, pp. 427–432.

[114] M. Du, K. Wang, Y. Chen, X. Wang, Y. Sun, Big data privacy preserving in multi-access edge computing for heterogeneous internet of things, IEEE Commun. Mag. 56 (8) (2018) 62–67.

[115] Z. Zhou, X. Chen, E. Li, L. Zeng, K. Luo, J. Zhang, Edge intelligence: paving the last mile of artificial intelligence with edge computing, Proc. IEEE 107 (8) (2019) 1738–1762.

Further reading

[116] J. Zhao, R. Mortier, J. Crowcroft, L. Wang, Privacy-preserving machine learning based data analytics on edge devices, in: Proc. AIES, 2018, pp. 341–346.

[117] Y. Li, et al., A network-centric hardware/algorithm co-design to accelerate distributed training of deep neural networks, in: Proc. 51st Annu. IEEE/ACM Int. Symp. Microarchitecture (MICRO), 2018, pp. 175–188.

[118] B. Reagen, et al., Minerva: enabling low-power highly-accurate deep neural network accelerators, ACM SIGARCH Comput. Archit. News 44 (3) (2016) 267–278.

[119] L. Zeng, E. Li, Z. Zhou, X. Chen, Boomerang: on-demand cooperative deep neural network inference for edge intelligence on industrial internet of things, in: IEEE Netw, 2019.

[120] H. Li, C. Hu, J. Jiang, Z. Wang, Y. Wen, W. Zhu, JALAD: Joint Accuracy-and Latency-Aware Deep Structure Decoupling for Edge-Cloud Execution [online], 2018, arXiv:1812.10027. Available https://arxiv.org/abs/1812.10027.

[121] S. Venugopal, M. Gazzetti, Y. Gkoufas, K. Katrinis, Shadow puppets: cloud-level accurate AI inference at the speed and economy of edge, in: Proc. USENIX workshop hot topics in edge Comput. (HotEdge), 2018.

[122] S. Jain, J. Jiang, Y. Shu, G. Ananthanarayanan, J. Gonzalez, ReXCam: Resource-Efficient Cross-Camera Video Analytics at Enterprise Scale [online], 2018, arXiv:1811.01268. Available https://arxiv.org/abs/1811.01268.

[123] Z. Fang, M. Luo, T. Yu, O.J. Mengshoel, M.B. Srivastava, R.K. Gupta, Mitigating multi-tenant interference in continuous mobile offloading, in: Proc. Int. Conf. Cloud Comput, 2018, pp. 20–36.

About the authors

Anubhav Singh, born on 1 January 2000 in Ballia, Uttar Pradesh, is pursuing MSc Forensic Science from Rashtriya Raksha University and has completed BSc (Hons.) Forensic Science from Galgotias University, Greater Noida, UP. He has a Diploma in Photography and PG Diploma in IT Fundamentals for Cybersecurity at IBM. He has completed several certificate courses. He is a certified graphics editor and designer. He has published more than book chapters in national and international research and review papers in peer-reviewed international journals. He has organized more than 4 National and international conferences and has participated and presented his work at more than 12 national and international conferences and workshops.

Kavita Saini is presently working as a professor at the School of Computing Science and Engineering, Galgotias University, Delhi NCR, India. She received her PhD degree from Banasthali Vidyapeeth, Banasthali. She has 18 years of teaching and research experience supervising masters degree and PhD students in emerging technologies. She has published more than 40 research papers in national and international journals and conferences. She has published 17 authored books for UG and PG courses for a number of universities including MD University, Rothak, and Punjab Technical University, Jallandhar, with national publishers. Kavita Saini has edited many books with international publishers, including IGI Global, CRC Press, IET Publisher, and Elsevier, and has published 15 book chapters with international publishers. Under her guidance, many MTech and PhD students are carrying out research work. She has also published various patents. Kavita Saini has also delivered technical talks on "Blockchain: An Emerging Technology," "Web to Deep Web," and other emerging areas and has handled many special sessions in international conferences and special issues in international journals. Her research interests include web-based instructional systems (WBIS), blockchain technology, Industry 4.0, and cloud computing.

Varad Nagar, born on 17 March 2002, in Varanasi, Uttar Pradesh, is currently pursuing BSc (Hons.) Forensic Science from Vivekananda Global University, Jaipur, Rajasthan, India, and is also pursuing foundation degree from IIT Madras in data science and programming. He has participated and presented his work at more than 10 national and international conferences and workshops. He has published more than 12 papers in Scopus indexed journals and 10 book chapters in various national and international publications, and several papers/chapters are under progress. He has hands-on experience on a variety of sophisticated instruments like UV-visible spectrophotometers, IR spectroscopy, SEM, etc.

Vinay Aseri, born on 28 December 2001, at Churu, Rajasthan, is currently pursuing BSc (Hons.) Forensic Science from Vivekananda Global University, Jaipur, Rajasthan, India. He has participated and presented his work at more than 15 national and international conferences and workshops. He has published more than 10 papers in Scopus indexed journals and 13 book chapters in various national and international publications, and several papers/chapters are under progress. He has hands-on experience on a variety of sophisticated instruments like UV-visible spectrophotometer, IR spectroscopy, SEM microscopy, etc.

Mahipal Singh Sankhla, born on 19 May 1994, in Udaipur, Rajasthan, is currently working as an assistant professor in the Department of Forensic Science, Vivekananda Global University, Jaipur, Rajasthan. Prior to this, he has served as an assistant professor in the Department of Forensic Science, Institute of Sciences, SAGE University, Indore, MP. He has completed BSc (Hons.) Forensic Science and MSc Forensic Science. Currently, he is pursuing PhD in Forensic Science from Galgotias University, Greater Noida, UP. He has undergone training at the Forensic Science Laboratory (FSL) Lucknow, CBI (CFSL) New Delhi, Codon Institute of Biotechnology, Noida, and Rajasthan State Mines & Minerals Limited (R&D Division), Udaipur. He was awarded Junior Research Fellowship-JRF, a DST-funded project at Malaviya National Institute of Technology—MNIT, Jaipur, the Young Scientists Award for Best Research Paper Presentation at the 2nd National Conference on Forensic Science and Criminalistics, and Excellence in Reviewing Award in the *International Journal for Innovative Research in Science & Technology (IJIRST)*. He has edited 4 books and published 10 book chapters in various national and international publications. He has published more than 120 research and review papers in peer-reviewed international and national journals. He has participated and presented his research work at more than 25 national and international conferences and workshops and organized more than 25 national and international conferences, workshops, and FDPs.

Pritam P. Pandit is currently pursuing his masters degree in forensic science from Vivekananda Global University, Jaipur, Rajasthan. He has completed Post-Graduate Diploma in Forensic Science and Related Laws from the Government Institute of Forensic Science, Aurangabad, with distinction and has graduated from Rajarshi Chhatrapati Shahu College, Kolhapur, Maharashtra, with first class. He has also completed Maharashtra State Certificate in Information Technology (MSCIT) with 90%. He has published more than

10 papers in Scopus indexed journals and more than 11 book chapters in various national and international publications, and several papers/chapters are under progress.

 **Rushikesh L. Chopade** is currently pursuing his masters degree in forensic science from Vivekananda Global University, Jaipur, Rajasthan, India. He has completed Post-Graduation Diploma in Forensic Science and Related Laws from the Government Institute of Forensic Science, Aurangabad, and BSc Chemistry, Botany, Zoology (CBZ), from GSG College, Umarkhed, with a first class. He has published more than 6 papers in Scopus indexed journals and more than 10 book chapters in various national and international publishers, and several papers/ chapters are under progress.

CHAPTER SEVENTEEN

5G—Communication in HealthCare applications

R. Satheeshkumar[a], Kavita Saini[b], A. Daniel[b], and Manju Khari[c]

[a]Department of Electronics and Communication Engineering, Galgotias College of Engineering and Technology, Noida, India
[b]School of Computing Science and Engineering (SCSE), Galgotias University, Delhi, Uttar Pradesh, India
[c]School of Computer and System Sciences, Jawaharlal Nehru University, New Delhi, India

Contents

Abstract

Every industry will benefit greatly from edge computing enabled by 5G. It moves computing and data storage closer to the point where data is generated, allowing for improved data control, lower costs, continuous operations, and faster insights and actions. It is a distributed system in which data is handled as close as feasible to the original data source. This architecture requires the efficient use of resources like as cell phones, tablets, sensors and laptops that are not always linked to a network.

Data transmission is prohibitively expensive. Latency is minimized and end users enjoy a better experience by bringing computation closer to the source of data. The Internet of Things (IoT) and Augmented Reality (AR) or Virtual Reality (VR) are two emerging Edge Computing application cases. The rush that consumers felt when playing an Augmented Reality-based Pokemon game, for example, would not have been conceivable if the game did not include "real-timeliness." Because the smart phone, rather than the central servers, was doing the AR, it was conceivable.

1. Introduction

The fifth-generation communication is referred to as 5G. After 1G, 2G, 3G, and 4G mobile communication networks, it is a new global wireless communication standard. 5G-allows for the creation of a new type of wireless network that connects nearly everyone and everything, including machines, objects, and gadgets.

The 5G of mobile infrastructures expected to built soon for commercial purpose, supporting to video familiarity via mobile internet, perk up the excellence of voice and a many of application state of affairs,. The healthcare business is transitioning to an epoch of information and quickness, and more than a few time-critical applications, such as telemedicine, require 5G connections at the same time [1].

Healthcare will be radically different in 10 or 20 years. Hospital operations will be more efficient, clinicians will have access to real-time data to aid in clinical decision-making, and patient outcomes will improve. Patients will be able to make better health decisions for themselves, resulting in fewer difficulties later in life.

The 5G network will be exemplify by towering speed of internet and smart networks. Presently, to download a full film over 4G gets roughly 8 min; with 5G, it will gets less than 5 s.

The video gaming, high definition and three dimensional (3-D) video, television interaction, virtual reality, driverless vehicles, robotics, and manufacturing advancement, will be possible thanks to the speed of internet. Because there will be billions of devices online, not all data will need to be transferred at the same time. Some applications necessitate real-time connectivity, while some can send data during off-peak hours.

Intelligent networks, which will manage travel and systems design in ways that promote well-organized direction-finding and decision making, are a major part of defining 5G. However, 5G will introduce intelligence across the network, and information analytics will be an integral element of the service. Real-time aggregation and analysis will be possible with 5G, allowing users to make sense of data and enhance and modify the capabilities of each application.

2. 5G—IOT for *E*-healthcare

The recent advancements in big data-oriented wireless technologies, such as forthcoming 5G, internet of things (IoT) connected devices, information analytics, and edge computing, as well as methodologies, has allowed the interconnected healthcare services at an advantage and healthier life.

There is an unprecedented need for linked healthcare, thanks to the rapid rise of interconnected devices, Internet of Things (IoT) services, apps and wireless technologies. In terms of size, rate, multiplicity, connected healthcare, and pace, together with linked strategy and sensors, produces a tremendous number of datasets [2]. The massiveness, complexity, and multidimensionality of the wireless big data generated by linked healthcare, handling it is a tough problem. With a very the highest accuracy, quick reply time, improved resource effectiveness, and next-version of big data cognizant wireless technology, such as the forthcoming 5G paired with IoT, recommend an important potential to improve the complexity. Although the development of 5G-aware big data analytics, the Internet of Things, cognitive sensors, and has permitted the factual vision of linked healthcare to be understand, human emotion recognition is serious for hold emotional care to create an enhanced life. The conventional methods of emotional care communiqué may not offer sufficient sensitive get in touch with. It's difficult to produce a structure that can gather, recognize, excavation, examine, and method successful information at the side of healthcare big data in an appropriate and precise manner outstanding to the intricacy, dynamicity, and multidimensionality of healthcare services [3].

A cognitive healthcare based on 5G grow to be a practicality with the increase of IoT and smart homes [4]. The healthcare industry is booming and has the potential to generate billions of dollars in income. It is as well supply to the improvement of human life by giving the most up-to-date healthcare amenities at any time and from anywhere. The growing linked healthcare has seen enormous growth thanks to the wireless network, edge computing, and cloud computing technology.

Many countries have constructed smart cities to provide people with cutting-edge technology. Embedding an emotion recognition module within the framework is a current trend in healthcare. Some components in the framework of cognitive are devoted to be familiar with the emotion of user or mind state; after be familiar with the emotion, a service supplier can distribute an suitable service to the user, or use the emotional reaction to perk up the quality of service in the future [1,5]. The framework also includes inside and outside placing method for locating the user as necessary.

Consider the case of a client who is over the age of 65. Assume the client be alive only in a smart home with numerous IoTs. The customer subside to the floor and is unable to move. The IoTs are constantly gathering signals, and a cloud server uses these data to detect a fall. Caregivers must be dispatched to the client by the service provider. A global positioning system (GPS) guides caretakers to the smart home where the customer lives; though, an accurate inner positioning system is necessary to pinpoint precisely where the customer collapsed. Once the exact site has been determined, the caregivers may hurry there devoid of invading other people's privacy.

E-health arose from a demand for better documenting and patient health tracking and treatments carry out on them, notably for insurance compensation principles. Usually, health care giver keep document report of their patient's medical histories and current conditions. Increasing healthcare expenses and technology advancements, on the other hand, prompted the creation of electronic following systems [6]. The e-health expertise development, a new area called telemedicine arose telecommunication expertise are used to deliver fitness treatment tenuously.

Consumers benefit from improvements in e-health by being able to obtain prescriptions over the Internet and have them delivered directly to their homes. Health care centers and acute care facilities maintain network pages detailing their proficiency and services for patients. The e-health have supported community-dwelling people with disabilities by allowing patient communication via text, audio, or video conferencing to track home-based development.

The new technologies (5G) are expected to explore the difficulties raised above, as well as performance criteria such as high speed, connection and capacity density, enhanced reliability, peak throughput per connection, low latency, low power consumption and system spectral efficiency [7,8].

5G wireless technology is designed to provide multi-gigabit per second hit the highest point information speeds, improved reliability, enormous network capacity, ultra-low latency, increased availability, and a more consistent user experience to a larger number of users. Higher performance and efficiency allow for new user experiences and industry connections.

Real-time multicasting, data throughput, ad hoc peer-to-peer, latency, and data encryption are just a small amount of the benefits of 5G above existing wireless network technologies. Healthcare, with its current IoMT infrastructure, is a vital area for possibly revolutionary 5G technologies.

The probable of 5G in the sanatorium workflow is already being demonstrated by current research trends. After the COVID-19 epidemic, Zhou et al. [9] documented the creation of small house hospitals' in China, and are now constructing a 5G network solution that fix each small house network area to the host via a VPN tunnel erected on pinnacle of a 5G network. Zhou emphasizes the success of the 5G hospital solution, the benefits that can only be obtained all the way through this network, and the prospect of joining numerous sanatorium.

Furthermore, Li and Wang [6], Stefano and Kream [10], and Anwar and Prasad [11], explain and show the emerging 5G infrastructures for telemedicine, which have a significant impact on both the speed of growth and the total reimbursement for healthcare. Furthermore, Alani and Saeed [5], and Mavrogiorgou et al. [12] highlight the BigData advantages associated with low-latency, resilient, ultra-fast, and protected data transmission throughout healthcare communications using a 5G network. Finally, Mattos and Gondim [13] discuss how a correctly developed 5G infrastructure can provide significant prospects for direct machine-to-machine (M2M) connections in healthcare.

A sequence of alters introduced by "Internet+" and "Smart+" assist the "intellectualization" of the social order, and this tendency is reproduced in the healthcare commerce [2]. innovative service-oriented system architecture, space interfaces, and end-to-end system slicing characterize 5G technology [4,14,15], which can become accustomed to the network supplies of a variety of applications. In addition, 5G offers powerful technological hold up for the expansion of 5G healthcare, and smart health checkup applications is one of the most important upright industry for 5G technology. In the medical industry, 5G can hold up enormous medical image information transport and giving out, real-time distant manage of smart devices, and ultrahigh-definition video communication [7,16], and thus can meet the network needs of multidisciplinary consultation, remote surgery, intelligent diagnosis of medical images, and other medical applications [17,18]. The medical manufacturing, on the other hand, is grappling with issues such as centralized medical resources, opaque people, multifaceted information systems, and a wide range of diverse medical apparatus. Because many medical appliance scenarios need the integration of applications, several medical devices, and services, 5G confronts numerous hurdles when used in the medical field.

One of the prospect orientations of the service model of healthcare is 5G smart healthcare, which slot in several knowledge such as 5G, the Internet of Things (IoT), fog computing, edge computing, and artificial intelligence. Innovations in 5G smart healthcare, such as new healthcare service models and resolution, are based on the idea of mobile healthcare, Internet healthcare and telemedicine. The administrator balance of the R16 edition of 5G will be exposed by the 3rd Generation Partnership Project (3GPP) in July 2020, representing that the technological answer for 5G smart healthcare will be particular further. Once the principles and safety supplies are met, all types of 5G smart healthcare manufacture will be hastened, and many of them will be more available to the general people. According to IHS Markit, the 5G-enabled physical condition and robustness sector will be worth more than a trillion dollars.

The medical manufacturing is being improve at a rapid pace, and 5G is pull towards you a lot of notice from all in excess of the globe. A lot of nation have leap into the use and extension of 5G smart healthcare, responsibility a slew of research and progressive carry out in the aim and completion of 5G healthcare system and apps [19]. Verizon, for instance, said in the second half of 2018 that it would roll out a 5G profitable wireless system and 5G core network in the United States. In October 2018, China Mobile and Zhengzhou University's First allied hospital exposed that doctors could do inaccessible ultrasound surgeries and real-time analysis via the 5G system. In February 2019, Vodafone and hospital linked on 5G-based distant surgical procedure examination mean to create Spain's first 5G smart hospital. In June 2019, the University hospital of Birmingham in the United Kingdom demonstrate the feasibility of 5G ambulance and secluded ultrasonography process in collaboration with British Telecom and West Midlands 5G (WM5G).

In October 2019, Huawei, China Mobile, and Zhengzhou University's first affiliated hospital issued a white paper on a 5G healthcare system based on flexible network segment. A COVID-19 immunity detection and analysis system based on 5G cloud computing in April 2020, released by Sichuan University and China Mobile where a 5G distant CT scan was employed in COVID-19 preventive system.

With 5G smart healthcare, a range of new or better medical appliance situation have emerged. Virtual surgery, remote consultation, emergency rescue, remote teaching, and monitoring, for example, have all been fast developed using 5G equipments [15,17,20]. While these application situation get better the quality of the patient experience and health services, they also introduce new security risks. The security element of 5G healthcare will cooperate key roles in national, and information security [21], because medical information contains a lot of user seclusion and some medical appliance situation damage patient protection.

5G technologies is a new network that be different from 4G and 3G and it enables mobile Internet of things by combining considerable processing capabilities with a virtual computing environment (mIoT). The severe supplies in healthcare appliances to offer flexible, cost-effective, and reliable services have been the subject of ongoing study. 5G promises to boost the healthcare industry by extending service options to a variety of healthcare domains, including healthcare professionals, caregivers, and patients.

According to analysts, the healthcare business is booming because it accounts for a considerable portion of national GDP, with figures ranging from 6% in China to 10% in Europe and 18% in the United States. E-Health curve on a common hype since the early 2000s, with high aspirations of rapid improvement in experience quality, service quality, and affordable savings in health care [22]. In its most current study on e-Health, the World Health Organization (WHO) believes that "there is a need for stronger political commitment for e-Health, underpinned by sustainable finance, and for effective policy implementation" [16].

This can be performed by employing millimeter wave, tiny cell, beamforming, full-duplex, and other techniques [23]. Unlike earlier network technologies, 5G uses three distinct spectrum bands: lower-range frequency band, middle range frequency band, and high range frequency band. This not only enhances the network's capacity but also allows even the tiny devices to execute high calculation and attach fast to the system's dispersed handing out control [24].

The number of linked devices it maintain, exceptionally low latency, and network speed are the features that set 5G apart from its predecessors.

With the help of new technology, IoT can give high-quality solutions. In the sphere of medicine, it becomes a new reality of an original idea that provides the best service to COVID-19 patients and executes precise surgeries. Complicated problems are easily managed and digitally controlled during the current epidemic [22]. In the field of medicine, the Internet of Things tackles new challenges in developing effective support systems for physicians, patients and surgeons. For effective IoT deployment, the various process phases are carefully identified.

Remote surveillance became possible in the health sector thanks to IoT-enabled technology, which opened up the possibility of healthy preserve patients, empowering clinicians to give superior care and security. It also enlarged patient contribution and approval by making interaction with clinicians more convenient and efficient. Furthermore, remote health monitoring helps to cut down on healthcare wait and re-admissions. In healthcare, IoT has a substantial impact on cost reduction and enhanced therapeutic benefits [16]. Without a doubt, the Internet of Things (IoT) alters the healthcare industry by expanding the reach of devices and people's connections to solutions. IoT recommend healthcare appliances that benefit families, doctors, patients, insurance businesses, and hospitals.

Robots that are connected to the internet of things are flattering further general. It's used to clean medical center, disinfect healthcare equipment, distribute medications, giving doctors and nurses more time to focus on their patients. China is the first country for example, to use UVD robots from a Danish business to clean its medical services during the emergency. These IoT-enabled robots [25] help disinfect treatment area and clean rooms.

The all knot of wires tied to a patient is not convenient for the patient, then results in limited movable, and greater than before nervousness but it is also difficult to handle for the staff. Purposeful disconnections of sensors by exhausted patients, failures to correctly reattach sensors, and as patients are encouraged around in a hospital are fairly common. Wireless sensing gear that is less apparent and has a constant network connection to back-end medical record systems helps to eliminate knots of wires with patient concern while also reduces the errors [25].

3. 5G—Industrial Internet of Thongs (IIoT)

To move beyond conventional networks of mobile, safe system access point must be disseminated, intelligence and allowing smooth the tiniest linked sensors to be empowered. With a mixture of applications assembled about a variety of types of "smart" devices with sensors, and machine learning models, the IIoT has the probable to make over omnipresent computing. The smart devices make bigger our understanding and awareness of the globe as an agitated-connected environment, posturing fresh challenges in preparing the IIoT for extensive deployments. By combining 5G networks for connectivity with external devices, data conversion, digital processing, and spectrum detection. 5G is projected to be more than just a latest mobile communications production. Instead, it is previously being referred to at the same time as the "unifying fabric" will join billions of procedure in a few of the most efficient, speedy ways possible, and dependable. Allowing technology is expected to have an innovatory influence.

The novel communication is intended to change industries and revolutionize the global of linked sensors. Of course, such a revolution would necessitate research and development to ensure sensor cohabitation and device interoperability with 5G networks. Integrated IIoT is a smart communication network that can learn about its users and their surroundings and utilize that information to assist them to attain their targets in a context-based way.

This significantly get better quality of user life while also assisting in the optimization and control of rapidly growing resource expenditure rates in IIoT framework. Services, or systems, people, devices that are involved with smart enabling technologies such as radio frequency identifications (RFIDs), sensors, 5G Smartphones, and other applications in our daily lives are considered inhabitants (users) of IIoT settings.

E-Healthcare is a significant application for 5G-based IIoT, as it attempts to keep a medical information patient in electronic environments such as the Cloud using cutting-edge communications paradigms. Wireless medicinal sensor networks (WMSNs) have become a popular technology for wireless sensor networks (WSNs) in healthcare application system [16].

For early diagnosis, hospitals have created a variety of therapeutic healthcare sensors that are used to sense the patient body and record fitness data such as temperature, heart rate, and blood pressure [25]. These healthcare data are also disseminated via professional handy devices, such as a personal data analyzer (PDA) and a Smartphone, for additional analysis. WMSNs have been used in patient health monitoring by several medical research communities [1,22–24]. As a result, a user authentication strategy is required to safeguard the medical information system from unauthorized access (Fig. 1).

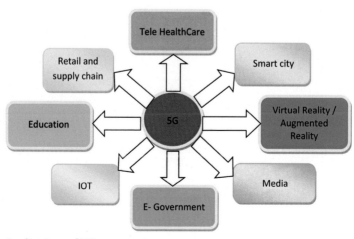

Fig. 1 Applications of 5G.

Technological advancements are transforming the globe about us, but they are also propelling the wireless sector to increase the next age band of communication network technology. Mobile technologies such as 1G, 2G, 3G, and 4G have concentrated on improving the rate and competence of wireless networks during the last 25 years, but there are still a few application areas anywhere recent wireless networks move violently to supply. Utility applications, wireless healthcare services, V2X communication (especially Vehicle-to-Vehicle (V2V) and Vehicle-to-Infrastructure (V2I) communication), and industrial automation, consumer, augmented reality and virtual reality services, and primary broadband access services, among others, are among these areas [26]. 5G applications are currently being used in a variety of industries, including energy [23–25], smart vehicle parking system [5], intellectual station area detection technology [21], medicare [15,25], and so on.

4. 5G—Network requirements for healthcare

Healthcare scenarios appear to have larger and more essential network requirements than scenarios such as highway traffic supervision, smart cultivation, and so on. The network's role in the healthcare environment is depicted in Fig. 2 using a conceptual diagram. The patient's ability to contact his caretaker, or any other emergency agency would necessitate continuous

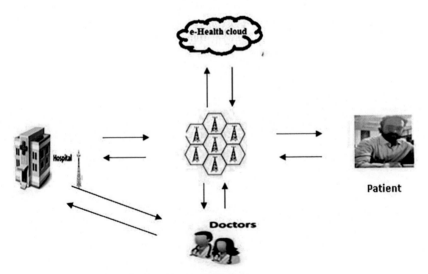

Fig. 2 Network's role in the healthcare environment.

connection and doctor. Doctors or caregivers on the other hand, would be required to respond to the appeal of client instantly, which would necessitate recover information from network servers.

The instance it takes to regain information and respond to the long-suffering is crucial. Monitoring in Mass-Casualty Disasters: While emergency medical triage techniques exist [20,21] their effectiveness might quickly deteriorate as the number of victims grows. Furthermore, during mass-casualty events, it is necessary to improve the assessment of first responders' health.

Wireless sensing systems' greater portability, scalability, and quick deployment capabilities can be utilized to more effectively report the triage levels of multiple victims and track the health status of first responders at the crisis scene [11].

There are two types of healthcare scenarios: "the one who is cared for" and "the one who is cared for by." The user or patient using the system would go under the "Cared for" group, while medical care professionals such as insurance providers, doctors and nurses, and so on would fall under the "Cared by" category [6].

Since it permits health information to be supply accumulated and common among many entities, electronic healthcare (e-health) has gotten a lot of attention [7]. E-health systems that use 5G networks may handle a wide range of applications while maintaining high information security, reliability, and seamless medical information transfer. Remote monitoring for telemedicine in real-time and the transfer of massive health information files are two prominent use cases. Healthcare practitioners may watch patients tenuously and meet real-time data without worrying about disconnections, network outages, or lag time, thanks to 5G's ultra-low latency [8,15]. It makes preventative care and other custom-tailored healthcare more accessible. Huge information files in the systems can be professionally transmitted between users and service providers for well-organized patient treatments thanks to the elevated crest information rates provided by 5G network. Huge picture files generated by Computerized Tomography (CT) or Magnetic Resonance Imaging (MRI) scans, for example, can be swiftly delivered to specialists for review [4,22]. Furthermore, the combination of 5G and e-health systems would allow for the effective construction of smart healthcare and the Internet of Medical Things (IoMT) [18,19,27]. Many members of a family unit can use general healthcare services, such as genetic testing [11], in addition to accessing personage services. They will be able to keep track of the health state of their family members in this manner.

The services will considerably improve the long-term benefits of medical treatments because families play an essential role in promoting health and minimizing the risk of illness [6]. Healthcare panel at sanatoriums, healthcare society, and crisis care centers can also provide group health services. The entire treatment would be improved as a result of the healthcare teams' joint efforts [8,9].

Patients, physicians, pharmacists, medical researchers, caretakers, and others are among the e-health users in a 5G-permitted healthcare location [9,27]. Healthcare suppliers (e.g., hospitals), data center administrators or medical experts who provide services for specific consumers should be the servers [8]. Because users and servers communicate over the open Internet, their sensitive information may be vulnerable to a variety of assaults. As a result, the security and privacy of e-health systems are critical [1]. It necessitates a comprehensive process capable of preventing potential security issues. Many studies have proposed protected password based or two-part (password plus smart tag) authentication solutions to solve security concerns [3,11]. These aren't, however, long-term solutions. For example, with two-factor authentication schemes, if the adversary knows password of the user, it can easily carry out assaults with a stolen smart card.

The entities must accumulate large qualifications (e.g., unique id, passwords) for receiving a rising numeral of medical care services with a single-server architecture defined in several works [28].

5. 5G—Virtual HealthCare

The "virtual visits" that get place between patients and healthcare expert via communications technology the Audio and video connectivity that permits "virtual" gathering to take position in real time, as of nearly every position are referred to as virtual healthcare.

A videoconference between a doctor and a patient at home can be considered a virtual visit. It might imply that instead of flying to another city, a patient can connect with an offshore health check expert passing through a high quality conference hookup at his or her local hospital. It may also make it easier for patients to locate experienced second observation online.

Virtual healthcare has thus far primarily been utilized for consultations, status reporting, check-ins, meeting and rather than in-depth diagnosis or treatment. Even Nevertheless, as expertise advances, extra severe disorders like diabetes are becoming more accessible through virtual healthcare.

Virtual healthcare also makes it easier for specialists to keep an eye on problems or operations from afar. Patient monitoring at home has also been found to be effective in the treatment of patients with continual sicknesses such as cardiac problem, hypertension, and diabetes where rehospitalizations are all too common due to a lack of communication or transparency regarding the patient's health.

6. TeleHealth vs. virtual health

Virtual healthcare, like remote patient management, is commonly confused with telemedicine, or telehealth although they are not the similar obsession [21]. Telehealth is a larger word that encompasses any distant and/or technology-driven healthcare. Virtual healthcare is a subset of telehealth.

Telehealth technology might take the form of a phone, videoconferencing capability, or an interactive voice response system.

Telemedicine is a term that refers to the management of various healthcare problem with no having to observe the patient in person. To address a concerns of patient and assess their state remotely, healthcare providers can use telehealth stages such as live vision, live audio, or instant messaging [23]. Getting health advices, guide patients via at-home workout, or referring them to a nearby professional or ability are all examples of this. Even more thrilling is the rise of telemedicine mobile apps, which allow patients to receive care directly from smartphones or tablets [28].

VirtualHealth has continuously been acknowledged as the leading provider of medical management and related technology on a national level [29]. We have strong expertise in value-based care and count the majority of the country's largest health plans among our clients, which matches exceptionally well with our objective to make healthcare more proactive.

Virtual healthcare is a wide expression that refers to all of how healthcare suppliers communicate with their patients over the internet. In adding up to care for patients via telemedicine, doctors can communicate with them remotely using live video, audio, and instant messaging.

Telemedicine has reached a critical juncture (also known as telehealth). Doctor-patient contacts have largely been face-to-face in health check workplace, houses, hospitals, and care facilities for hundreds of years. Medical technology, as well as speedier networks such as 5G, are transforming the features of healthcare.

Telemedicine has been progressing at a steady pace. The emergence of wearable medical check devices to observe essential signs has been particularly striking [12]. Although the concept of doctors visiting patients through the Internet is enticing, it has yet to be widely implemented.

The COVID-19 epidemic, however, has changed that. Health institutions are progressively more implementation tele medical care technology to help patients to keep patients and medical workers safe [1]. As a result, the telemedicine sector is expanding rapidly over the world. Medical equipment manufacturers are scrambling to keep up with demand. Furthermore, the rise of telehealth and other healthcare activities is being fueled like never before by today's 4G LTE and upcoming 5G wireless technology [19]. Extensive cellular system availability, easiness and pace of deployment, and greatly better bandwidth are all central technical portion.

7. 5G—Remote HealthCare monitoring

In the last few years, medical electronics have advanced significantly. A silicon platform is available from Maxim Integrated for device makers that are developing products to monitor common vital signs. Stress, Pulse/heart rate, electrocardiogram (EKG/ECG), blood oxygen (SpO2), body temperature, and UV light are all examples of these (skin exposure) [27]. With built-in GPS and cellular, Apple's new 6 Series smartwatch can measure blood oxygen and ECG. Fitbit has unveiled its Sense wristwatch, which uses a Wi-Fi connection to measure skin temperature, blood oxygen, ECG, and stress.

To save money on healthcare, sanatorium are attempting to release patients as soon as their condition has stabilized, such as those suffering from heart attacks. Hospitals will use wireless technologies to tenuously verify health of patients health problems after they have been discharged. Smartwatches are valuable for monitoring the health of wearer regularly.

8. 5G—Remote surgery

While the majority hospitals have experienced surgeons on staff, they rarely have specialists in every surgical area of expertise [18]. Frequently, they will seek the advice of doctors with specific knowledge. Video and audio have been used by experts in specialized medical specialties to assist the surgical team throughout the procedure.

As medical technology progresses, many and more surgeries are reliant on it. A little surgical microscope, for example, can be put into the body and, with the help of a high-definition 4K monitor like the Olympus Visera, vividly depict the specifics of the area to be operated on.

With a 5G network connection, the same data that the surgical team sees may be live-streamed in real-time to the behind specialist team, and in the future, in even greater 8K quality. The experts will be intelligent to advise the team in real-time as the surgery takes place thanks to this real-time connection.

The first 5G remote brain surgery has been completed by doctors. The patient had Parkinson's disease and was over 1500km distant. During a three-hour procedure, the patient received a cavernous brain stimulation embed. Dr. Ling Zhipei performed the pioneering surgery using a computer powered by China Mobile and Huawei's 5G network to manipulate instruments in Beijing from his position in Sanya City.

9. 5G—Futures and robotics in healthcare

We have merely scratched the surface of what 5G can accomplish for telehealth and healthcare. It will enhance patient remote monitoring, home medical care, and surgical procedures. 5G will improve the way emergencies are handled in smart cities. The dangerous information will be available in near real-time, allowing team to react much faster and provide better treatment.

Many of these choices and functions are likely to be undertaken by AI in the future. To increase precision, machine learning, artificial intelligence, and robotics can play a larger part in procedures [12].

In addition, carers and emergency teams will have more time and will be better prepared to deal with unforeseen emergencies such as sickness, accidents, and natural catastrophes [17]. Fig. 3 shows 5G enabled robotic healthcare environment.

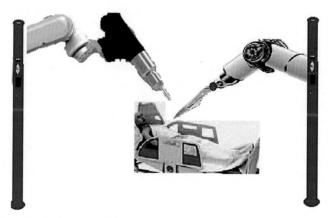

Fig. 3 5G enabled robotic healthcare environment.

The concept of remote surgery is based on the exchange of medical data. Medical data is digitized and transmitted via cable or wireless telecommunication networks, including images, audio, and video. Surgeons can use the networks to control the surgical robot and execute surgeries from afar [11,22]. The main barriers to real-time remote surgery have been system slowness and network instability. The latest revolution of the 5th generation wireless infrastructure (5G) has made remote surgery a reality. The 5G network provides exceptional high-speed, low-latency, and high-bandwidth capabilities [30].

Many experts believe that 5G will transform healthcare by giving firms more connection power and higher broadband speeds. Remote robotic-assisted surgery is one area where 5G could be very revolutionary.

According to a paper by Fitch Solutions Macro Research, a unit of Fitch Group, doctors might potentially undertake remote, global treatments that are currently impossible. Robot-assisted surgery is still a popular alternative to traditional surgery because it offers more precision and accuracy [2].

10. 5G—Impact on HealthCare

Furthermore, the utilization of 5G technology has the potential to protect quality while also lowering overall medical expenses.

1. The use of various medical related sensors and remote patient health monitoring equipment to aid patients living in rural places have access to top medical support are just a few instances. Using video conferencing or telemedicine to bridge the geographic distance and provide high-quality care to underserved populations can help bridge the gap.

2. Point-of-care testing (POCT), a relatively new technology, can save money by keep away from the costly hospital visits. Patients can use mhealth technology, digital platforms, or remote monitoring equipment instead of visiting a huge medical center. By 2018, the POCT market is expected to be worth $27.5 billion. 36 These devices improve patient accessibility by bringing technology to the patient's bedside or home.

3. Health therapies from home are a technique of providing larger quality health care to patients without requiring them to travel long remoteness to healthcare of hospitals facilities. It can send medical information electronically and have health center offer diagnosis and treatment suggestions from afar.

4. In many areas, diabetes is a big issue. The condition of Mississippi exposed that 13% of its people have diabetes, with 54% of those populace

living in rural areas with inadequate access to appropriate care. Health check authorities, on the other hand, saw cost savings of $339,184 for around 100 patients participating in the program and predicted Medicaid savings of $189 million per year after establishing a Diabetes Telehealth Network with remote healthcare services.

5. Health IoT can keep expenses down and save money while maintaining high-quality treatment by keeping people out of hospitals. Not every medical issue necessitates a trip to the hospital. The routine concerns can be diagnosed remotely, providing patients with more options than traditional treatment.

6. Using voice recognition software to automate administrative tasks can save time and money. According to a study of technology in sanatoriums, it allowed doctors to "provide care without being stopped by data inflowing and querying tasks." 41 People may record medical information without stop and enter data thanks to the program.

In a nutshell, 5G technologies connect devices so that smarter and faster decisions can be made. These data enable caretakers and policymakers to have real-time awareness of people, diseases, and symptoms, allowing them to create new insights. Interoperable gadgets operate in tandem with intelligent notification systems to guarantee that each patient receives the best possible care.

11. Conclusion

Cost savings alone can be a motivator for many businesses to implement edge computing. Companies that used the cloud for many of their apps may have discovered that bandwidth expenses were greater than anticipated and are looking for a less expensive alternative. Edge computing could be a good option.

Carriers all around the world are introducing 5G wireless technologies, which promise tremendous bandwidth and low latency for apps, allowing businesses to scale their data capacity from a garden hose to a firehose. Rather than simply providing faster bandwidth and instructing businesses to process data in the cloud, many carriers are including edge-computing technologies into their 5G installations to provide speedier real-time processing.

While the original purpose of edge computing was to lower bandwidth costs for IoT devices across long distances, it's apparent that the emergence of real-time applications that demand local processing and storage will continue to propel the technology ahead in the future years.

References

[1] K. Saini, P. Raj, "Handbook of Research on Smarter and Secure Industrial Applications Using AI, IoT, and Blockchain Technology", IGI Global, 2021, ISBN13: 9781799883678, ISBN10: 1799883671, EISBN13: 9781799883685.

[2] K. Saini, V. Agarwal, A. Varshney, A. Gupta, E2EE for data security for hybrid cloud services: a novel approach, in: IEEE International Conference on Advances in Computing, Communication Control and Networking (IEEE ICACCCN 2018) Organized by Galgotias College of Engineering & Technology, Greater Noida, 2018, pp. 12–13, https://doi.org/10.1109/ICACCCN.2018.8748782.

[3] D.N. Le, R. Kumar, B.K. Mishra, J.M. Chatterjee, M. Khari, (Eds.), Cyber Security in Parallel and Distributed Computing: Concepts, Techniques, Applications and Case Studies, John Wiley & Sons, 2019.

[4] M. Chen, J. Yang, Y. Hao, S. Mao, K. Hwang, A 5G cognitive system for healthcare, Big Data Cogn. Comput. 1 (1) (2017) 2.

[5] M.M. Alani, H. Tawfik, M. Saeed, O. Anya, Applications of Big Data Analytics: Trends, Issues, and Challenges, vol. 219, Springer, 2018.

[6] E.-L. Li, W.-J. Wang, 5G will drive the development of health care, Chin. Med. J. 132 (23) (2019) 2895–2896, https://doi.org/10.1097/CM9.0000000000000534.

[7] S.A.A. Shah, E. Ahmed, M. Imran, S. Zeadally, 5G for vehicular communications, IEEE Commun. Mag. 56 (1) (2018) 111–117.

[8] S. Ullah et al., "UAV-enabled healthcare architecture: issues and challenges," Futur. Gener. Comput. Syst., vol. 97, pp. 425–432, 2019. [Online]. Available: http://www.sciencedirect.com/science/article/pii/S0167739X18318247.

[9] B. Zhou, Q. Wu, X. Zhao, W. Zhang, W. Wu, Z. Guo, Construction of 5G all-wireless network and information system for cabin hospitals, J. Am. Med. Inform. Assoc. 27 (6) (2020) 934–938, https://doi.org/10.1093/jamia/ocaa045.

[10] G.B. Stefano, R.M. Kream, The micro-hospital: 5G telemedicine-based care, Med. Sci. Monit. Basic Res. 24 (2018) 103.

[11] S. Anwar, R. Prasad, Framework for future telemedicine planning and infrastructure using 5G technology, Wirel. Pers. Commun. 100 (1) (2018) 193–208, https://doi.org/10.1007/s11277-018-5622-8.

[12] A. Mavrogiorgou, A. Kiourtis, M. Touloupou, E. Kapassa, D. Kyriazis, M. Themistocleous, The road to the future of healthcare: Transmitting interoperable healthcare data through a 5G based communication platform, in: Proc. Eur., Medit., Middle Eastern Conf. Inf. Syst., in Lecture Notes in Computer Science, 341, 2019, pp. 383–401, https://doi.org/10.1007/978-3-030-11395-7_30.

[13] W.D. de Mattos, P.R.L. Gondim, M-health solutions using 5G networks and M2M communications, IT Prof. 18 (3) (2016) 24–29, https://doi.org/10.1109/MITP.2016.52.

[14] K. Ahmad, A. Kamal, K.A.B. Ahmad, M. Khari, R.G. Crespo, Fast hybrid-MixNet for security and privacy using NTRU algorithm, J. Inform. Security Appl. 60 (2021), 102872.

[15] A. Ksentini, P.A. Frangoudis, Toward slicing-enabled multi-access edge computing in 5G, IEEE Netw. 34 (2) (2020) 99–105.

[16] C. Qingping, H. Yan, C. Zhang, Z. Pang, L. Xu, A reconfigurable smart sensor interface for industrial WSN in IoT environment, IEEE Trans. Ind. Informat. 10 (2) (2014) 1417–1425.

[17] J. Ordonez-Lucena, P. Ameigeiras, D. Lopez, J.J. Ramos-Munoz, J. Lorca, J. Folgueira, Network slicing for 5G with SDN/NFV: concepts, architectures, and challenges, IEEE Commun. Mag. 55 (5) (2017) 80–87.

[18] A. Vergutz, G. Noubir, M. Nogueira, Reliability for smart healthcare: a network slicing perspective, IEEE Netw. 34 (4) (2020) 91–97.

[19] R. Khan, P. Kumar, D.N.K. Jayakody, M. Liyanage, A survey on security and privacy of 5G technologies: potential solutions, recent advancements, and future directions, IEEE Commun. Surveys Tuts. 22 (1) (2020) 196–248. 1st Quart.

[20] D. Rupprecht, A. Dabrowski, T. Holz, E. Weippl, C. Pöpper, On security research towards future mobile network generations, IEEE Commun. Surveys Tuts. 20 (3) (2018) 2518–2542. 3rd Quart.

[21] A. Akhunzada, S.U. Islam, S. Zeadally, Securing cyberspace of future smart cities with 5G technologies, IEEE Netw. 34 (4) (2020) 336–342.

[22] B. Karschnia, "Industrial Internet of Things (IIoT) benefits, examples control engineering," Control Engineering 2017. Accessed on Aug. 31, 2017. [Online]. Available: http://www.controleng.com/single-article/industrial-internet-of-things-iiot-benefits-examples/a2fdb5aced1d779991d91ec3066cff 40.html.

[23] S.R. Jena, R. Shanmugam, R. Dhanaraj, K. Saini, Recent advances and future research directions in edge cloud framework, Int. J. Eng. Adv. Technol. (IJEAT) 2249-8958, 9 (2) (2019), https://doi.org/10.35940/ijeat.B3090.129219.

[24] W. Chung, C. Yau, K. Shin, A cell phone based health monitoring system with self-analysis processing using wireless sensor network technology, in: Proc. Int. Conf. IEEE Eng. Med. Biol. Soc., 2007, pp. 3705–3708.

[25] R. Satheeshkumar, R. Arivoli, Real time virtual human hand for diagnostic robot (DiagBot) arm using IOT, J. Adv. Res. Dyn. Control Syst. 12 (Special Issue 1) (2020).

[26] A. Keller, The Road to 5G: Drivers, Applications, Requirements and Technical Development, Ericsson, Huawei, and Qualcomm, Global Mobile Suppliers Assoc., Washington, DC, USA, 2015.

[27] R. Gupta, S. Tanwar, S. Tyagi, N. Kumar, Tactile-internet-based telesurgery system for healthcare 4.0: an architecture, research challenges, and future directions, IEEE Netw. 33 (6) (2019) 22–29.

[28] E. Gilman, D. Barth, Zero Trust Networks: Building Secure Systems in Untrusted Networks, O'Reilly Media, Sebastopol, CA, USA, 2017.

[29] I. Ahmad, S. Shahabuddin, T. Kumar, J. Okwuibe, A. Gurtov, M. Ylianttila, Security for 5G and beyond, IEEE Commun. Surveys Tuts. 21 (4) (2019) 3682–3722. 4th Quart.

[30] A. Banchs, D.M. Gutierrez-Estevez, M. Fuentes, M. Boldi, S. Provvedi, A 5G mobile network architecture to support vertical industries, IEEE Commun. Mag. 57 (12) (2019) 38–44.

About the authors

R. Satheeshkumar (1984) was born in manalmedu town, Nagapattinam, Tamilnadu, India. He received Diploma in Computer Technology (2003) from the Directorate of Technical Education, Tamilnadu, India. He received a B.E. degree in Electronics and Communication Engineering (2007) from Anna University, Chennai, India. He received M.E. degrees in Electronics and Control Engineering (2013) from the Sathyabama University, Chennai, India. He was an

Embedded Software Engineer with the Orbit Controls and Services from 2007 to 2010. He was an Embedded Engineer with the Lumisense Technologies from 2010 to 2011. He was an Embedded Engineer with Porus Technologies from 2013 to 2015. At present Ph.D. thesis submitted in Robotics under the Department of Electrical Engineering in Annamalai University, Annamalai Nagar, India. His area of interest is embedded systems, RTOS, and Robotics.

Kavita Saini is presently working as Professor, School of Computing Science and Engineering, Galgotias University, Delhi NCR, India. She received her Ph.D. degree from Banasthali Vidyapeeth, Banasthali. She has 18 years of teaching and research experience supervising Masters and Ph.D. scholars in emerging technologies.

She has published more than 40 research papers in national and international journals and conferences. She has published 17 authored books for UG and PG courses for a number of universities including MD University, Rothak, and Punjab Technical University, Jallandhar with National Publishers. Kavita Saini has edited many books with International Publishers including IGI Global, CRC Press, IET Publisher Elsevier and published 15 book chapters with International publishers. Under her guidance many M.Tech and Ph.D. scholars are carrying out research work. (kavitasaini_2000@yahoo.com)

She has also published various patents. Kavita Saini has also delivered technical talks on Blockchain: An Emerging Technology, Web to Deep Web and other emerging Areas and Handled many Special Sessions in International Conferences and Special Issues in International Journals. Her research interests include Web-Based Instructional Systems (WBIS), Blockchain Technology, Industry 4.O, and Cloud Computing.

Dr. A. Daniel is currently working as an Associate Professor in School of Computing Science and Engineering in Galgotias University, Greater Noida, Uttar Pradesh. He completed his B.E and M.E both in Anna University. He has completed his Ph. D. in Computer Science and Engineering at Shri Venkateshwara University, Uttar Pradesh. His research interests are Cloud Computing, Networking etc. He has published several articles in reputed international journals. He has membership in IEEE, ACM, IFERP, IAENG and CSTA. (danielarockiam@gmail.com)

Dr. Manju Khari is an Associate Professor in Jawaharlal Nehru University, New Delhi, prior to the university she worked with Netaji Subhas University of Technology, East Campus, formerly Ambedkar Institute of Advanced Communication Technology and Research, Under Govt. Of NCT Delhi. Her Ph.D. in Computer Science and Engineering from National Institute of Technology Patna and She received her master's degree in Information Security from Ambedkar Institute of Advanced Communication Technology and Research, affiliated with Guru Gobind Singh Indraprastha University, Delhi, India. She has 80 published papers in refereed National/International Journals and Conferences (viz. IEEE, ACM, Springer, Inderscience, and Elsevier), 10 book chapters in a Springer, CRC press, IGI Global, Auerbach. She is also co-author of two books published by NCERT of XI and XII and co-editor in 10 edited books. She has also organized 05 International conference sessions, 03 Faculty development Programme, 01 workshop, 01 industrial meet in her experience. She delivered an expert talk, guest lecturers in International Conference and a member of reviewer/technical program committee in various International Conferences. Besides this, she associated with many International research organizations as Associate Editor/ Guest Editor of Springer, Wiley and Elsevier books, and a reviewer for various International Journals.

CHAPTER EIGHTEEN

The integration of blockchain and IoT edge devices for smart agriculture: Challenges and use cases

Swati Nigam[a] (ID)**, Urvashi Sugandh[a]** (ID)**, and Manju Khari[b]** (ID)
[a]Department of Computer Science, Faculty of Mathematics and Computing, Banasthali Vidyapith, Banasthali, India
[b]School of computer and System Sciences, Jawaharlal Nehru University, New Delhi, India

Contents

Abstract

Growth of IoT (Internet of Things) has proved its importance in the various sector. But due to some limitations of security, privacy, etc. it is not possible to use IoT devices in the field of agriculture at its fullest. To overcome these limitations Blockchain is used as it provides security, privacy and it helps in monitoring, examining, to authenticate the agriculture data. With the help of Blockchain, traditional methods of collection, rearranging and distribution of Agri-products can be replaced by more trust-worthy, decentralized, vitreous, and immutable style. In agriculture sector, Blockchain and Internet of things can be amalgamated to have better results which leads us to one level up in the field of agriculture and may control or improve the supply chain in proper

Advances in Computers, Volume 127
ISSN 0065-2458
https://doi.org/10.1016/bs.adcom.2022.02.015
507

manner. The consequences of using blockchain and IoT in combination will result in better understanding to supervise and managing the agriculture effectively. This chapter will illustrate the importance of using blockchain and IoT collectively to develop smart agriculture from traditional agriculture. A model is also proposed to overcome the challenges encountered in agriculture sector, based on IoT applications with the help of blockchain. Also, a review is mentioned about the main characteristics and functions of blockchain used in agriculture sector such as livestock grazing, crops and food supply chain. Finally, some of the open issues Blockchain and security challenges are elaborate.

1. Introduction

Many limitations have been overcome due to the continuous growth and development of Internet of things (IoT). With the implementation of IoT, various devices can share the data or information to other devices without using wired network. IoT may be defined as the interconnection of the devices for collecting and exchanging the information among themselves [1]. IoT may be identified as an infrastructure consists of gateways and sensors. With the help of sensors and gateways it, IoT devices create a network and can communicate with the surroundings any time to give rise the mass production of various data resources such as audio–video, images, etc. IoT has find out its application in various fields which belongs to our day-to-day activities such as in hospitals, households, universities, and agriculture [2].

Agriculture sector uses IoT to convert the farming techniques into smart farming techniques. Basically, IoT devices operated by various organizations through which data is first collected, then distributed as per need, data is stored, analyzed by the experts and then proper action will be taken [3,4]. In smart agriculture large number of IoT nodes are placed in wide range resulting in large amount of data sharing, consequently, it gives rise to network traffic, latency issues, high energy consumption, etc. Due to reasons only limited or essential data could be collected instead of the complete data. Also, the mentioned limitations signifies that there is no existence of communication in between the nodes, excepting for some defined and conditions.

Internet of things technologies also help farmers with precision farming. It used to be that farmers had to map the entire landscape of an area and then plot their plots into precise dimensions, which required using sophisticated equipment [5,6]. Now, even the most intricate forms of farming can be monitored digitally by modern devices. Irrigation systems can map precise areas where water conservation is most critical, for example. High-resistance grass seed used in place of traditional turf, or modified tillage techniques, can be easily managed with the assistance of livestock monitoring tools.

1.1 Requirement of IoT in agriculture

IoT devices holds the capability to collect and send the important data via internet to the server where the storage of data is done to use it in future by the user. This is done by the constant monitoring of the data receive by the IoT enabled devices. Monitoring activity may help the user to take appropriate action or decisions.

Devices enabled with the IoT make use of sensors to collect and transmit the data to another device or a server and vice-versa. After analyzing the received data, IoT devices may take appropriate action to improve the environment and further again, consequences of the actions taken are to be analyzed to avoid risks or to further improved the surroundings [7,8].

In current scenario, IoT enabled devices are needed in the agricultural sector for the better understanding of the actions to be taken for the better results.

Smart sensing systems are used in Precision agriculture to manage the agricultural produces, animals, fields, automatically [9].

Internet of things' technologies help in precision agriculture because they let farmers precisely measure and map specific areas. This accuracy is particularly significant when it comes to irrigation systems that rely on water levels and flows [10]. With precise measurements and maps, farmers can plan on the quantity of water that their fields will need. They no longer have to guess how much water a field needs, which is often incorrect or settle for a lower rate than they want to pay for.

IoT has various applications in agriculture field but due to some limitations it does not fulfill the requirements needed to upgrade the farming. But use of IoT in combination with the blockchain has overcome from its limitations. Also, there is lacking trust in centralized system due to which it needs the decentralization [11].

Livestock production is another aspect of smart agriculture that relies heavily on the Internet of things. In the past, farmers had to send out their entire fleet of horses, pigs, and cattle in order to keep up with the demand of the marketplace. Nowadays, tractors, combine harvesters, and livestock-delivery systems can provide farmers with an efficient way of maintaining and harvesting their produce [12]. Drones are making this possible at a fraction of the cost of traditional farming, since they allow for greater precision in handling and transporting massive quantities of produce.

Internet of things technologies enable farmers to get access to their own data as well, through things like GPS. Modern devices have onboard computers that store information, and software enables them to present this information in graphical formats such as dashboards or charts. Agricultural researchers can view historical data or conduct statistical analysis of current farming practices. Internet of things technologies allow you to easily access and share this information with other people who work in the industry, allowing collaboration and communication between team members [13,14].

One application of Internet of things technology for agriculture is food production. Modern farming practices are geared more toward conserving resources than increasing production. Thanks to the Internet, farmers can easily see how things are done in another location and apply the same principles in their own fields. With this knowledge, they can increase their yield and reduce water consumption. With the improved food quality, greater value for money and reduced environmental impact, an Internet of things' technology is a key part of the future of agriculture [15].

The Internet of things (IoT), is revolutionizing the way farming is done today by allowing farming equipment to communicate with one another to coordinate activities, improve efficiency, and to make a better, more environmentally friendly crop or product. With the Internet of things, farmers can access information about their crops or soil conditions easily and quickly. They can also receive visual feeds of their crops or soil from a mobile phone or web cam [16]. These feeds provide critical data on what crops need to be improved upon so that they can be delivered more quickly to consumers. With all of the information available at the touch of a button, farmers can make better decisions about how to grow their crops and increase profitability.

Blockchain is a distributed ledger system where the data is not controlled by single or centralized authority. It is a decentralized system and give permission to numerous systems to keep a copy of all the transaction records taken place in the network. Basically, blockchain had been noticed in 2009 and first, implemented in Bitcoin as cryptocurrency [17,18]. In this system, manipulation of information is quite impossible as the copy of all the transaction distributed among the specific nodes. Currently, Blockchain has been implemented in various fields such as healthcare, finance, banking, agriculture, etc.

After IoT, blockchain brings next step upgradation in the field of agriculture. It has the superb capability to convert traditional farming into smart farming, shares the data in a transparent manner and results in effective use of shared data. Use of blockchain in combination with IoT leads to secure data routing done by IoT devices and can avoid the attacks on the system [19].

1.2 Requirement of blockchain in agriculture

Concepts of blockchain enables to transform the current agriculture into smart agriculture. Fundamental property of transparency ensures to resist any kind of fraud in the collection or sharing of the information. Also, it is helpful in tracking of the agricultural products to avoid the wastage of the foods and blocking of the products. With the implementation of blockchain, users keep an eye on the food supply chain [20].

Using the Blockchain in agriculture is one of the most unique uses for this emerging technology. Traditional methods of securing private data such as credit card information or sensitive financial transactions are often susceptible to hacks. By decentralizing data entry and communication, the Blockchain allows for a more secure and faster transaction processing by multiple participants in the agricultural industry [21]. The supply chain, for example, is very long, involving many different parties using their own distinct siloed monitoring systems. However, using the Blockchain provides the solution of an extended version of the traditional supply chain, using the same technology and methodologies but applied to a much larger portion of the agricultural market.

The benefits of the use of the Blockchain in agriculture is multi-faceted, not only for the farmers, and producers they employ, but also for the public at large. The applications for this technology are practically endless. With the ability to track each phase of the food supply chain, from growing through to marketing, there are fewer risks of trade-offs between speed and safety. For instance, if a manufacturer senses a customer demand for lower cost or lower quality food, they can easily adjust production or reduce processing times to meet those standards without incurring financial loss [22,23]. If a farmer senses a trend toward lean meat or lower volume, they can also adjust accordingly. This provides the public with a clearer picture of the food supply chain, providing an accurate depiction of where the waste is being generated and which assets require the most attention to improve quality and efficiency.

In addition to using the Blockchain for agriculture, it has been used as an innovation platform in other sectors. Distributed ledger technology has been increasingly used by hedge funds and capital markets, to reduce the opportunities for corruption and strengthen the integrity of the transactions and portfolios. Similarly, big data has been used to improve traceability of natural resources across different industries. Transparency and traceability are two important tenants of the Blockchain ecosystem, and its use in agriculture shows how its application can streamline and optimize the agricultural sector and help create a more transparent public space [24].

2. Blockchain technology: An overview

Blockchain is a robust technology gained attention after the implementation of cryptocurrency named as "Bitcoin." This cryptocurrency comes in notice after publishing the white paper of Bitcoin, with contribution of number of experts under the guidance of "Satoshi Nakamoto" in 2008. Blockchain name suggests itself a "chain of blocks." Therefore, blockchain consists of verified blocks connected with each other and each blocks contains the information about transactional data, timestamp, and the hash address of the last block. In blockchain data is stored on the distributed ledger and works in a decentralized manner. The very first block in the blockchain is known as "genesis block" [25].

Components in the blockchain architecture are as explained below:

Block: Block is the main and basic element in the blockchain structure which is used to store the complete set of transaction data and is further shared among the existing nodes in the network.

Chain: This chain is to be referred as the series of the block in a particular order.

Transaction: It a very basic unit of information in this architecture which serves its purpose.

Nodes: In blockchain architecture, nodes are defined as the user s or the computer system who has the complete information about Blockchain ledger.

Miners: Miners are the special type of nodes and performing the function of verification of the block before they get added in the blockchain structure.

Consensus: Consensus are defined as a procedure which helps in the smooth operation of Blockchain.

All the above, mentioned components belongs to the architecture of the blockchain are the major units. Table 1 shows the fields of blocks and their size.

Table 1 Fields of blocks and size.

Sr.no.	Fields of block	Size (bytes)	Illustration
1	Block	4	Size hold by block itself
2	Transaction counter	1–9	Holds the count of transaction taken
3	Block Header	80	Consists of several field
4	transaction	Not fixed	Details of the transaction are to be recorded

As it is mentioned that Genesis block is the first block existing in the blockchain and having no parent in the blockchain. Every connected in the chain has its own specific address known hash value considered as the identity of that block. Due to the presence of components in the block, it is considered as a unit. Basically, a block is consisting of block header and block identifiers. First, all the field of a block will be explained based on its size. Components of a block header are as follows (Fig. 1):

Version: it defines the upgradation level of the software or protocol which are in use. This field occupies 4 bytes.

Timestamp: It provides the time of creation and updating of the block to avoid any kind of fraud. This field is of 4 bytes.

Previous Block Hash: This is also known as parent block hash and point to the previous last block. It is a main component to build a linking in between the chain due to which blockchain is formed in proper manner. This field is of 32 bytes.

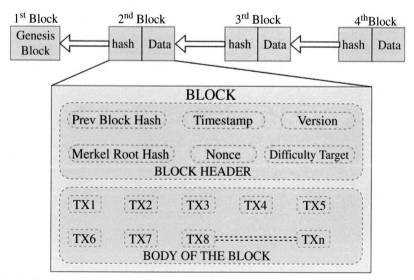

Fig. 1 Structure of a block in blockchain.

Nonce: It is a random number and generally starts with 0. This number increases on the progression of the hash computation. This field is of 4 bytes.

Markle root Hash: It is represented as a binary tree and each hash codes serves as a node. Markle root hash is used for the identification of transaction takes place in the block. This field is of 32 bytes.

N-bits: it is also known as difficulty target. It defines the difficulty level to find the targeted hash of the current block. This field is of 4 bytes.

3. Working of blockchain

Concept of blockchain is based on distributed ledger technology collecting the records of digital transactions in an immutable and transparent manner. Therefore, Blockchain may be defined as a decentralized and distributed or shared ledger holds the information of digital transactions performed in the network without including any third party or any intermediaries. Blockchain works in an efficient manner where the data is immutable and secure. At very first, a transaction request in initiated by the user in network. in response of the request, a block is created to represent the creation of transaction. After the creation of the block, transaction request is broadcasted to every node existing in the network will verify and validate the transaction. Being a verified transaction, it may get contracts, cryptocurrency, records as a reward and then transaction is added to a new block for the ledger. Further on this block is added to the blockchain in a permanent and immutable way [26]. Fig. 2 shows the working of blockchain.

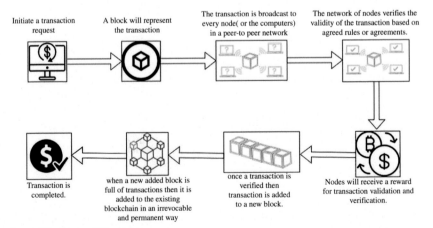

Fig. 2 Working of blockchain.

4. IoT: An overview

The Internet of things (IoT) has played an important role in our day-to-day needs. This technology is also known as Internet of Everything and make a link in between the devices to connect with each other. It is acknowledged in a very short duration by the organizations as one of the essential fields for future use. As a result, existence of communication in between the devices represents importance of this technology for the industries. The concept of this technology has been proposed in 2009 by Ashton. He makes it possible to establish communication in between human-to-things, human-to-human and things-to-things. By using IoT, each object gained its unique identity. IoT is implemented in combination with other technologies to bring revolution in existing fields [27,28].

Some prerequisites are needed to implement the Internet of Things (IoT) in a successful manner. These are follows as:

Hardware: Importance of hardware or devices is explained in its name by itself. Hardware needed in IoT are sensors, cameras, actuators, CCTV, and embedded hardware require to establish communication in between devices.

Middleware: These are the computing tools and storage devices used to store the data and for data analysis with big data and cloud analytics.

Presentation: At every representation of data is quite important due to the reason need of interpretation and visualization are required. Better the usage of tools may lead to better understanding of the data.

5. Working of IoT

1. **Sensors**

   With the help of sensors, a large data can be collected as it is capable to capture every minute data, and this collected can have complexities at different level. This data may be temperature monitoring data, video, etc.

2. **Linking**

   Data collected with the help of sensors is to be stored on the cloud infrastructure, which is done by using Wi-Fi, satellite networks, Bluetooth, etc.

3. **Data processing**

   Data collected on the cloud will be processed by using the tools or software. It may be processed in terms of monitoring a particular data whether it resides in defined limits or not. This activity will be performed by the user to take the appropriate actions.

4. User Interface

Now the collected information must be represented to the user, and it is done by using interface. It might be mobile phone, Laptop, computer system or PLC.

IoT has many applications in many areas but now in this chapter main concerned is about the use of IoT to convert classical agriculture into smart agriculture. Fig. 3 shows the working of IoT.

Smart agriculture is a strategic system used to enhance crop production and reduce expenses. By using Internet of things technologies, agricultural experts can collect, collate, analyze, and distribute real-time information about weather patterns, soil composition, and climate conditions. The Internet of things is expanding the reach of such technology to not only improve the quality of farming but also make the process of farming easier and more accurate [29,30]. With the Internet of things, farmers no longer have to wait for information, weather conditions, and other critical data during their field work. With the help of the Internet of things, they can now conduct research on weather patterns, crop production, and other aspects of the land with just a few clicks of a mouse.

In the past, agriculture relied largely on land surveys to accurately map the physical structure of the land and to determine the optimum spacing and nutrients needed for different crops [31]. With the advent of modern computer software, the process has been made even more precise and accessible to the point where precise and complete data about crops can be analyzed at the push of a button. This allows farmers to know what crops should be planted where, when they should be planted, and how much space each one must occupy. Through the Internet of things, land surveys are becoming

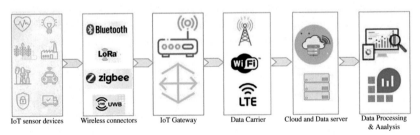

Fig. 3 Working of IoT.

obsolete because the actual physical structure of the land can now be viewed in real time through digital imaging. Computer programs that provide accurate topographical and soil moisture mapping can be accessed instantly from any location.

6. Edge computing: An overview

Edge computing is an example of distributed computing that pulls together data and computation closer to the real-world sources of data, bringing real-time computing and information storage closer to the actual devices that need access to it. This is said to dramatically increase response times, reduce bandwidth and boost productivity. In the case of vehicle fleets, this means that fleet managers can make informed decisions about vehicle maintenance and scheduling. Foreseen failures could be mitigated through predictive maintenance procedures [32,33]. On the other hand, existing vehicles can still be kept in good running order, as long as they receive regular maintenance, which most truck and van fleets already do. The combination of improving efficiency and reducing costs has real-world value for both current as well as future generations of vehicles.

In case of agriculture, it is reshaping the future of agriculture. Autonomous tractors and robots have been developed and can be operated automatically without human interference. Edge computing technology is capable to transform the agriculture area. In IoT system or IoT based devices performs some essential jobs and are represented in the figure as a pyramid including the three layers (1) at the bottom is edge computing layer (2) above that fog computing layer is present and at the top is (3) Cloud computing layer. Fig. 4 shows the working model of edge computing.

Edge computing: This layer performs the processing on the edge servers which are in direct contact of millions of controllers and sensors. These edge servers have potential of analyzing and take decisions on real time site.

Fog Computing: This layer acts as an intermediate in between the edge computing layer and cloud computing layer. It can perform the computing on the proliferation of the data. Moreover, it has the capability to perform the additional analytics and filtrating.

Cloud computing: Basically, this layer is used to store the data. Important data is stored in this layer collected from the edge nodes and fog nodes. After then multiple analytics is done to make decisions.

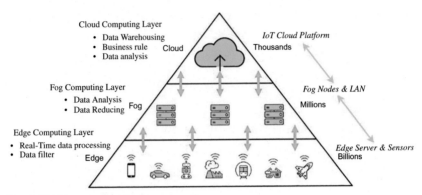

Fig. 4 Working model of edge computing.

7. A proposed model for smart agriculture using blockchain and IoT

Smart agriculture using blockchain and IoT comprising of some of the activities and data including the agricultural parameters to sense, knowledge about location where it has to be identified, tracking of data from source to destination to take decision, to view results by using an application [34,35]. Our proposed model defined on the basis of six layers after the employment of blockchain and IoT are as follows (Fig. 5):

- *Sensor layer and edge computing layer:* This layer consists of sensors which are used to sense the environment, analyze and after analyzing proper actions have to be taken. In market there are various types of sensors, and each sensor has its specific application [36,37]. These sensors can be planted under the soil, i.e., underground, above the crops. The sensors put down under the soil are water resistant generally used to take measurement of moisture existing in the soil, chemical properties of the soil, pH level, etc. While edge computing increases the speed of data collection activity and hence its efficiency [38,39].

- *Link Layer:* Under this layer, currently existing routing and networking technologies are implemented to make information exchange in between the sensors. Generally, IoT uses WSNs (Wireless Sensors networks) for the efficient management of the crop and field. Use of WSN provides various benefits to monitor the fields, to optimize the quality of crops [40,41].

- *Fog Computing layer:* With the help of fog layer information can be shared with the user or farmers will be on time [42]. This layer is in direct

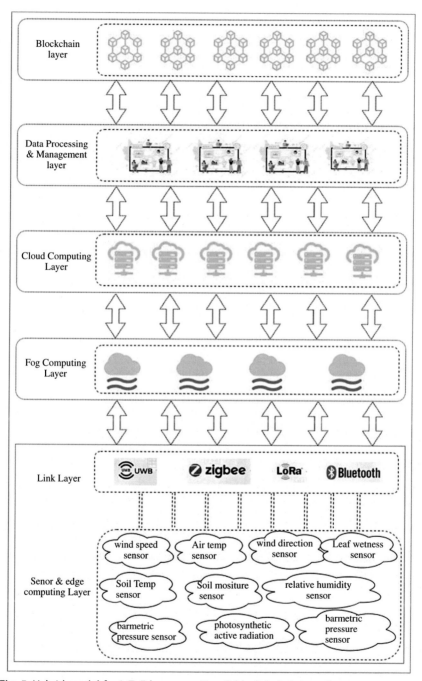

Fig. 5 Hybrid model for IoT, Edge computing & blockchain integration.

contact of cloud computing layer and works on the time-sensitive information. Due to which the farmers can get the information on time and may reduce the loss or risk factors [43,44].

- *Cloud computing layer.* This layer lies in between the fog computing layer and data processing and management layer. Cloud computing layer may generate the distributed pool of It resources such as storage, software information, network with the help of a suitable infrastructure ensuring the good level of food security [45–47].
- *Data processing and management layer.* Basic function of this layer is to process the collected data and do analysis of the same. This layer manages the data and mine the data related to interset by using efficient data mining techniques [48,49]. In every field data analysis is most important activity to make prediction and decision to avoid risks and losses. This is the reason this layer plays an important role in the proposed model [50,51].
- *Blockchain layer.* After releasing of processed data, this data will enter into the the blockchain layer where it will become advantageous for the farmers. This data will be verified by the special nodes. After verification data will be added in the block and futher this block would be added in the chain hence become a part of the bockchain [52,53].

As per the above Fig. 6 a model is proposed for the for the monitoring of food supply chain. This model consists of three layers named as (1) Physical data layer (2) Logical data layer and (3) Web interface layer. By introducing these three layers in the proposed model it consists of extendable, efficient and scalable implementation. Physical layer consists of IoT nodes, IoT gateway [54,55]. IoT nodes keep the track of the farm to monitor the environment and conditions of the growth. IoT nodes gather the data from the various farms which is then forward to the base station with the help of gateways and routers. Also, another gateway known as RFG gateway is used to synchronize the two irrelevant data and to maintain the management. To store the data, receive from physical data layer, in the database or server, logic data layer is used [56,57]. As per requirement data is used to analyze or to make decision could be received from the logic data layer in appropriate form. Some codes are generated for the linking of the data to the database. As a result, web application layer will identify them. With the help of this model data can be converted to IoT networks for the smooth sharing of data and operation in between the IoT devices [58,59].

As you may be aware, food traceability is critical in this market. Tracking items is a major concern, and most corporations seem to be hesitant to do

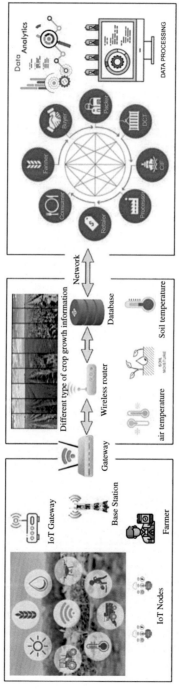

Fig. 6 Agriculture monitoring and supply chain.

so. Consumers have no idea how or in what method the food was prepared if sufficient monitoring is in place. As a result, the industry needs assistance at this time. As a result, blockchain in food traceability has the potential to change the situation for everyone. The blockchain is well-equipped to track food from its suppliers to the customer who purchases it. As a result, businesses may use this power to improve their visibility on the manufacturing line and provide higher-quality food to the market. In the field of food safety, blockchain can put customers at rest and allow them to acquire all of their goods without fear of contamination or other issues. Because the food business is currently ill-equipped to identify the real source of contamination. It may even go unnoticed until its too late in most situations. As a result, organizations may use blockchain to track how they handle raw materials and final goods. Furthermore, they may treat their byproducts with greater care in order to preserve the quality and grade of their delicacies. Fig. 7 shows the proposed architecture of food safety using blockchain technology.

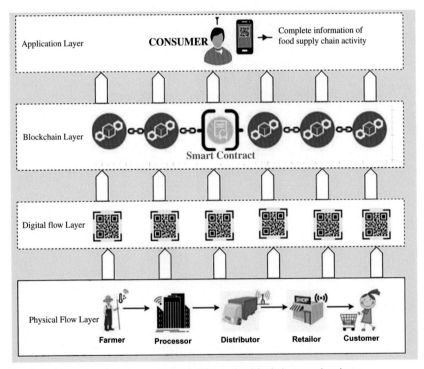

Fig. 7 Proposed architecture of food safety using blockchain technology.

8. Advantages of blockchain, edge computing and IoT based agriculture

Due to benefits of implementing blockchain in agriculture, it is in high demand. Blockchain is considered as a disruptive technology because of its significance. Advantages of blockchain are as follows:

Agriculture supply chain: Implementation of blockchain in agriculture supply chain leads to convert the industry into more efficient industry which was not before due to unavailability of automation and the proper use of technology. Therefore, use of blockchain in food supply chain deduct the costing of farming activities. This may increase the overall efficiency of the process [60,61].

Transparency: Concept of blockchain also provides the transparency regarding the data. The subsides generated for the farmers by the government. By this tracing of the actual facts of the subsidies can be followed [62]. User can check clearly about the flow of money.

Fair prices: Use of Blockchain can update the prices to be fair. Many brokers are involved in the process of buying of crops and not provide the fair prices to the farmers. By using blockchain, scenario will become totally different [63,64]. Farmers can sell their crops or products to genuine buyers from whom they can negotiate.

Inventory management: Condition of Farm inventory could get better by reducing the wastage in them. And farmers to paid for the wastages in direct or undirect way. With the help of blockchain, storage condition or surroundings could be tracked so that the propre action will be taken [65–67].

Updating of farm management software: In current scenario, software is used in agriculture are based on client server architecture only not working at their maximum efficiency. But these software's can be updated on the implementation of blockchain [68,69].

Secure for IoT optimization: IoT devices are very efficient in keeping track of their products and provides safety to the devices also. These devices can collect real time data could analyze it and take appropriate action. On the other hand, sometimes these devices become so vulnerable about prediction of natural calamities [70,71]. Cloud services required for storing the data or information may get easily prone to cyber-attacks. All these issues can be solved by using blockchain.

9. Summary of the research for applying blockchain and IoT in agriculture (Table 2)

Table 2 Summary of the research for applying blockchain and IoT in agriculture.

Ref. no	Author name	Use case	Year	Blockchain contribution
[1]	C. Xie, Y. Sun, and H. Luo	Agricultural Products Tracking	2017	Double-chain storage structure and build a highly secured method for monitoring agricultural goods based on blockchain. Double-chain storage assures that farm produce data cannot be intentionally tampered with or destructed by storing it in a blockchain transaction hash
[2]	Y. P. Lin et al.	ICT e-agriculture	2017	Blockchain-based e-agriculture platform for regional and local application is suggested in this paper. Evaluating blockchain technology for use in ICT e-agriculture applications is made easier with the introduction of this tool
[4]	M. Tripoli and J. Schmidhuber	Supply Chain	2018	DLTs in agri-foods: Possibilities, Advantages, and Strategies is the goal of this article. Institutional and technical limitations and obstacles to its implementation are also identified. Agricultural supply chains and rural development initiatives may both benefit from digital records, cryptography, and disintermediation of transaction processing and data storage, thanks to DLTs
[5]	M. P. Caro, M. S. Ali, M. Vecchio	Agri-Food supply chain	2018	IoT devices that produce and consume digital data throughout the supply chain may be seamlessly integrated into the AgriBlockIoT blockchain-based traceability system presented in this study
[6]	K. Leng, Y. Bi, L. Jing.	Supply Chain	2018	In this research, we present a public blockchain of agricultural supply chain network based on the double chain architecture, focusing on the dual chain structure and its storage mode, resource rent-seeking and matching mechanism, and consensus algorithm. Openness and security in transaction information may be taken into consideration by using a double-chain structure for agricultural supply chains, according to these findings
[7]	J. Hua, X. Wang, M. Kang	Supply Chain	2018	Here, we propose the use of blockchain methods to create an agricultural provenance system with decentralized, community maintenance and consensus trust in order to alleviate the trust issue inside the supply chain. Management activities like (fertilizing, irrigation, etc.) as well as a specific data structure are stored

Ref	Authors	Topic	Year	Description
[8]	J. T. Hao, Y. Sun, and H. Luo	Agricultural Products Tracking	2018	Based on IPFS and the blockchain, we provide a new data storage mechanism in this article. As a starting point, the sensors' real-time data is stored in IPFS and accessed through web browsers. The IPFS hash address of the origin data is then stored on the blockchain to prevent a hostile user from performing a data faking attack. A blockchain-based authentication process is then designed. It is able to authenticate the data and protect it effectively
[9]	J. Lin, Z. Shen, A. Zhang	Food Traceability	2018	As part of this research, researchers suggest a blockchain-based food traceability system that includes all stakeholders in a smart agricultural ecosystem, even if they don't trust each other
[13]	A. Croxson, R. S. Sharma	Agri-Food supply chain	2019	Q-methodology was used to analyze whether or not blockchain technology may be the answer of problems associated with agro-food supply chain and give advice for organizations on how they can implement this technology. Four unique groups were found in the sector, each with a different take on the possibilities of blockchain
[14]	Z. Wang and P. Liu	Agricultural products tracking	2019	Blockchain technology is used to store the traceability data of farm commodities securely, and this paper proposes an agricultural product traceability model that can covering the full industrial chain of agricultural goods. Buyers can retrieve the data from the truthful source of traceability for agricultural products through this model, which uses blockchain technology to store traceability data
[19]	M. Kim, B. Hilton, Z. Burks	Food Traceability	2019	End-to-end food traceability, "farm to fork" in this context, may be achieved using the Blockchain platform and Internet of things devices exchanging GS1 messaging standards, according to this article. Supply chain stakeholders will have full access to the distributed ledger
[23]	K. Salah, N. Nizamuddin	Supply Chain	2019	With the suggested method, there is no longer a requirement for a trusted central authority, intermediaries and transaction records, which improves efficiency and safety. The suggested approach is based on the use of smart contracts to control and manage all exchanges and transactions among the supply chain network players
[24]	A. Kamilaris, A. Fonts	Agri-Food supply chain	2019	Examines the influence of blockchain technology on agricultural and food supply chain, provides active projects and efforts, evaluates the overall implications, problems and possibilities with a critical perspective of the development of these projects in this article

Continued

Table 2 Summary of the research for applying blockchain and IoT in agriculture.—cont'd

Ref. no	Author name	Use case	Year	Blockchain contribution
[27]	T. Surasak, N. Wattanavichean	agricultural products tracking	2019	Our traceability solution benefits greatly from the combination of the blockchain database and the Internet of Things since all of the data is collected in real time and stored in a highly secure database. Supply chain management and Food traceability might become more dependable and public awareness of monitoring and quality control in Thailand could be resurrected with our approach
[28]	S. Madumidha, P. Siva Ranjani	Agri-Food supply chain	2019	Decentralizing blockchain-based traceability for agriculture is presented in this article, which allows the construction of blocks that are constantly integrated with IoT devices, from providers to consumers
[32]	X. Li and D. Huang	Agri-Food supply chain	2020	From the source through manufacturing, service, security, and sales, the article examines how Internet technology has evolved over time in relation to the agriculture industry's supply chain. The paper also proposed the development of a blockchain-based system for tracking agricultural products
[33]	W. Lin et al	Farm Overseeing	2020	In this article, we conduct a survey to learn more about the agricultural uses of blockchain technology. Technical aspects, including data model, cryptographic algorithms and consensus procedures are presented in great depth first. Second, current agricultural solutions are classified and examined to show the utilization of blockchain methods
[34]	Y. Chen, Y. Li, and C. Li	Farm Overseeing	2020	This approach includes the complete ecological farm's circular agriculture cycle onto the blockchain. Data may now be shared more widely thanks to the adoption of numerous smart gadgets that automatically gather and upload data
[35]	M. D. Borah, V. B. Naik, R. Patgiri	Supply Chain	2020	Blockchain technology incorporated into the supply chain to increase product traceability and use. In order to accomplish these objectives, authors plan to use BCT to construct a value chain that is transparent from the farm to the fork
[36]	P. Chun-Ting, L. Meng-Ju	Agricultural Traceability	2020	This project uses IoT sensors and a blockchain-based agricultural service platform to track food from farm to fork

Ref	Focus	Year	Description
[38]	Supply Chain	2020	Authors want to find out what the biggest obstacles are to Indian ASC using blockchain. A thorough literature search and the opinion of experts from five agribusiness, academia, and agro-stakeholders were used to identify the constructions
[42]	Farm Overseeing	2020	A Blockchain-enabled Smart IoT Framework with Artificial Intelligence is proposed in this study, which offers an effective means of integrating blockchain and AI for IoT with existing IoT applications and approaches
[43]	Agricultural Traceability	2021	According to an AgriChain system, there are anticipated to be a large number of transactions that are waiting in queues before they can be entered into blocks, and these transactions are evaluated in detail in this chapter using a queuing model that gauges the performance of this system
[45]	Farm Overseeing	2021	In order to save energy, we've developed the Improved LEACH protocol. A new energy efficiency threshold limit is implemented in this protocol. Blockchain with the ILEACH protocol may be used to create an intelligent agricultural system
[49]	Agricultural products tracking	2021	Agricultural product supply chains under blockchain technology are examined, and the existing state of agricultural supply chain networks is examined
[50]	Trust Management	2021	It is our goal to build a sustainable supply chain strategy that incorporates non-cooperative game theory, in which the Bayesian formula is used to integrate previous experience and present conditions
[53]	Supply Chain	2021	For agro merchants, we have developed a model that is both effective, efficient, and satisfying, as well as a food traceability system which uses blockchain and IoT to make their company smarter and wealthier
[54]	Supply Chain	2021	An increase in the global desirability index for a blockchain–enabled supply network over the conventional supply chain indicates that the supply chain's sustainability may be improved by using blockchain technology. Practitioners will benefit from the findings of this research, which attempts to provide actionable advice on how they may use this technology

Continued

Author columns:
Ref	Authors
[38]	V. S. Yadav, A. R. Singh
[42]	S. K. Singh, S. Rathore, and J. H. Park
[43]	S. S. Patra, C. Misra
[45]	S. J. Anand, K. Priyadarsini
[49]	Y. Zhang, J. Li, and L. Ge
[50]	Y. Bai, K. Fan, K. Zhang
[53]	S. Al-Amin, S. R. Sharkar
[54]	A. A. Mukherjee, R. K. Singh

Table 2 Summary of the research for applying blockchain and IoT in agriculture.—cont'd

Ref. no	Author name	Use case	Year	Blockchain contribution
[56]	T. H. Pranto, A. A. Noman	Farm Overseeing	2021	This study demonstrated that blockchain is irreversible, accessible and transparent in agriculture while also underlining the robust mechanism presented by its partnership with smart contracts and IoT
[57]	M. Biswas, T. M. N. U. Akhund	Supply Chain	2021	The suggested model integrates Artificial intelligence technology, Internet-of-things, and Blockchain to build a smart and future agricultural system that gives farmers with a safe and open transaction method to rich, novel, and effective decision assistance
[58]	I. Eluubek kyzy, H. Song, A. Vajdi, Y. Wang	Supply Chain	2021	Our design considers trustworthiness, scalability, and share allocation. We use a cyber-physical system to assure product quantity and quality. Scalability is addressed with a novel consensus mechanism and a public service platform model
[62]	S. Hu, S. Huang, J. Huang	Supply Chain	2021	Author use the data integrity of blockchain and the edge computing paradigm to build an OASC trust architecture that has a substantially higher cost-efficiency ratio
[63]	W. Liu, X. F. Shao, C. H. Wu, and P. Qiao	Farm Overseeing	2021	This review lays the groundwork for ongoing research into the use of ICTs and BTs in agriculture, with implications for technical advancement and agricultural sustainability
[64]	M. H. Ronaghi	Supply Chain	2021	The peculiarity of the study is that it identifies the relevance of blockchain aspects in agriculture and provides the management framework of blockchain in supply chain operations from an applied perspective
[65]	G. S. Sajja, K. P. Rane, K. Phasinam	Farm Overseeing	2021	This article examines the use of blockchain technology in food supply chains, crop insurance, smart agriculture, and agricultural products transactions
[67]	S. Saurabh and K. Dey	Agri-Food supply chain	2021	According to the findings of the research, supply chain players' adoption-intention decision processes may be influenced by dis-intermediation, traceability, pricing, trust, compliance and coordination and control. Additional considerations for a sustainable and scalable supply chain collaboration and sustainability architecture are provided by the adoption factors

10. Challenges and open issues

Amalgamation of blockchain and IoT may overcome the challenges of IoT in very efficient manner. But still some of the new challenges and issues comes in notice the are considered as strong obstruction. The challenges and issues are defined as below (Fig. 8):

1. *Handling of big data in the blockchain*: In the blockchain, on confirming of the new block is broadcasted over the entire network. Being a peer-to-peer network, this new will be appended with each node. And this procedure will consume very amount of storage which approximate of 720 GB.
2. *Trade-off between power consumption and performance*: Blockchain algorithms requires a very high computational power which acts as a restriction in the applications based on this technology. Due to the reasons user are not confirmed about the efficiency of Blockchain.
3. *Transparency and privacy*: In certain applications, such as banking, transparency is critical, and this is where blockchain comes in. However, when storing and retrieving IoT data from specific IoT applications such

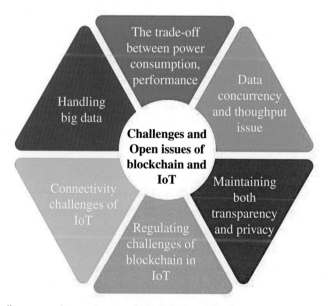

Fig. 8 Challenges and open issues of blockchain and IoT.

as eHealth on the blockchain, user confidentiality may be significantly damaged. IoT access control must be cost-effective in order to retain a reasonable level of openness and privacy.

4. *Regulating challenges of blockchain in IoT*: Decentralization, anonymity, immutability and automation are all properties of blockchain technology that have the potential to improve the security of a wide range of IoT applications. Immutability refers to the fact that data is permanently stored in the distributed transaction ledger (DTL) on the P2P network and cannot be erased or updated. Additionally, records cannot be vetted for privacy before being published on the blockchain owing to a lack of governance. Code such as smart contracts on a DTL might result in illegal action. Because of the DTL's anonymity, it is difficult to identify the persons involved in unlawful service transactions.

5. *Data concurrency and throughput issue*: In blockchain the throughput is to a limited amount due to existing protocol and consensus mechanism used in this technology while in IoT, high data concurrency is existed due to the non-stoppable streaming of data. Hence, blockchain technology is required to increase the throughput to synchronize with IoT devices.

6. *Connectivity challenges in IoT*: IoT devices are connected with highly efficient devices and networks for the sharing of the data with the user. To connect the devices with blockchain, IoT is restricted up to a limit for the execution of applications.

11. Conclusion

The goal of this research was to find out how critical it is to include IoT and blockchain technology with edge computing into the development of smart systems and precision agricultural applications. Due to this technical integration, it has been shown that blockchain may provide new solutions to chronic security and performance concerns in IoT-based precision agricultural systems. With the help of Blockchain, traditional methods of collection, rearranging and distribution of Agri-products can be replaced by more trust-worthy, decentralized, vitreous, and immutable style. In agriculture sector, Blockchain and Internet of things can be amalgamated to have better results which leads us to one level up in the field of agriculture and may control or improve the supply chain in proper manner. This chapter presents an overview of blockchain technology, IoT with edge computing and how these three technologies used in smart agriculture.

Chapter proposed a hybrid architecture for IoT, edge computing and blockchain technology integration. Two framework is also proposed for smart agriculture, one for monitoring and supply chain management, and second for food safety. In the last, a tabular summary is also present for different research carried out in the related area from 2017 to 2021.

References

[1] C. Xie, Y. Sun, H. Luo, Secured data storage scheme based on block chain for agricultural products tracking, in: *Proc. - 2017 3rd Int. Conf. Big Data Comput. Commun. BigCom 2017*, 2017, pp. 45–50, https://doi.org/10.1109/BIGCOM.2017.43.

[2] Y.P. Lin, et al., Blockchain: the evolutionary next step for ICT e-agriculture, Environ. - MDPI 4 (3) (2017) 1–13, https://doi.org/10.3390/environments4030050.

[3] E.Y.T. Adesta, D. Agusman, A. Avicenna, Internet of things (IoT) in agriculture industries, Indones. J. Electr. Eng. Informatics 5 (4) (2017) 376–382, https://doi.org/10.11591/ijeei.v5i4.373.

[4] M. Tripoli and J. Schmidhuber, "Emerging Opportunities for the Application of Blockchain in the Agri-Food Industry Agriculture," FAO, United Nations, http://www.fao.org/3/CA1335EN/ca1335en.pdf.

[5] M.P. Caro, M.S. Ali, M. Vecchio, R. Giaffreda, Blockchain-based traceability in Agri-food supply chain management: a practical implementation, in: 2018 IoT Vert. Top. Summit Agric. - Tuscany, IOT Tuscany 2018, 2018, pp. 1–4, https://doi.org/10.1109/IOT-TUSCANY.2018.8373021.

[6] K. Leng, Y. Bi, L. Jing, H.C. Fu, I. Van Nieuwenhuyse, Research on agricultural supply chain system with double chain architecture based on blockchain technology, Futur. Gener. Comput. Syst. 86 (2018) 641–649, https://doi.org/10.1016/j.future.2018.04.061.

[7] J. Hua, X. Wang, M. Kang, H. Wang, F.Y. Wang, Blockchain based provenance for agricultural products: a distributed platform with duplicated and shared bookkeeping, in: *IEEE Intell. Veh. Symp. Proc.*, vols. 2018-June, 2018, pp. 97–101, https://doi.org/10.1109/IVS.2018.8500647.

[8] J.T. Hao, Y. Sun, H. Luo, A safe and efficient storage scheme based on blockchain and IPFs for agricultural products tracking, J. Comput. 29 (6) (2018) 158–167, https://doi.org/10.3966/199115992018122906015.

[9] J. Lin, Z. Shen, A. Zhang, Y. Chai, Blockchain and IoT based food traceability for smart agriculture, in: ICCSE'18: Proceedings of the 3rd International Conference on Crowd Science and Engineering, 2018, pp. 1–6, *dl.acm.org.* https://doi.org/10.1145/3265689.3265692.

[10] O. Bermeo-Almeida, M. Cardenas-Rodriguez, T. Samaniego-Cobo, E. Ferruzola-Gómez, R. Cabezas-Cabezas, W. Bazán-Vera, Blockchain in agriculture: a systematic literature review, Commun. Comput. Inf. Sci. 883 (2018) 44–56, https://doi.org/10.1007/978-3-030-00940-3_4.

[11] D. Kos, S. Kloppenburg, Digital technologies, hyper-transparency and smallholder farmer inclusion in global value chains, Curr. Opin. Environ. Sustain. 41 (2019) 56–63, https://doi.org/10.1016/j.cosust.2019.10.011.

[12] O. Lamtzidis, D. Pettas, J. Gialelis, A novel combination of distributed ledger technologies on internet of things: use case on precision agriculture, Appl. Syst. Innov. 2 (3) (2019) 1–31, https://doi.org/10.3390/asi2030030.

[13] A. Croxson, R.S. Sharma, S. Wingreen, Making sense of blockchain in food supply-chains, in: *Australas. Conf. Inf. Syst*, 2019, pp. 97–107. Accessed: Nov. 15, 2021. [Online]. Available: https://aisel.aisnet.org/acis2019/10/.

[14] Z. Wang, P. Liu, Application of blockchain technology in agricultural product traceability system, in: Lect. Notes Comput. Sci. (Including Subser. Lect. Notes Artif. Intell. Lect. Notes Bioinformatics), 2019, pp. 81–90, https://doi.org/10.1007/978-3-030-24271-8_8. vol. 11634 LNCS.

[15] M.C. Aldag, The use of blockchain Technology in Agriculture, Zesz. Nauk. Uniw. Ekon. w Krakowie 4 (982) (2019) 7–17, https://doi.org/10.15678/znuek. 2019.0982.0401.

[16] X. Shi, et al., State-of-the-art internet of things in protected agriculture, Sensors (Switzerland) 19 (8) (2019), https://doi.org/10.3390/s19081833.

[17] R. Mavilia, R. Pisani, Scaling blockchain for agricultural sector: the Agridigital case, in: 3rd Int. Sci. Conf. ITEMA Recent Adv. Inf. Technol. Tour. Econ. Manag. Agric, 2019, pp. 55–60, https://doi.org/10.31410/itema.2019.55.

[18] S. Umamaheswari, S. Sreeram, N. Kritika, D.R. Jyothi Prasanth, BIoT: blockchain based IoT for agriculture, in: Proc. 11th Int. Conf. Adv. Comput. ICoAC 2019, 2019, pp. 324–327, https://doi.org/10.1109/ICoAC48765.2019.246860.

[19] M. Kim, B. Hilton, Z. Burks, J. Reyes, Integrating blockchain, smart contract-tokens, and IoT to design a food traceability solution, in: 2018 IEEE 9th Annu. Inf. Technol. Electron. Mob. Commun. Conf. IEMCON 2018, 2019, pp. 335–340, https://doi.org/10.1109/IEMCON.2018.8615007.

[20] F. Antonucci, S. Figorilli, C. Costa, F. Pallottino, L. Raso, P. Menesatti, A review on blockchain applications in the Agri-food sector, J. Sci. Food Agric. 99 (14) (2019) 6129–6138, https://doi.org/10.1002/jsfa.9912.

[21] V.S. Yadav, A.R. Singh, Use of blockchain to solve select issues of Indian farmers, in: AIP Conf. Proc., vol. 2148, 2019, https://doi.org/10.1063/1.5123972.

[22] S. Wingreen, R. Sharma, P. Jahanbin, S. Wingreen, R. Sharma, A blockchain traceability information system for trust improvement in agricultural supply chain, in: 2019 European Conference on Information Systems At: Stockholm-Uppsala, Sweden, 2019, pp. 5–15. Accessed: Nov. 15, 2021. [Online]. Available: https://aisel.aisnet. org/ecis2019_rip/10.

[23] K. Salah, N. Nizamuddin, R. Jayaraman, M. Omar, Blockchain-based soybean traceability in agricultural supply chain, IEEE Access 7 (2019) 73295–73305, https://doi.org/10.1109/ACCESS.2019.2918000.

[24] A. Kamilaris, A. Fonts, F.X. Prenafeta-Boldú, The rise of blockchain technology in agriculture and food supply chains, Trends Food Sci. Technol. 91 (2019) 640–652, https://doi.org/10.1016/j.tifs.2019.07.034.

[25] M. Shyamala Devi, R. Suguna, A.S. Joshi, R.A. Bagate, Design of IoT blockchain based smart agriculture for enlightening safety and security, Commun. Comput. Inf. Sci. 985 (2019) 7–19, https://doi.org/10.1007/978-981-13-8300-7_2.

[26] V.S. Yadav, A.R. Singh, A systematic literature review of blockchain technology in agriculture, in: Proc. Int. Conf. Ind. Eng. Oper. Manag, 2019, pp. 973–981. Accessed: Nov. 15, 2021. [Online]. Available: http://ieomsociety.org/pilsen2019/ papers/256.pdf.

[27] T. Surasak, N. Wattanavichean, C. Preuksakarn, S.C.H. Huang, Thai agriculture products traceability system using blockchain and internet of things, Int. J. Adv. Comput. Sci. Appl. 10 (9) (2019) 578–583, https://doi.org/10.14569/ijacsa.2019.0100976.

[28] S. Madumidha, P. Siva Ranjani, U. Vandhana, B. Venmuhilan, A theoretical implementation: agriculture-food supply chain management using blockchain technology, in: Proc. 2019 TEQIP - III Spons. Int. Conf. Microw. Integr. Circuits, Photonics Wirel. Networks, IMICPW 2019, 2019, pp. 174–178, https://doi.org/10.1109/IMICPW. 2019.8933270.

[29] J. Potts, Blockchain in agriculture, SSRN Electron. J. (2019), https://doi.org/10.2139/ ssrn.3397786.

[30] T. Alam, Blockchain and its role in the internet of things (IoT), IJSRCSEIT 5 (1) (2019) 151–157, https://doi.org/10.32628/cseit195137.

[31] P. Singh, N. Singh, G.C. Deka, Prospects of Machine Learning with Blockchain in Healthcare and Agriculture, igi-global.com, 2020, pp. 178–208.

[32] X. Li, D. Huang, Research on value integration mode of agricultural E-commerce industry chain based on internet of things and blockchain technology, *Wirel. Commun. Mob. Comput.*, vol. (2020) 2020, https://doi.org/10.1155/2020/8889148.

[33] W. Lin, et al., Blockchain Technology in Current Agricultural Systems: from techniques to applications, IEEE Access 8 (2020) 143920–143937, https://doi.org/10.1109/ACCESS.2020.3014522.

[34] Y. Chen, Y. Li, C. Li, Electronic agriculture, blockchain and digital agricultural democratization: origin, theory and application, J. Clean. Prod. 268 (2020), https://doi.org/10.1016/j.jclepro.2020.122071.

[35] M.D. Borah, V.B. Naik, R. Patgiri, A. Bhargav, B. Phukan, S.G.M. Basani, Supply Chain Management in Agriculture Using Blockchain and IoT, Springer, 2020, pp. 227–242, https://doi.org/10.1007/978-981-13-8775-3_11.

[36] P. Chun-Ting, L. Meng-Ju, H. Nen-Fu, L. Jhong-Ting, S. Jia-Jung, Agriculture blockchain service platform for farm-to-fork traceability with IoT sensors, in: *International Conference on Information Networking*, vol. 2020, 2020, pp. 158–163, https://doi.org/10.1109/ICOIN48656.2020.9016535.

[37] G. Mirabelli, V. Solina, Blockchain and agricultural supply chains traceability: research trends and future challenges, Procedia Manuf. 42 (2020) 414–421, https://doi.org/10.1016/j.promfg.2020.02.054.

[38] V.S. Yadav, A.R. Singh, R.D. Raut, U.H. Govindarajan, Blockchain technology adoption barriers in the Indian agricultural supply chain: an integrated approach, Resour. Conserv. Recycl. 161 (2020), https://doi.org/10.1016/j.resconrec.2020.104877.

[39] L. Hang, I. Ullah, D.H. Kim, A secure fish farm platform based on blockchain for agriculture data integrity, Comput. Electron. Agric. 170 (2020), https://doi.org/10.1016/j.compag.2020.105251.

[40] X. Li, D. Wang, M. Li, Convenience analysis of sustainable E-agriculture based on blockchain technology, J. Clean. Prod. 271 (2020), https://doi.org/10.1016/j.jclepro.2020.122503.

[41] S. Benedict, Serverless blockchain-enabled architecture for IoT societal applications, IEEE Trans. Comput. Soc. Syst. 7 (5) (2020) 1146–1158, https://doi.org/10.1109/TCSS.2020.3008995.

[42] S.K. Singh, S. Rathore, J.H. Park, BlockIoTIntelligence: a blockchain-enabled intelligent IoT architecture with artificial intelligence, Futur. Gener. Comput. Syst. 110 (2020) 721–743, https://doi.org/10.1016/j.future.2019.09.002.

[43] S.S. Patra, C. Misra, K.N. Singh, M.K. Gourisaria, S. Choudhury, S. Sahu, qIoTAgriChain: IoT blockchain traceability using queueing model in smart agriculture, in: EAI/Springer Innovations in Communication and Computing, 2021, pp. 203–223.

[44] Q.N. Tran, B.P. Turnbull, H.-T. Wu, A.J.S. de Silva, K. Kormusheva, J. Hu, A survey on privacy-preserving blockchain systems (PPBS) and a novel PPBS-based framework for smart agriculture, IEEE Open J. Comput. Soc. 2 (2021) 72–84, https://doi.org/10.1109/ojcs.2021.3053032.

[45] S.J. Anand, K. Priyadarsini, G.A. Selvi, D. Poornima, Iot-based secure and energy efficient scheme for precision agriculture using blockchain and improved Leach algorithm, Turk. J. Comput. Math. Educ. 12 (10) (2021) 2466–2475. Accessed: Nov. 15, 2021. [Online]. Available: https://www.turcomat.org/index.php/turkbilmat/article/view/4857.

[46] M. Sandeep Kumar, V. Maheshwari, J. Prabhu, M. Prasanna, R. Jothikumar, Applying blockchain in agriculture: a study on blockchain technology, benefits, and challenges, in: EAI/Springer Innovations in Communication and Computing, Springer Science and Business Media Deutschland GmbH, 2021, pp. 167–181.

[47] H. Patel, B. Shrimali, AgriOnBlock: Secured Data Harvesting for Agriculture Sector Using Blockchain Technology, ICT Express, 2021, https://doi.org/10.1016/j.icte. 2021.07.003.

[48] U. Sengupta, H.M. Kim, Meeting changing customer requirements in food and agriculture through the application of blockchain technology, Front. Blockchain 4 (2021), https://doi.org/10.3389/fbloc.2021.613346.

[49] Y. Zhang, J. Li, L. Ge, Research on agricultural product supply chain based on internet of things and blockchain technology, Adv. Intell. Syst. Comput. 1343 (2021) 11–17, https://doi.org/10.1007/978-3-030-69999-4_2.

[50] Y. Bai, K. Fan, K. Zhang, X. Cheng, H. Li, Y. Yang, Blockchain-based trust management for agricultural green supply: a game theoretic approach, J. Clean. Prod. 310 (2021), https://doi.org/10.1016/j.jclepro.2021.127407.

[51] S. Dong, L. Yang, X. Shao, Y. Zhong, Y. Li, P. Qiao, How can channel information strategy promote sales by combining ICT and blockchain? Evidence from the agricultural sector, J. Clean. Prod. 299 (2021), https://doi.org/10.1016/j.jclepro.2021.126857.

[52] C.S. Bhusal, Blockchain Technology in Agriculture: a case study of blockchain Start-up Companies, Int. J. Comput. Sci.Iinf. Technol. 13 (5) (2021), https://doi.org/10.5121/ ijcsit.2021.13503.

[53] S. Al-Amin, S.R. Sharkar, M.S. Kaiser, M. Biswas, Towards a blockchain-based supply chain management for e-agro business system, Adv. Intell. Syst. Comput. 1309 (2021) 329–339, https://doi.org/10.1007/978-981-33-4673-4_26.

[54] A.A. Mukherjee, R.K. Singh, R. Mishra, S. Bag, Application of blockchain technology for sustainability development in agricultural supply chain: justification framework, Oper. Manag. Res. (2021), https://doi.org/10.1007/s12063-021-00180-5.

[55] G. da Silva Ribeiro Rocha, L. de Oliveira, E. Talamini, Blockchain applications in agribusiness: a systematic review, Future Internet 13 (4) (2021), https://doi.org/10.3390/ fi13040095.

[56] T.H. Pranto, A.A. Noman, A. Mahmud, A.B. Haque, Blockchain and smart contract for IoT enabled smart agriculture, PeerJ Comput. Sci. 7 (2021) 1–29, https://doi.org/ 10.7717/PEERJ-CS.407.

[57] M. Biswas, T.M.N.U. Akhund, M.J. Ferdous, S. Kar, A. Anis, S.A. Shanto, BIoT: blockchain based smart agriculture with internet of thing, in: *Proceedings of the 2021 5th World Conference on Smart Trends in Systems Security and Sustainability, WorldS4 2021,* 2021, pp. 75–80, https://doi.org/10.1109/WorldS451998.2021.9513998.

[58] I.E. Kyzy, H. Song, A. Vajdi, Y. Wang, J. Zhou, Blockchain for consortium: a practical paradigm in agricultural supply chain system, Expert Syst. Appl. 184 (2021), https://doi. org/10.1016/j.eswa.2021.115425.

[59] R. Mavilia, R. Pisani, Blockchain for Agricultural Sector: The Case of South Africa, African J. Sci. Technol. Innov. Dev, 2021, https://doi.org/10.1080/20421338.2021. 1908660.

[60] M. Verma, Smart contract model for trust based agriculture using blockchain technology, Int. J. Res. Anal. Rev. 344 (2) (2021) 2348–2349. Accessed: Nov. 15, 2021. [Online]. Available: www.ijrar.org.

[61] N. Niknejad, W. Ismail, M. Bahari, R. Hendradi, A.Z. Salleh, Mapping the research trends on blockchain technology in food and agriculture industry: a bibliometric analysis, Environ. Technol. Innov. 21 (2021), https://doi.org/10.1016/j.eti.2020.101272.

[62] S. Hu, S. Huang, J. Huang, J. Su, Blockchain and edge computing technology enabling organic agricultural supply chain: a framework solution to trust crisis, Comput. Ind. Eng. 153 (2021), https://doi.org/10.1016/j.cie.2020.107079.

[63] W. Liu, X.F. Shao, C.H. Wu, P. Qiao, A systematic literature review on applications of information and communication technologies and blockchain technologies for precision agriculture development, J. Clean. Prod. 298 (2021), https://doi.org/10.1016/j.jclepro.2021.126763.

[64] M.H. Ronaghi, A blockchain maturity model in agricultural supply chain, Inf. Process. Agric. 8 (3) (2021) 398–408, https://doi.org/10.1016/j.inpa.2020.10.004.

[65] G.S. Sajja, K.P. Rane, K. Phasinam, T. Kassanuk, E. Okoronkwo, P. Prabhu, Towards Applicability of Blockchain in Agriculture Sector, Mater. Today Proc, 2021, https://doi.org/10.1016/j.matpr.2021.07.366.

[66] W. Ren, X. Wan, P. Gan, A double-blockchain solution for agricultural sampled data security in internet of things network, Futur. Gener. Comput. Syst. 117 (2021) 453–461, https://doi.org/10.1016/j.future.2020.12.007.

[67] S. Saurabh, K. Dey, Blockchain technology adoption, architecture, and sustainable Agri-food supply chains, J. Clean. Prod. 284 (2021), https://doi.org/10.1016/j.jclepro.2020.124731.

[68] P.R. Srivastava, J.Z. Zhang, P. Eachempati, Blockchain technology and its applications in agriculture and supply chain management: a retrospective overview and analysis, Enterp. Inf. Syst. (Oct. 2021) 1–24, https://doi.org/10.1080/17517575.2021.1995783.

[69] T. Narayanaswamy, P. Karthika, K. Balasubramanian, Blockchain Enterprise: use cases on multiple industries, in: EAI/Springer Innovations in Communication and Computing, Springer Science and Business Media Deutschland GmbH, 2022, pp. 125–137.

[70] B. Bera, A. Vangala, A.K. Das, P. Lorenz, M.K. Khan, Private blockchain-envisioned drones-assisted authentication scheme in IoT-enabled agricultural environment, Comput. Stand. Interfaces 80 (2022), https://doi.org/10.1016/j.csi.2021.103567.

[71] N. Kamalakshi, Naganna, Role of blockchain in agriculture and food sector: a summary, EAI/Springer Innov. Commun. Comput (2022) 93–107, no. 978-3-030-76215-5. https://doi.org/10.1007/978-3-030-76216-2_6.

About the authors

Urvashi Sugandh is an assistant professor at the Department of Computer Science Engineering, HMR Institute of technology and management, affiliated to Guru Gobind Singh Indraprastha University, Delhi by Govt. of NCT Delhi. Currently, she is pursuing Ph.D. from Banasthali Vidyapith, Rajasthan. She has been awarded master's degree in Master of Technology from Department of Information Technology of Banasthali Vidyapith, Rajasthan in 2014. She has published two patents and four research papers in International/National Conferences. Her Research area is Blockchain, Data Mining and software engineering.

Manju Khari is an associate professor in Jawaharlal Nehru University, New Delhi, prior to the university she worked with Netaji Subhas University of Technology, East Campus, formerly Ambedkar Institute of Advanced Communication Technology and Research, Under Govt. Of NCT Delhi. Her PhD in Computer Science and Engineering from National Institute of Technology Patna and she received her master's degree in Information Security from Ambedkar Institute of Advanced Communication Technology and Research, affiliated with Guru Gobind Singh Indraprastha University, Delhi, India. She has 80 published papers in refereed National/International Journals and Conferences (viz. IEEE, ACM, Springer, Inderscience, and Elsevier), 10 book chapters in a Springer, CRC Press, IGI Global, Auerbach. She is also co-author of two books published by NCERT of XI and XII and co-editor in 10 edited books. She has also organized five international conference sessions, three faculty development programme, one workshop, one industrial meet in her experience. She delivered an expert talk, guest lecturers in International Conference and a member of reviewer/technical program committee in various International Conferences. Besides this, she associated with many International research organizations as Associate Editor/Guest Editor of Springer, Wiley and Elsevier books, and a reviewer for various International Journal.

Swati Nigam is currently an assistant professor at the Department of Computer Science, Banasthali Vidyapith, Rajasthan, India. She has been awarded PhD degree in Computer Science from Department of Electronics and Communication, University of Allahabad, India in 2015. She has been a post-doctoral fellow under National Post-Doctoral Fellowship scheme of Science and Engineering Research Board, Department of Science and Technology, Government of India. Earlier she has been awarded Senior Research Fellowship by Council of Scientific and Industrial Research,

Government of India. She has authored more than 20 articles in peer-reviewed journals, book chapters and conference proceedings. She has also published an authored book in Springer publications. She is a designated reviewer of several SCI journals like IEEE Access, Computer Vision and Image Understanding, Journal of Electronic Imaging, Multimedia Tools and Applications, etc. She has been publication chair, publicity chair, TPC member and reviewer of various reputed conferences. She is a professional member of IEEE and ACM. Her research interests include object detection, object tracking and human behavior analysis.

Printed in the United States
by Baker & Taylor Publisher Services